Soil and Rock Construction Materials

For Maria

Soil and Rock Construction Materials

G.H. McNally

BSc, BA, DIC, MAppSc, MIE Aust, CPEng
Senior Lecturer in Engineering Geology
Department of Applied Geology,
University of New South Wales

Routledge
Taylor & Francis Group

LONDON AND NEW YORK

First published 1998 by E & FN Spon

This edition published 2013 by Routledge
2 Park Square, Milton Park, Abingdon, Oxon OX14 4RN
711 Third Avenue, New York, NY 10017, USA

Routledge is an imprint of the Taylor & Francis Group, an informa business

© 1998 G.H. McNally

Typeset in 10/12pt Palatino by
Cambrian Typesetters, Frimley

British Library Cataloguing in Publication Data
A catalogue record for this book is available from the British Library

ISBN 0 419 21420 8

Publisher's Note
The publisher has gone to great lengths to ensure the quality of this reprint but points out that some imperfections in the original may be apparent.

Contents

Preface

The origins of this book go back to 1969, to my first job as a geology graduate. I was put in charge of a laboratory that tested a variety of road construction materials, including natural gravel roadbase, aggregate, concrete and earthfill. These were topics that my university training had scarcely touched upon, yet the small amount of reference literature available at the time was aimed at people already well acquainted with testing procedures and specifications. There was little in print to inform a new practitioner as to why materials behaved in the way they did, how test results could be interpreted, why specifications were strict on particular aspects, and so on. Fortunately, my colleagues and the laboratory staff gave generously of their time and experience to bring me up to scratch, and I had the opportunity of visiting dozens of construction sites (and of learning the hard way, by making mistakes!)

The second starting point towards this book came in 1990, when I was asked to prepare a short course on soil and rock materials for the Key Centre for Mines at the University of New South Wales, in conjunction with the materials branch of the NSW Roads and Traffic Authority. This one-week course has now been given five times, in Australia and Malaysia, and its content has been steadily expanded. The idea of presenting the course material in book form came about because there is still no introductory text in this field, although some topics (such as concrete materials) are the subject of many volumes. Other topics, such as pavement design and recycled materials, are dealt with extensively in journals but in few books.

Although this book has been written from a geological viewpoint, it is also directed towards geotechnical engineers, environmental consultants, and land-use planners and regulators. Only an elementary knowledge of geology is presumed; indeed, most of the subject matter has been drawn from mining and civil engineering, mineral processing, soil and rock mechanics, ceramics and several other disciplines. This is not a state-of-the-art volume, and I have especially avoided going into detail about testing procedures and specification requirements wherever possible, to avoid the discussion becoming tedious. What I have tried to draw out are the reasons why some material qualities are sought and how they may be modified.

One of the problems in writing about natural construction materials is that the literature is voluminous, but poorly accessible – most of it is not only unpublished, but is available only in limited-circulation reports. Furthermore, many of the comments in these pages are based on my own experience and – even more so – that of my professional colleagues. The latter are busy people, not given to writing academic papers about matters that are, in their eyes, everyday knowledge. In reaching out to an international readership, I have tried to keep in mind the difficulties of access to this literature, and differences in engineering practice and terminology. Fortunately, my own country, Australia, contains a wide variety of terrains and construction problems, from those of heavily-trafficked freeways down to unsealed rural roads. Engineering construction here is based on American, British and European practice, though some aspects are drawn from countries like South Africa, which have similar terrain and climate. In citing references I have preferred to quote books rather than journals, and journals rather than limited-circulation conference proceedings, even though the latter are often the most informative.

I am grateful to many people for helping me with this book, but two deserve special thanks. The first is my good friend and long-time mentor, Jim Williams (Manager, Scientific Services for the NSW Roads and Traffic Authority), who jointly presented the short courses around which this book is based. The second is the late L.G. (Gerry) Wearne, who was my senior laboratory technician in 1969–70 and who began my education in materials technology. Others who contributed to the short courses include Joe Whitehead, Ed Haber, Greg Won, Ian Stewart and Gary Rigozzi. A number of my former students, especially Ian Wilson, contributed reference material and photographs, and are acknowledged in the text.

Greg McNally
Sydney, June 1997

CHAPTER 1

Introduction to soil and rock materials

This book is intended to provide an introduction to the assessment, extraction, processing and environmental management of soil and rock construction materials, from a geological perspective. These *geomaterials* are sometimes referred to as low-cost resources, or as a category of industrial minerals. In Australia the producers of quarried products are known collectively as the *extractive industries*. Geomaterials differ from other mineral commodities in many ways: their widespread, though uneven, distribution; the large volumetric demand for them, concentrated around urban areas; their low ex-pit value, because substitutes can always be found; and, consequently, their sensitivity to transport costs. Large and high-quality geomaterial deposits far from cities are worthless, while inferior materials close to customers are often upgraded.

The traditional task of geologists in this industry was to locate and prove up deposits meeting specified criteria within acceptable hauling distance of the construction site. With the increasing scale of quarry operations and rising public concern over landscape degradation, this role has expanded. Geologists are now generally part of a project team, in which they may be asked to give advice on the geological aspects of product quality control, blast design, crusher selection, quarry restoration and so on. They will be expected to know enough about downstream processing to prepare a preliminary pit layout and to suggest how the properties of the raw material are likely to influence this.

The geologist should certainly take these things into account when planning a quarry investigation, and attempt to map, sample and log drillholes with a wider view than simply evaluating quality and reserves. At a later stage the person responsible for laboratory testing and quality control in a large quarry is also, more often than not, likely to have been originally trained as a geologist. His or her duties may also include those of the site environmental officer – responsible for matters such as blast vibration monitoring, dust control and pit restoration.

A number of significant trends in construction-materials technology have emerged over the past 20 years or so, as outlined below:

- *Fewer but larger sites*   The first of these is towards fewer but larger extractive sites, at the expense of many small ones. Big quarries offer economies of scale: a variety of products, more elaborate processing facilities, large reserves, continuity of operation and access to specialist skills. These operations are better equipped, financially and technically, to meet increasingly stringent environmental and specification requirements, and to accommodate fluctuating market demand. There is a price to be paid for this producer concentration, of course: big quarries are more intrusive within a landscape than small ones; haulage to distant construction sites is more expensive; and price competition is diminished.
- *Maximum use of existing quarries*   The second development is towards maximum use of existing quarries, even where only inferior materials remain. The aim of this, from the regulator's viewpoint, is to minimize further landscape degradation at replacement sites. 'High grading' and early abandonment of partly worked resources are nowadays discouraged. From the operator's viewpoint this makes best use of

Figure 1.1   Quarry reclamation, the old way. This alluvial gravel pit near Urana, New South Wales, has been simply abandoned. Note that some natural revegetation has occurred, despite a poor soil and gully erosion. On a positive note, this quarry offers a wildlife refuge in an otherwise monotonous farmland plain, and is the only surface exposure of these Late Tertiary gravels in the State.

fixed capital assets, retains favourably located operating pits, and avoids expensive environmental investigations and legal disputes over new sites.

- *Greater use of upgraded materials*   The third trend, which is a corollary of the above, is greater use of upgraded marginal materials (or simply non-standard materials, and even wastes) to extend the life of the limited reserves of first-class rock. This can be achieved by blending, stabilization and careful quality control. In effect, the trend is away from natural deposits that happen to have suitable properties, and towards materials that are processed or even manufactured to meet a specification. It is also desirable that a material not be overspecified, to a standard above what the end-use demands, because this is wasteful of resources.

- *Temporary land-use*   Finally, it is emphasized that quarrying is only a temporary use of the land, and that reclamation of mined-out land should be planned well before abandonment (Figures 1.1 and 1.2). Because old quarries are commonly enveloped by urban growth, this may be self-financing in two ways: first, because the voids can be used as waste disposal sites; and, secondly, the land may be later sold for

Figure 1.2   Preparing for quarry restoration in a north Queensland ferricrete (laterite) roadbase pit. At this stage the ferricrete surface has just been exposed; topsoil and vegetation mulch are stockpiled on the left, thin subsoil overburden in the other two mounds. Low soil fertility and low rainfall inhibit site rehabilitation, and the shallow depth (about 1.5 m) of the ferricrete means that a large area must be stripped to obtain the required volume of gravel. Usually the time interval between stripping and restoration is 6–12 months.

industrial or other higher-value end-uses. In other cases, restoration may be paid for by a levy on existing producers, who benefit from restrictions on competitors entering the market, before being returned to public ownership as recreational facilities.

1.1   SCOPE OF THE BOOK

Chapters 2–4 discuss the investigation and assessment of aggregate and roadbase deposits. Emphasis is placed on methods of evaluating rather than searching for these materials, since in most cases their general location will be known beforehand. There appears to be a worldwide trend towards increased production from hard rock quarries at the expense of sand and gravel pits, partly because these are better suited to large-scale extraction and partly because they cause less environmental damage per tonne mined. However, crushed rock sources are more likely to be weathered or altered, and hence to produce stone of suspect durability. They also require blasting, which increases extraction costs and creates vibration nuisances.

Chapter 3 also discusses two important developing sources of sand and gravel – marine dredging, and weakly consolidated sandstone and conglomerate. These offer very large reserves, but present problems of salt content and fines disposal, respectively. Chapter 4 deals with natural road gravels, which are a heterogeneous collection of gravel–sand–fines mixtures, such as 'laterite' gravel, which are used for pavements in secondary roads. Though formerly favoured because they occurred close to construction sites and required minimal processing, their use in this form is declining. Instead, they are increasingly being upgraded by crushing and stabilization to meet heavier traffic requirements (Figure 1.3).

Chapters 5 and 6 discuss the extraction and processing of hard rock and granular materials. Processing is taken to include crushing, particle shaping, screening, sizing, beneficiation and dewatering. These techniques – mostly borrowed from mineral processing – are becoming more important as premium sources are exhausted and inferior deposits have to be worked. The chief technical issues here are: in blasting, the replacement of nitroglycerine-based explosives by ANFO, watergels and, most recently, emulsions; in crushing, the wider use of impact crushers, even in moderately strong and abrasive rock, and especially for product shaping; and in sand workings, more extensive application of size classification techniques (Figure 1.4).

Chapter 7 describes the testing and specification of aggregate and prepared roadbase. The principal categories of laboratory test methods, and their applicability, are summarized. There have been few major changes in materials testing over the past 30 years, but one has been the introduction of repeated-load triaxial testing to simulate the fatigue

Figure 1.3 Calcrete pavement beneath a thin sprayed seal, Morgan, South Australia. This is typical of duricrust road gravels still used with minimal processing (in this case grid rolling, but elsewhere single-stage crushing) for secondary road pavements. The coarse fragments in the foreground have been moved to the edge of the road shoulder during construction, and the roadbase material beneath the seal is somewhat finer.

effects of traffic on road pavement courses. Another innovation has been the move from prescriptive specifications to ones based on quality assurance and more refined sampling techniques.

The use of coarse and fine aggregate in Portland cement concrete, asphaltic concrete and bituminous surfacing – and their differing requirements – are the subject of Chapters 8 and 9. In concrete aggregates (Chapter 8), intact rock strength is now thought to be less important than the tenacity of the cement–aggregate bond, especially where tensile strength is at a premium. In concrete mix design, there is now a wider acceptance of slag and fly ash as cement extenders and fillers, and not just because they are cheap. In bituminous surfacing, asphaltic veneers are progressively displacing sprayed seals, and more emphasis is being placed on skid resistance, low tyre noise and fatigue life in mix designs.

Earth embankment materials, compaction and road pavement design are discussed in Chapters 10 and 11. In Australia at least, earthfill compaction standards have not improved over the past 30 years, though lift thicknesses and roller sizes have both doubled. Embankment technology has, nevertheless, greatly profited from the application of

Figure 1.4   Small sand processing plant, near Renmark, South Australia, which includes a lump breaker, screens, a spiral and bucketwheel classifier, and a hydraulic classifier. A tailings pond allows process water to clarify before discharge to the Murray River (just beyond top edge of photo). The feed is very silty but well-graded Parilla Sand (Pliocene).

geosynthetics, particularly for tensile reinforcement (as in 'reinforced earth'; Figure 1.5). Modern deep cuttings produce less-weathered rock and hence a requirement that rockfill, with its attendant problem of settlement, be accepted in highway embankments. This in turn has dictated that a form of fill zoning be adopted to encompass common fill, select fill, transition zones and rockfill materials.

The double requirements of allowing for settlement of high fills and heavier traffic loadings have dictated that pavement thickness and stiffness be greatly increased in freeway-standard roads. This has been achieved partly by the wider use of crushed and bound roadbases, including deep lift asphalt, and partly through the re-introduction of concrete pavements. The effects of cyclic loading, especially by heavy trucks, are now fully appreciated in road design, where pavement life is expressed in *equivalent standard axles* (ESAs) rather than years. Analytical design methods based on triaxial testing and numerical modelling are gradually supplanting the former empirical approach.

Railway ballast, rockfill and macadam pavements are discussed in Chapter 12. These are all open-graded, free-draining broken rock materials, but of different maximum particle size and durability. The chief development in this area has been the widespread adoption of

Figure 1.5 Reinforced earth bridge abutments and underpass beneath a freeway under construction at Ourimbah, New South Wales. The solidity of the abutments is deceptive: the cladding slabs are simply slotted together and held in place by steel tendons embedded in the dune sand backfill.

weak rolled rockfill for embankment dams. This replaces strong end-dumped rockfill, which was sluiced but otherwise compacted under self-weight, resulting in large and long-term settlements.

Chapter 13 deals with the even coarser stone used in shoreline protection works. The main geotechnical challenge here is devising blasting procedures for generating armour blocks around 10 tonnes weight and zoning the breakwater to use this stone most effectively. Another problem lies in testing such unwieldy materials; ultrasonic velocity measurements appear to offer much potential.

Rock cut and split for use as dimension stone is the topic of Chapter 14. Dimension stone is experiencing a worldwide architectural revival, and indeed it is the only extractive industry with a significant export component, though facing stone has long displaced block masonry as the dominant application. The use of stone veneers as thin as 20–40 mm in facing panels places considerable emphasis on flexural tensile strength and durability, favouring granites over sandstones or limestones. Another important development in this area has been the growth of the stonework preservation and restoration sector, and the emphasis on stone salvage and matching.

Chapters 15 and 16 deal with limestone and brick clays, two commodities that are important industrial raw materials as well as construction materials. This is especially true of limestone, which, though a versatile source of aggregate, crushed roadbase, armourstone and dimension stone, is most valuable as a raw material in cement-making and other chemical processes. Carbonate rock terrains also give rise to unique landforms, soils and faunal assemblages – and therefore to a disproportionate share of environmental opposition to quarrying (Figure 1.6). The continued use of carbonate rock for aggregates, where

Figure 1.6   Quarrying heritage – a limestone kiln, now abandoned, built into a rock face near Mount Gambier, South Australia. The kiln was built out of offcuts from Gambier Limestone used primarily as sawn dimension stone.

other rock types would do as well or better, and of high-calcium limestone for cement-making, is not easy to defend.

Brick clays (Chapter 16) nowadays serve two markets: as a raw material in the mass production of extruded bricks in tunnel kilns; and, to a much smaller extent, as an ingredient in batch-produced pressed 'boutique bricks' for facing and restoration. In both applications the natural materials are heavily modified and blended, becoming less dependent on the source rock mineralogy and particle composition.

Environmental aspects of the construction-materials industry are the themes of the last five chapters. In Chapter 17 alternatives to traditional 'mineral aggregates' are discussed. These include wastes, whose exploitation can both remove landscape eyesores and reduce pressure on natural resources. However, they are not equally prized: most are inhomogeneous and therefore only usable as common fill close to the dump site. By far the most valuable are iron and steel slags, which are today classified as by-products rather than wastes and are produced to specification. Indeed, blast furnace slag in granulated or pelletized form is more valuable as a cement extender or lightweight aggregate than as roadbase. Demolition materials (crushed bricks, concrete masonry and asphalt pavements) are now being widely recycled, partly in response to Government directives and partly as a means of avoiding tipping fees.

The upgrading of inferior pavement materials by means of granular or chemical stabilization is the subject of Chapter 18. In the past, stabilization always performed better in the laboratory than on the road, but modern purpose-built equipment has improved the mixing consistency. Cement and lime–fly ash pozzolans are the most common additives, though foamed bitumen is making a comeback. One important application is the *in situ* renovation of over-age natural gravel pavements, which in Australia alone could be needed for thousands of lane kilometres on rural highways over the next 20 years.

Chapter 19 deals with the principal environmental problems of establishing new quarries, with particular emphasis on their geological justification – to counter the argument that, although the need has been established, a better site is available elsewhere. Minimization of blasting nuisances (Chapter 20) is an important issue during the working life of a quarry, especially since statutory limits for air and ground vibration are now being set far below damaging levels, to where the blast is scarcely noticeable. Ground vibrations are satisfactorily controlled by interhole delays, but airblast suppression presents more difficulties. Face reorientation is one palliative, along with more effective stemming. Blast monitoring is now routine, and the results can be used to improve fragmentation as well as meeting regulatory conditions.

The final chapter is concerned with reclamation, with the theme that quarries can no longer be simply abandoned at the end of their economic life. Nor can site rehabilitation be left until last; it must be integrated with extraction throughout the working life of the pit. However, most pits can be profitably backfilled with waste prior to final reclamation, though this must take into account the prevailing hydrogeological conditions.

1.2   MATERIAL CATEGORIES

A simple classification scheme for the construction materials discussed in this book is presented in Table 1.1. The materials are categorized in terms of product size, method of processing, end-use and whether they are soil (i.e. they can be dug without blasting) or rock in origin. Many of the material classes are produced in parallel, such as washed sand being sold for concrete fine aggregate and unwashed sand from the same pit

Table 1.1 Classification of construction materials

Cut and broken rock
   Dimension stone (structural and cladding)
   Breakwater armourstone and rock core
   Dam rip-rap, embankment slope protection blocks
   Pitching and beaching stone
   Rubble, dam rockfill
   Rock slag

Coarse crushed rock
   Coarse aggregate for concrete and asphaltic concrete
   Surfacing aggregate ('chippings')
   Railway ballast, macadam pavements and gabion stone
   Free-draining sub-base, drainage layers
   Recycled concrete aggregate (RCA) and recycled asphalt (RAP)

Fine crushed rock
   Prepared roadbase and sub-base
   Processed granular filters
   Bedding material, grit, crusher dust

Sand and gravel
   Crushed and screened river gravel
   Washed fine aggregate and sand filters
   Mortar sand ('fat' sand), plastering sand
   Sand fill, stabilizing grit
   Granulated slag

Soil materials
   Brick, tile and pipe clay
   Natural roadbase
   Stabilized soils
   Common fill
   Select fill, sub-ballast, capping layers
   Pulverized fly ash (PFA) and furnace bottom ash (FBA)

used for bricklaying mortars. Specification requirements vary with the proposed end-use; hence rock crushed for use as sealing aggregate is required to meet more stringent test criteria than that accepted as road sub-base, even though both may come from the same quarry and even from the same parent rock. The amount of selection, winning, processing and quality control required to meet the specification is naturally reflected in the price demanded.

Broken rock and cut stone

Broken rock refers to material that has been ripped or blasted and possibly passed over a bar screen (or 'grizzly') to remove oversize, but which is otherwise unprocessed. In general this product is cobble-sized to boulder-sized, with minimal fines. Very durable rock is preferred for breakwater stone and rip-rap, but lower-quality stone is acceptable for rubble and rockfill. *Cut stone* is rock that has been split, sawn, wire-cut or otherwise extracted in block form. These blocks are further sawn, dressed and finished for sale as dimension stone.

Crushed rock

Coarse crushed rock ballast is of similar quality to concrete aggregate, but made up of coarser particle sizes and devoid of fines. Slightly inferior source rock is acceptable for *fine crushed rock* roadbase, which is subjected to less intense loading and for which a proportion of void-filling fines is required to ensure high compacted density ('dense grading'). Crusher grit and dust are by-products from aggregate production used as a source of 'sharp' (i.e. angular and coarse) sand in concrete and as a filler in asphalt.

Sands and gravels

Sands and gravels are primarily of alluvial origin, although other sources such as dunes, marine deposits and weathered sandstone are also worked. Processing includes crushing coarse gravel (to create rough-faced particles as well as to reduce their size), washing (to remove fines), screening and 'classification' (sand size separation by settling velocity). Granulated slag is an artificial sharp sand produced by rapid water cooling.

Soil materials

The soil materials listed in Table 1.1 are a diverse lot, including weathered rock and boiler ash, whose only common characteristic is that they are relatively easy to dig. Most naturally occurring roadbase materials and select fill are used with little or no processing. Brick clay, on the other hand, is an industrial raw material and may be ground, blended, moistened and chemically modified. Between these extremes are the stabilized soils, to which sand, cement, lime or bitumen are added to improve their strength, stiffness and durability.

1.3   AUSTRALIAN PRODUCTION AND RESOURCES

Production of soil and rock construction materials in Australia currently amounts to about 150 million tonnes annually, with an ex-quarry value of around A$1000 million. The uncertainty is due to different methods of reporting and categorization in different states, and to under-reporting of some materials such as unprocessed roadbase. The official figures indicate that production of crushed rock aggregate is increasing faster than that of alluvial gravel, and that aggregate prices rose 160% during the 1980s, twice the consumer price index increase.

A compilation of Australia's total resources of construction materials would be meaningless, since their location close to markets is much more important than their geological reserves. However, a brief survey of the main aggregate sources in four capital cities (Sydney, Melbourne, Adelaide and Brisbane), which account for about 70% of Australian demand, may throw some light on supply trends.

- *Sydney*   The largest hard rock quarries in the Sydney region work basalts and dolerite for aggregate, and volcanic breccia for roadbase. The volcanic breccia is a marginal-quality material, but it is relatively abundant and closer to markets than the basalt. River gravels and coarse sand are currently extracted from the Nepean–Hawkesbury River floodplain and blended with dune sand for concrete-making. Reserves of both coarse and fine aggregate are declining, and future supplies of the former are being sought from microsyenite intrusions and basalts more than 100 km distant from the city centre (where superquarries capable of producing 5–10 million tonnes per annum (Mtpa) have been proposed). Future sources of fine aggregate are from crushed sandstone and offshore sand bodies.
- *Melbourne*   About 67% of Melbourne's aggregate and roadbase requirements are obtained from basalts, the Older Volcanics (east and north of the city) and Younger Volcanics (west and north). Although the total volume of the basalts is very much greater than that in the vicinity of Sydney or Brisbane, there are problems due to secondary mineralization (chloritized 'green basalts') and the tendency of glassy varieties to become polished in bituminous surfacing. Most of the non-basalt production is supplied from acid volcanic and hornfels quarries. Construction sand requirements are met mainly from

fluviatile deposits and dunes; although these require washing, large reserves are available.

- *Adelaide* Coarse aggregate for the Adelaide metropolitan area is mostly obtained from seven quartzite and six carbonate (limestone, dolomite) quarries, which together produce about 5 Mtpa. The carbonate rocks are considered to be the best available aggregate sources here, and are mainly used for sealing and concrete batching. The quartzites, which vary between hard sandstones and true silicified arenites, are used for roadbase. Reserves are large and adequate for 70–80 years, but as in other cities these are not well distributed, with quartzite predominating to the city's east whereas carbonate rock is more abundant to the south. Large reserves of construction sand are contained within Tertiary basins to the south and northwest.
- *Brisbane* Brisbane derives most of its aggregate from river gravels dredged from the Brisbane River and the North and South Pine Rivers, but the former is rapidly being depleted and has to be supplemented with crushed basalt, acid volcanics, granite and metamorphics. Roadbase is also obtained from the hard rock quarries. Sand comes mainly from alluvial sources, supplemented by dune mining on North Stradbroke Island.

Although Sydney is undoubtedly the worst-served in geological terms, all four cities have problems in meeting the demand for soil and rock construction materials. Much of Melbourne's abundant basalt is weathered or altered, Adelaide lacks any igneous rock sources, and Brisbane has to depend on metamorphic rocks and some marginal acid and intermediate lavas. All have environmental problems arising from quarrying, such as the conspicuous Hills Face quarry sites in Adelaide or the presumed connection between dredging and flood damage along the Brisbane River. The expansion of each city has in the past engulfed operating or potential quarry sites, particularly sand deposits. There has been considerable concentration of ownership in the Australian quarrying industry over the past decade or so, although the trend towards very large quarries evident in the UK, where one ballast quarry and seven brick-pits supply most of the country's needs, has not yet taken hold.

1.4 GENERAL REQUIREMENTS FOR ROCK MATERIALS

These may be divided into the requirements of the rock material itself, and those of the source.

Rock material

Though they vary somewhat, the sought-after qualities of the rock material used for aggregate, ballast, roadbase and armourstone are as follows:

- *Strength*   Adequate strength is required to resist crushing, shearing and flexural failure. Strong rocks also tend to have very low porosity (hence superior durability), high elastic modulus (hence toughness) and high specific gravity. Although high intact strength is generally desirable in aggregates, particularly surfacing aggregates, it may result in 'harsh' roadbase mixtures that require the addition of fines.
- *Durability*   This must be sufficient to resist breakdown during the working life of the road pavement or concrete structure. Durable rocks are generally composed of tightly interlocking grains of stable minerals (i.e. those which are unweathered, unaltered and insensitive to water) and consequently are almost devoid of porosity.
- *Shape and size*   The material must possess satisfactory particle shape and size distribution. These qualities influence the workability of concrete mixes (the ease with which they can be poured, pumped and compacted), their cement demand and durability. In the case of roadbase, poor grading (i.e. an uneven mixture of particle sizes) will reduce the maximum compacted density and hence pavement strength. Unsatisfactory particle shape and grading can increase bitumen requirements for sprayed road seals and asphaltic mixtures, or cause excessive aggregate stripping under traffic.
- *Surface properties*   The material must also possess satisfactory surface properties, which include bitumen and cement adhesion, chemical non-reactivity and polishing resistance. These characteristics are related to particle mineralogy, surface roughness, adhering moisture films and dust coatings. Unfortunately, good adhesion is often best displayed by porous lithologies, while wear resistance has to be sacrificed to polishing resistance.

Rock material sources

The general requirements of rock material sources include the following:

- *Reserves*   There should be adequate proven (i.e. closely drilled, sampled and tested) reserves to last through the working life of the quarry (1–100 years, though 20–30 years is usual), with allowances for wastage.
- *Economic hauling distance*   The source should be located within economic hauling distance of potential users and capable of being worked without undue mining or processing costs. 'Economic' hauling distances are extremely variable, but 40–50 km is common in Australian and British urban areas for quality aggregates.
- *Uniformity*   A source must be capable of producing a uniform product (or, better still, a range of products), which is within specification or can easily be upgraded. This is often as much a function of the processing equipment as of the *in situ* rock.

- *Acceptability* The exploitation of a source must be environmentally acceptable to prevailing community values or (more commonly) capable of meeting environmental constraints imposed by regulatory bodies.

The requirements for quarried rock products will be discussed further in subsequent chapters, but are included here to highlight differences between them. It should be emphasized that, although similar general properties (for example, strength, durability and near-cubic shaped particles) are required from most of the material categories, the specification limits imposed for these can vary with the proposed end-use and are not applied with the same rigour in all cases.

1.5 THE SUPERQUARRY CONCEPT

One response to the parallel trends of rising *per capita* consumption of aggregates and increasing public hostility to new or expanded quarries has been the development of very large quarries capable of producing 5–10 Mtpa of high-quality crushed stone. Though these might be located at relatively isolated inland sites or in remote coastal areas, it is the latter, typified by Glensanda in western Scotland, which have received most attention. Such a pit can replace about 20 medium-sized quarries – neatly increasing production efficiency through economies of scale, while greatly decreasing the overall environmental impact. By using large self-loading bulk carriers, stone can be transported 1000 km by sea at less cost than for 50 km of road haulage, making an international trade in aggregates feasible.

The geological characteristics of such a quarry site would include:

- A source of homogeneous hard rock sufficient for up to 100 years of production, say 1000 Mt of probable reserves and 100 Mt of proven (drilled) reserves. Assuming a nominal quarry depth of 100 m, a deposit of this size would occupy about 4 km². This would most likely be a large unaltered intrusive igneous body capable of being crushed with minimum waste and fines generation.
- A near-shore location adjacent to deep water, preferably screened from view and with a sheltered anchorage capable of handling ocean-going ships in the 50 000 to 100 000 tonne class. Although remote from residences, the site would have to be easily accessible to its small workforce.

The success of such a venture would also require suitable terminal depot facilities (for unloading, temporary stockpiling and possible further processing) at coastal ports close to urban markets. The combination of deep-draught wharves, a sufficiently large waterside storage area and good access to trunk rail and road connections might prove difficult to obtain. It could also incur opposition from nearby residents, not to mention predatory pricing from established local producers! Furthermore, the demand for first-class aggregate (for road sealing, high-strength concrete and railway ballast) is only a small proportion of the total aggregate and roadbase market, most of which would continue to be met by abundant but somewhat inferior local materials. When the superquarry concept was first mooted, it was suggested that up to 40 such pits could be developed in Scotland alone, but only one has eventuated, and more realistic projections suggest perhaps 10 worldwide by the end of the century (though the number of inland 5–10 Mtpa quarries could be much greater).

REFERENCES AND FURTHER READING

The following books are general references covering much of the content of this book. More specific references on individual topics are listed at the end of each chapter.

American Society for Testing Materials (ASTM) (1948) *Symposium on Mineral Aggregates.* ASTM Special Technical Publication No. 83.

Collis, L. and Fox, R.A. (eds) (1985) *Aggregates: Sand, Gravel and Crushed Rock Aggregates for Construction Purposes.* Geological Society of London, Engineering Geology Special Publication No. 1.

Dutton, A.H. (1993) *Handbook on Quarrying,* 5th edn. South Australia Department of Mines and Energy, Handbook No. 1.

Fookes, P.G. (1991) Geomaterials. *Quarterly Journal of Engineering Geology,* **24**, 3–15. [Introduction to a thematic issue dealing with many quarrying topics.]

Institution of Mining and Metallurgy (IMM) (1965) *Opencast Mining, Quarrying and Alluvial Mining.* Institution of Mining and Metallurgy, London.

Institution of Mining and Metallurgy (IMM) (1983) *Surface Mining and Quarrying.* Institution of Mining and Metallurgy, London.

International Association of Engineering Geology (IAEG) (1984) International Symposium on Aggregates, Nice, May. *IAEG Bulletin,* Nos 29 and 30.

Kirk, M. (1993) Coastal quarry or superquarry? A review of strategy for hard-rock coastal quarries. *Quarry Management,* August, pp. 19–27.

Knill, J.L. (ed.) (1978) *Industrial Geology.* Oxford University Press, Oxford.

Lay, M.G. (1990) *Handbook of Road Technology,* vol. I, 2nd edn, Chs 8, 11. Gordon and Breach, New York.

Manning, D.A.C. (1995) *Introduction to Industrial Minerals.* Chapman and Hall, London.

Orchard, D.F. (1976) *Concrete Technology,* vol. 3: *Properties and Testing of Aggregates,* 3rd edn. Applied Science, London.

Prentice, J.E. (1990) *Geology of Construction Materials.* Spon/Chapman and Hall, London.

Sand and Gravel Association of Great Britain (SAGA) (1967) *Pit and Quarry Textbook.* MacDonald, London.

Selby, J. (1984) *Geology and the Adelaide Environment.* South Australia Department of Mines and Energy, Handbook No. 8.

Smith, M.R. and Collis, L. (eds) (1993) *Aggregates: Sand, Gravel and Crushed Rock Aggregates for Construction Purposes*, 2nd edn. Geological Society of London, Special Publication No. 9.

Weinert, H.H. (1981) *The Natural Road Construction Materials of Southern Africa.* Academia, Pretoria and Cape Town.

Wylde, L.J. (ed.) (1978) *Workshop on the Geology of Quarries.* Australian Road Research Board, Report ARR 80.

Wylde, L.J. (1979) *Marginal Quality Aggregates Used in Australia.* Australian Road Research Board, Report ARR 97.

CHAPTER 2

Hard rock materials

The term 'hard rock' construction materials is used here to mean quarried or shot stone, which must be blasted, crushed and screened prior to use, as distinct from alluvial gravels (Chapter 3) or broken rock and duricrust materials (Chapter 4), which are simply ripped and blended. Though more costly to extract than the naturally occurring gravels, quarried hard rock can be processed to deliver a variety of consistent-quality products. These include breakwater stone, ballast, aggregate and roadbase. Inferior materials can also be upgraded by careful blending of screened fractions, using rock either selectively mined from different parts of the quarry or imported from distant pits.

Although the geological and economic aspects of quarry site selection and evaluation are emphasized in this chapter, environmental factors are

Table 2.1 Siting criteria for hard rock quarries

Economic criteria
  Product in present or future demand
  Acceptable production, processing and haulage costs
  Likelihood of local authority/regulatory authority approval
  No excess quarrying capacity available in region
  Close access to trunk roads and/or main-line rail
  Minimal upgrading of existing local roads and bridges required
  Low scenic- and agricultural-value land, and sufficient available for purchase
    or lease
  Few or no residences within 500 m of pit boundary
  No other obvious environmental constraints or infrastructure costs

Geological criteria
  Sufficient reserves for initial 10–20 years' operation
  Quality adequate to meet specification requirements
  Overburden and weathering depths not excessive and extent known
  Possibility of upgrading some overburden/waste to a saleable product
  Jointing closely spaced, but predictably oriented and tight
  Water table below proposed pit floor
  Topographic barriers to sight, sound and dust available

nowadays of equal or greater importance. The first two are summarized in Table 2.1, while environmental requirements are dealt with in Chapter 19. Hard rock quarry sites tend to occupy hilly land of little agricultural value – though not necessarily of low scenic value. Three common types are illustrated in Figure 2.1.

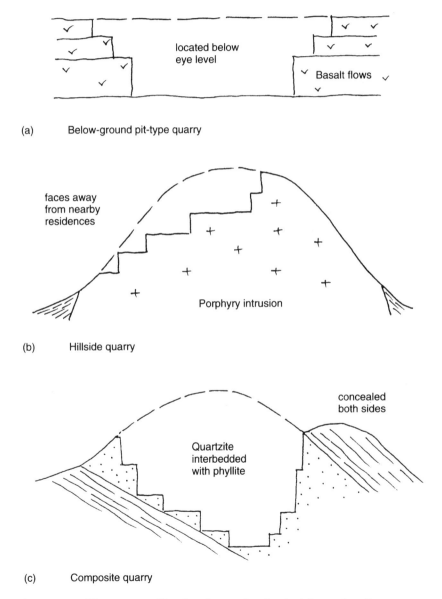

(a)   Below-ground pit-type quarry

(b)   Hillside quarry

(c)   Composite quarry

Figure 2.1   Three types of hard rock quarries. See text for explanation.

- *Below-ground category* This is characteristic of thick basalt flows, or horizontally bedded rocks such as limestone, situated on level plains. The main problems are concealment, particularly in the early stages of development, and groundwater inflows below the water table (which may also cause nearby wells to dry up). Irregular weathering profiles and altered zones between flows also cause quality control difficulties. Furthermore, the unweathered flows themselves may vary in quality; some are glassy and some contain more deleterious secondary minerals than others.
- *Hillside type* This is cheap to develop, weathering depths are minimal and rock quality – in boldly outcropping and homogeneous intrusives – is generally very good. Blasting costs may be minimized by 'snakeholing' or shooting inclined holes at the base of the slope. However, this sort of quarry is most conspicuous and would rarely be approved for development these days unless the faces could be concealed behind ridges, or at least oriented away from the public gaze.
- *Composite type* This represents a quarry where extraction is confined to a particular layer; for instance, a quartzite bed, a major sill, a limestone layer, or a very wide dyke. In this situation the quarry shape is dictated by the geology to a greater degree than the previous two categories. Extraction has to be carried out along-strike, and because of the narrow pit width available careful slope design is required to ensure maximum product yield. Rock quality in layered deposits is usually variable – more so vertically than horizontally – hence different products are mined on different benches, and blending may be desirable to make best use of the deposit. With bed strike lengths of many kilometres and generally rugged surrounding topography, opportunities for concealment are good. One visual screening technique that is widely used in both composite and hillside quarries is to work downwards from the top, leaving a facade of undisturbed rock around the pit and only an inconspicuous slot for road access to the workings inside the hill.

Relative to their output of crushed stone, hard rock quarries are smaller consumers of land than sand and gravel pits, but generate the additional nuisance of blast vibration. They also tend to be longer-lived, over 100 years in some cases, and require more capital. Among the economic site selection factors listed in Table 2.1, access to trunk roads and remoteness from residents are tending to become more important, while distance to markets is becoming less so. The likelihood of consent from planning and environmental authorities also looms large in the site selection process; sites likely to be the cause of administrative delays or litigation will be discarded wherever possible.

2.1   LOCATION OF QUARRY SITES

The best guide to prospective hard rock quarry sites, as indeed to all extractive mineral sites, is the nearby presence of existing or abandoned workings in similar geological formations. Old pit faces can provide a good idea of the rock types present, their structure and weathering patterns, and bulk sampling locations. Even where old workings are no longer accessible – where, for example, they have been backfilled or built over – local geological knowledge based on their records may be helpful.

Background geological data

The background geological data on a particular area may be published or unpublished; the more general investigations tend to be published, but more specific information usually remains in geological survey open-file reports. Construction-materials resources on the outskirts of major cities have been the subject of planning studies by public authorities worldwide over the past decades. These investigations have included geological mapping, scout drilling at selected sites and limited suites of testing. Their aims were to identify and classify the principal deposits in a region, and to preserve these as long-term extractive resources by restrictive zoning before they could be sterilized by urban expansion.

The primary source in a new area will usually be the regional geological map. In Australia these are mostly at 1:250 000 scale, with selected areas mapped at 1:100 000; in the UK resource maps are at 1:50 000 or larger scales. The larger-scale maps will not necessarily be in published form – many are available only as dyeline plans from mapping authorities. In many undeveloped areas good-quality regional geological maps are a legacy from colonial administrations keen to develop mineral resources, or from previous aid projects.

Because most of these maps were not compiled with construction-materials exploration as their principal aim, they have to be carefully interpreted. Generally they show bedrock geology well but ignore superficial deposits unless these are thicker than (say) 3–5 m. The legend and accompanying notes (if any) can be used to identify potentially quarryable formations, but very often hard and soft units are lumped together within these formations on the basis of their stratigraphic continuity rather than their economic potential. Formations with definite quarrying potential include limestone and quartzite beds, granite stocks, lava flows and intrusions wider than (say) 50 m. Maps will probably be much less useful for pinpointing thin dykes, minor intrusions and hard quartzite beds or lava flows in otherwise non-prospective sedimentary rock formations. Even where these features are shown, they may be much more numerous than the map indicates, or areally more extensive. At the very least the map will suggest large areas where further investigation is not warranted and a few sites worth a closer look.

Topographic maps

Although the regional geological map may suggest prospective hard rock formations, it will not give any indication of their state of weathering (Figure 2.2) or the extent of individual deposits. These will have to be resolved by inspection, initially by driving around the area and subsequently on foot at selected sites. At this stage medium-scale topographic maps (say 1:25 000 or 1:50 000) may be useful for supplementing the geological map. Orthophotomaps – cheaply prepared topographic overlays on enlarged and tilt-corrected airphotos – are likely to be even more help. These large-scale (1:2500 to 1:10 000) maps, in conjunction with airphotos, are more likely to show abandoned quarries, excavations, scrapes and large outcrops – all signposts to prospective sites. Airphoto interpretation is also helpful for locating access roads, drilling water points, fence lines and other vehicular barriers and (not least important!) nearby residences that could be affected by the quarry. Airphotos are usually flown every few years, so their information is more up-to-date than maps.

2.2   QUARRYING LITHOLOGIES

Almost any unweathered and unaltered igneous or high-grade metamorphic rock, plus some indurated sedimentary rocks, may be suitable as a source of crushed stone. Lithology is only a rough guide to suitability; dolomitic shale, for example, would be considered an unlikely aggregate source in Canada (where some varieties are reactive in concrete) but it supplies the best-quality stone in Adelaide. Worldwide, the most common source of aggregate is undoubtedly carbonate rock, which is discussed in Chapter 15. In the USA, the largest single market for aggregates, limestone and dolomite make up about two-thirds of the quarried stone.

Basic lavas

Basic lavas are probably second only to limestone as a hard rock source, owing to their wide distribution and volume (sometimes hundreds of metres thick and covering thousands of square kilometres). They can be variable in quality due to weathering along joints and flow surfaces, because of a vesiculate or brecciated fabric, and as a consequence of high olivine or secondary mineral content. The 'green' basalts of Melbourne

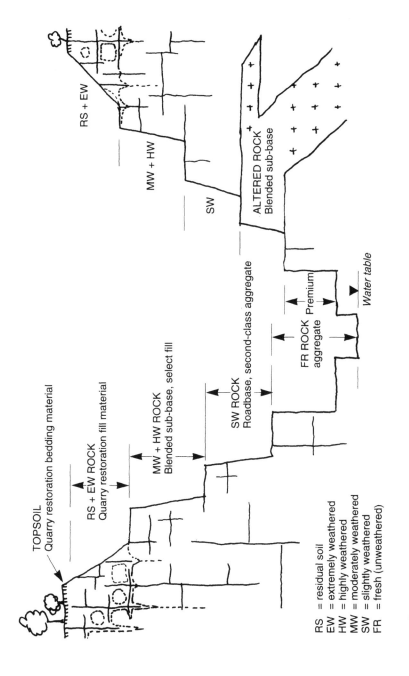

Figure 2.2 Cross-section through a quarry in weathered and altered hard rock, showing possible uses for the different material classes present. Note that, whereas weathering intensity decreases with depth, the alteration state may remain constant in affected rock.

Figure 2.3   Columnar structure of basalt lava distorted by flow down hill, Peats Ridge, New South Wales. Each column is about 0.2 m wide, making the rockmass easy to fragment using heave-energy explosive. Note faint bleaching due to alteration at the top of the flow.

are good examples of the latter problem, where a significant percentage of chloritic and smectitic alteration products cause rapid deterioration on exposure. Similar non-durable basalts are common in the northwest of the USA.

Thick non-glassy basalt flows are preferred for crushing, although scoria may be useful as a low-grade sub-base. Two significant quarrying benefits found in basalt flows are their non-abrasive mineralogy and the presence of closely spaced columnar joints (Figure 2.3), which reduce crushing and blasting costs respectively.

Basic intrusives

Basic intrusives such as dolerite are less abundant, but are often superior in quality due to slower cooling and less severe weathering effects. However, deuteric alteration, magmatic differentiation and multiple intrusions can result in several different lithologies of varying quality within a large quarry. Gravity settling of olivine-rich phases can give rise to smectitic alteration products, rendering much of the rockmass useless for concrete aggregate and suitable only as lower-quality roadbase. Thick sills and dykes, say 20 m or more in width, can provide good quarry sites where their strike length is sufficient. Small gabbro bodies may be better still, since they are less subject to altered or weathered margins and mineral layering. Dolerites are the most widely quarried rocks in South Africa, where their properties have been described by Weinert (1968).

Granites

Acid igneous rocks are generally less suitable for quarrying than basic lithologies. Granites and related plutonic rocks tend to be more pervasively weathered due to kaolinization within feldspar grains, and to be very abrasive because of their quartz content. Very small increases in porosity, say from 0.5 to 1% and presumably due to microfracturing, can result in large decreases in elastic modulus and other key quality indicators such as aggregate crushing value (ACV) and Los Angeles abrasion (LAA) loss. This is often due to hydrothermal alteration rather than weathering, and the distinction can have important practical consequences: weathered rock quality improves with depth, but altered rock remains inferior. The wide spacing between joints, compared with columnar-jointed basalt, greatly increases the required powder factor and hence blasting costs. Mica and clay liberated during crushing may coat particle surfaces and resist bitumen adhesion. Nonetheless, granites have been quarried successfully in Hong Kong, where they are the main sources of aggregate (Irfan, 1994).

Porphyries and microgranites

Medium-grained porphyries and microgranites (Figure 2.4) are more satisfactory as aggregate sources. This is possibly due to the presence of late-crystallizing potassic feldspars, which are more stable than early-stage calcic plagioclases, but may also be related to tighter grain-to-grain bonding than is the case with coarser-grained granite and granodiorite. Fine-grained acid extrusives (rhyolite, dacite) tend to vary in strength and durability, and are sometimes flaky due to flow banding or reactive because of the presence of glass. Good polishing resistance is a bonus in some cases. Intermediate igneous rocks (diorite, syenite, trachyte and andesite) are similar to acid rocks in their aggregate-making properties, but are less common.

Metamorphic rocks

Of the metamorphic rocks, the most attractive quarrying propositions are high-grade gneisses, greenstones (metabasalts) and some hornfelses. Gneiss is widely distributed in high-grade metamorphic terrains, generally Precambrian in age, and is favoured because of its relatively stable mineralogy and subdued foliation. Its geomechanical properties are similar to those of granite. Greenstones are similar, possibly less abrasive

Figure 2.4　Jointed porphyry, Canberra, ACT, Australia. This face is only about 6 m high, but the 40° forward-dipping joints would present a hazard in a higher bench; the joint set dipping at 80° to the right would be a more satisfactory face orientation.

but denser. Some varieties contain a proportion of fibrous minerals, the most objectionable of which is asbestos. Hornfels is quarried simply because of its hardness, which may be such that the crushed aggregate is too harsh for roadbase without the addition of fines.

Quartzite

Quartzite is widely exploited for roadbase, aggregate and breakwater stone, despite its variable properties. The degree of cementation, reflected in its particle specific gravity, can be quite inhomogeneous. Some 'quartzites' are really only moderately indurated sandstones. Other varieties are thinly interbedded with phyllite and metasiltstone, or are actually silicified greywackes and arkoses. Patchy silicification may be due to groundwater movement rather than metamorphism, producing silcretes rather than true quartzites. Even massive and well-cemented quartzites can be difficult to blast without generating oversize or excessive dust (which is a health hazard), and will be the cause of abrasive wear in crushers. Finally, quartzites tend to be either too brittle or too soft for surfacing aggregate, though some gritstones are very skid-resistant.

Other metamorphic rocks tend to be too foliated, too micaceous or too weathered to be acceptable, though even slate has been crushed for roadbase where no better rock was available. Apart from the obvious problem of platy particles, these rocks often contain small quantities of sulphide minerals, which oxidize to sulphuric acid (proportions as low as 0.5% can be troublesome).

Indurated sedimentary rocks

Some indurated sedimentary rocks are quarried for roadbase and aggregate. By far the most important of these is limestone, which is discussed in Chapter 15, and others include orthoquartzite and silicified greywackes, shales and mudstones (argillites). None is entirely satisfactory as high-quality aggregate, due to variable cementation (and hence variable strength and durability), a tendency to polish, and the risk of alkali–silica reaction (ASR) in concrete. They perform better as roadbase, because more fines are produced in crushing and their lack of durability is less critical in this application.

2.3 INVESTIGATION TECHNIQUES

The principal means of exploration in hard rock quarry sites is cored drilling, mostly supplemented by rotary percussion (tophammer or downhole hammer) drilling. Compared to foundation investigation and mineral exploration, the amount of drilling performed at quarry sites is remarkably small – only about one cored drillhole per million tonnes of resource in Australia. Some augering and backhoe pitting is generally carried out, mainly to probe the depth of weathered overburden and to fill in detail between outcrops. Geophysical techniques, chiefly seismic refraction and magnetometry, are occasionally used.

Core drilling

Core drilling is usually performed by truck-mounted wireline rigs using triple tube barrels and water flush. Skid-mounted drills are used on steep slopes and for angled holes. The largest possible core diameter is desirable, both to ensure maximum recovery and to supply the largest possible volume of rock sample per metre drilled. In Australian practice this means HQ core (62 mm diameter), but in the UK core diameters up to 110 mm (P and S sizes) may be possible in weaker and non-abrasive rock. Vertical holes are most common, except in steeply dipping beds or where joint spacing and orientation information is required. Ideally, quarry sites should be drilled on a systematic grid, at about 200 m centres. Once the overall geological structure of the deposit is established by coring, fill-in holes can be drilled by cheaper open hole methods. In the latter case the holes are chip-logged and sometimes geophysically logged (using density, sonic and gamma tools) as well. Drilling monitors are available to record penetration rate, machine torque and thrust and hence rock hardness, but do not appear to be widely used.

Core logging

Core logging is based largely on visual impressions and semi-quantitative test data. A checklist for quarry drillhole logging is presented in Table 2.2, showing the features to be recorded. The main

Table 2.2 Checklist for quarry drillhole logging

Cored drillholes

Intact rock properties
(by hand specimen and hand lens examination)
 Lithology, visible mineralogy (especially quartz percentage)
 Grain size, rock fabric
 State of weathering, alteration of mineral grains

Rockmass properties
 Joints, shears, fractures, seams visible in core
 Weathering/alteration along joints and layering
 Bedding, foliation, layering (dip, spacing)
 Groundwater inflows and drilling water loss zones (plus water salinity and
  static level fluctuations)

Drillcore properties
 Core recovery, location and causes of losses
 Fracture spacing indices (RQD[a], fractures/metre)
 Core state (solid sticks, fractured, fragmented)
 Location and results of strength index tests
 Location and sample numbers of core sticks submitted for laboratory testing

Geomechanical test results
 Uniaxial compressive strength (UCS)
 Indirect tensile strength (ITS)
 Point load strength index (PLSI)
 Schmidt hardness
 Sonic velocity (V_p)
 Bulk density
 Water absorption

Uncored drillholes

(samples at approximately 1 m intervals)
 Chip lithology, weathering state
 Maximum chip size, chip/dust ratio
 Dust colour and whether dry, damp or wet
 Rate of drill penetration, location of hard and soft bands
 Groundwater inflows and standing water level after drilling
 Geophysical logs (if run)

[a] RQD = rock quality designation.

purpose is to develop a three-dimensional geological model of the proposed quarry site, with emphasis on the classes of rock material present and the distribution of weathering and alteration within the rockmass. Core samples will be taken for petrographic examination and for limited geomechanical testing, but the quantity of drillcore available will usually be insufficient for routine aggregate testing. Sometimes additional holes are drilled alongside the first if extra rock is needed for this purpose. In any case the laboratory-crushed samples may bear little relationship to the shape and grading of the eventual quarry product, though they should be a good indicator of its durability.

Geological mapping

Geological mapping, typically at 1:500 or 1:1000, may be used to supplement the drilling data, although it must be remembered that surface outcrops are likely to be both more weathered and more closely jointed than the rockmass at depth. On the other hand, the visible rock exposures may represent the only really strong and unweathered material present within the deposit! This is a common trap with interbedded limestone/shale formations and granites. Similarly, some hill-capping basalt remnants may appear to be much larger in volume than in reality because of downslope movement of their copious scree.

At the reconnaissance stage this mapping may be done using airphoto and topographic map enlargements as a base, but by the time detailed drilling is under way contoured photogrammetric plans should be available. With these, plus supplementary compass-and-tape, stadia and electronic distance-measuring (EDM) surveys, the geologist can prepare a 'mud map' of the site. At a later stage a preliminary quarry plan showing the bench and face layout, haul roads, processing plant and stockpile locations will be drawn up. Interactive PC-based mine planning software packages allow a variety of layouts to be tried out. This computer modelling should identify possible mining and environmental problems, and help to plan further investigations to solve these. The main features that might be recorded on site investigation maps are listed in Table 2.3.

Trial blasting and crushing

Occasionally, trial blasting and trial crushing will be carried out as a final investigation stage. This would usually only be done – because of expense – when there is some doubt about marginal-quality material, to investigate blast vibration magnitudes, or where the quarry product has to meet a particularly strict specification. Normally production blasting patterns and loading, as well as crusher settings, are adjusted on a trial-

Table 2.3 Checklist for quarry site mapping

Map scales
   1:500 to 1:1000

Base maps
   Enlarged airphotos, photogrammetric plans (1 m contours), plane table or
      EDM[a] plans, compass-and-tape or compass-and-pace sketch maps
   Enlarged standard topographic or cadastral (property) maps and plans
   Low-altitude airphoto contact prints, for stereographic examination

Topographic detail
   Stream lines, permanent or non-perennial, and groundwater seepages
   Ridge lines (especially screening ridges) and watersheds
   Cliff lines, escarpments, natural benches and other geomorphic features
   Spot heights, benchmarks and survey control points

Rock exposures
   Continuous outcrops, scattered outcrops, bedrock beneath thin cover, floaters,
      scree
   Bedrock depths, location of probing holes, pits, trenches and costeans
      (bulldozer trenches)

Geology
   Lithology and layering/bedding/foliation
   Major structures (faults, dykes, etc.)
   Joint orientation, spacing, continuity, dip
   Weathering and alteration grades and distribution

Sampling data
   Drillhole locations and drilling method
   Surface and channel sample locations
   Pit and trench sampling locations

Other data
   Land ownership and boundaries, extractive lease boundaries
   Locations of existing dwellings and other nearby buildings
   Locations of existing/abandoned pits, faces, excavations, scrapes
   Roads (widths, surfacing, capacity), tracks, fences, gates, waterpoints
   Man-made features, present land-use, zoning classifications

---

[a] EDM = electronic distance measuring.

and-error basis to suit the rock coming out of the face and to satisfy
environmental regulations.

2.4   INVESTIGATION AIMS

Few quarries close because they run out of rock, but many have
exhausted their prime material or been forced to close prematurely
because haphazard working has depleted this. Therefore the main
objective of drilling and mapping at quarry sites is to determine the

quantity and quality of *all* the rock materials present, with a view to planning systematic extraction and correctly matching product quality to end-use. In other words, fresh rock should be used as first-class aggregate and not as rubble, and lesser rock grades designated as upper course roadbase, sub-base and select fill as their quality decreases (see Figure 2.2). This not only makes economic sense, it is becoming a condition of consent by environmental regulatory authorities.

Previous to drilling, the site would have been regarded as a hard rock *resource*, in other words an area believed on good geological evidence (such as outcrop patterns, presence of nearby quarries in the same rock formation, regional mapping and so on) to be underlain by workable rock in economical quantities. It is now necessary to quantify the resource – by drilling and detailed mapping – into *reserves*, of which there are three main categories:

- *Proved or measured reserves* are those which are either so closely drilled (at, say, 100 m centres) or so well exposed that there is no doubt about their continuity in three dimensions. At a new site reserves sufficient for 10–20 years of production will usually be proved up by drilling, although much larger indicated reserves are generally present (in some cases more than 100 years' worth).
- *Indicated or probable reserves* are semi-proven volumes of rock whose grade and tonnage are computed from widely and/or irregularly spaced drillholes. Although these cannot be regarded with the same confidence as proved reserves, it is expected that they will be upgraded once additional drilling is carried out.
- *Inferred or prospective reserves,* or simply 'resources', are those presumed to exist because of favourable geology or proximity to nearby operating quarry sites, but in which little or no drilling has yet been performed.

Although the determination of rock quality and quantity will always be the principal aim of quarry exploration programmes, much additional information for pit design and for planning can be obtained at very little extra cost during site investigation. Joint orientation and spacing measured during outcrop mapping, for example, is useful for blast design and face slope stability studies (see Figure 2.4). Groundwater depth and quality measured in exploratory drillholes, and rockmass permeability estimated from falling or rising head tests in the same drillholes, can influence the choice of explosive, pump capacity required in deep pits and the likelihood of future aquifer pollution.

Much of this sort of information will be required in any case as input for the *Environmental Impact Statement* (EIS), which must be submitted before developmental approval for the quarry can be obtained from the regulatory authorities (see also Chapter 19). In the case of a large quarry development proposal, a Public Enquiry may be convened, at which the quarrying company will be expected to answer a wide variety of objections. To do this successfully, a large amount of geological and geotechnical information may be required, much more than would be necessary simply to prove up the quarry's 20 year reserves. In fact, the requirements for EIS preparation have greatly increased the need for geological and engineering investigations prior to quarry opening, since a proposed operating and restoration plan is now obligatory and the old 'suck-it-and-see' approach to pit development is no longer acceptable.

2.5   CASE HISTORY: PROPOSED MOUNT FLORA QUARRY NEW SOUTH WALES

The Mount Flora intrusion is a 2.5 km wide microsyenite and gabbro body intruding Triassic sandstone and shale of the Sydney Basin about 7 km NNE of Mittagong, New South Wales. The intrusion covers 770 ha and is one of about a dozen in the area, not all of which break the surface. It is a complex igneous body, the greater part of which consists of a microsyenite laccolith about 200 m thick. This has been intruded by gabbro plugs and faulted, and is overlain in places by extrusive basalt. At this stage only the microsyenite is proposed for quarrying, and its importance lies in the fact that it is the closest unexploited large source (more than 1000 Mt inferred reserves) of aggregate-quality hard rock for the developing southwestern fringe of Sydney. It is also located a few kilometres from a freeway and a main-line railway, and in a sparsely populated area of no particular scenic significance (though close to a proposed national park). These factors have caused it to be regarded as a potential quarry site for at least 20 years, and three phases of investigation had been completed prior to 1990, when development was finally approved. These investigations comprised the following:

- *Diamond drillholes* – 36 diamond drillholes totalling 1670 m of coring across the whole of the intrusion, to evaluate geological structure and petrological variation, and to obtain test specimens.
- *Seismic refraction* – 18.2 line kilometres of seismic refraction to assess depth of weathering and rock quality in the upper portions of the igneous body.
- *Percussion drilling* – 900 m of percussion drilling to supplement coring within the selected initial quarry area of 40 ha.
- *Geological mapping* and airphoto interpretation of outcropping microsyenite.
- *Trial blasts* – two trial blasts, from which 300 t and 120 t of spalls were recovered for full-scale crushing tests.

The drilling outlined 32 Mt of probable reserves within the proposed quarry area, itself divided into two pits by a faulted and weathered zone. Much larger inferred reserves, totalling about 1200 Mt, are available for future development if required. The most unusual feature of the investigation was the extensive use of seismic refraction. Although this technique only penetrates to a depth of about 10 m, it was considered cost-effective for rapidly assessing what is by quarrying standards a very large site, and for estimating depth of weathered overburden and the likely quality of the near-surface rock.

The main problem that the investigations revealed was the extent of deuteric alteration in the upper layers of the intrusion. This has caused the naturally grey microsyenite to turn first pink and then white because of the prevalence of secondary minerals, and also increased the water absorption of the crushed rock up to around 5% in places. The economic result of this is that the upper benches of the proposed quarry will be used mainly for roadbase and concrete aggregate, while better-quality sealing aggregate will be obtained at depth as the proportion of secondary minerals diminishes.

2.6 CASE HISTORY: PROSPECT QUARRY, NEW SOUTH WALES

Prospect Hill is a prominent topographic feature rising about 90 m above the surrounding plain at the centre of the Sydney Basin, about 7 km west of Parramatta. The hill owes its form to the presence of a dish-shaped layered basic intrusion about 2 km long by 1 km wide and covering 280 ha. Its average thickness is about 100 m, and quarry development has been carried out down to about half this depth. The intrusion forced its way between flat-lying shales and sandstone, both uplifting and thermally metamorphosing these. A geological section through the deposit is shown in Figure 2.5.

The Prospect laccolith is the largest body of igneous rock in the Sydney metropolitan area and the site of the largest hard rock quarry in the region. This location has become even more strategic to Sydney's construction-materials needs over the past 20 years as the suburbs have expanded westwards and surrounded it. Indeed, though quarrying has been carried out at Prospect Hill since the 1870s, more than half the present pit – actually two quarries operated by rival companies – has been excavated since 1970. Production is now in decline due to depletion of the best materials, and Prospect quarry is due to be superseded by Mount Flora and other new hard rock sources over the next decade. In fact, this decline was one of the main arguments put forward in the Mount Flora EIS and at the environmental enquiry in order to justify approval of that project.

Nonetheless, Prospect quarry is an interesting example of the effect of geology and location on the development of a hard rock site in a region

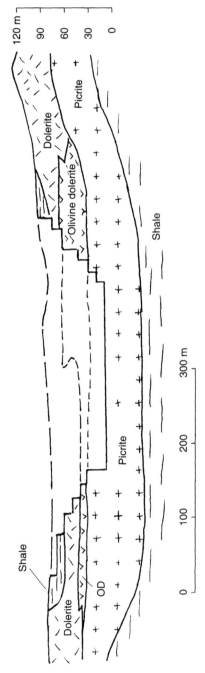

Figure 2.5 Geological section through Prospect quarry, west of Sydney, New South Wales.

where these are few and widely spaced. Though large, about $3\,km^3$, much of the intrusion is very poor-quality rock by aggregate standards, and almost all of the superior upper material has been extracted. Despite these limitations, the site is centrally located in a growing city of three million people, adjacent to a freeway, and has the most extensive quarrying facilities in the state. With selective mining and careful blending, a variety of roadbase mixes, sand and select fill are produced. Furthermore the alternative near-city sources, mainly volcanic breccia pipes, are smaller and also considered to be of only marginal aggregate quality.

The geological section (Figure 2.5) shows the igneous layering present in simplified form. The upper 30% of the laccolith is dolerite, grading downwards through olivine dolerite to picrite, which makes up more than half the volume of the intrusion. A thin chilled margin of basalt plus small amounts of late-stage differentiates (pegmatite, aplite and syenite) are also present. The main effect of *magmatic differentiation* within this intrusion has been to concentrate unstable olivine at depth, where it constitutes 25–40% of the picrite and causes that rock to disintegrate rapidly on exposure (Figure 2.6). The laccolith has also been subjected to extensive *deuteric alteration*, which has caused all rock types present to

Figure 2.6 Typical bench face at Prospect quarry. The dark rock is olivine-rich picrite, which disintegrates within weeks of exposure, hence the smooth, rounded exposure. The light-coloured rock is better-quality syenitic 'pegmatite'; note the sharper joint intersections in this more durable material.

contain between 20 and 70% of hydrous secondary minerals (mainly prehnite, zeolites, analcite and calcite).

In practical terms, this means that much of the Prospect rock is potentially non-durable and the quality becomes worse with depth. Only small quantities of first-class aggregate are now produced from the remnants of the chilled margin basalt and the less olivine-rich dolerite. With hindsight, the quarry might have been developed in a different fashion, the better-quality rock reserved for use as aggregate and the middlings for roadbase. This would have required extensive drilling and quarry planning, which is scarcely an option in a century-old quarry that has passed through the hands of many operators.

REFERENCES AND FURTHER READING

Beaumont, T.E. (1979) Remote sensing for location and mapping of engineering construction materials in developing countries. *Quarterly Journal of Engineering Geology*, **12**, 147–58.

Fookes, P.G. (1980) An introduction to the influence of natural aggregates on the performance and durability of concrete. *Quarterly Journal of Engineering Geology*, **13**, 207–29.

Haines, A. (1993) Mining geotechnical investigation in quarry design practice. *Quarry Management*, April, pp. 15–19.

Hawkins, A.B. (1986) *Site Investigation Practice*. Engineering Geology Special Publication No. 2, Geological Society, London.

Hoek, E. and Bray, J.W. (1977) *Rock Slope Engineering*, 2nd edn. Institution of Mining and Metallurgy, London.

Irfan, T.Y. (1994) Aggregate properties and resources of granitic rock for use in concrete in Hong Kong. *Quarterly Journal of Engineering Geology*, **27**, 25–38.

Oberholsen, R.E. (1982) Aggregates in South Africa: production, properties and utilization. *Quarry Management and Products*, **9** (April), 253–6.

Van Schalkwyk, A. (1981) Geology and selection of quarry sites. *Quarry Management and Products*, **8** (June), 414–18.

Weinert, H.H. (1968) Engineering petrology for roads in South Africa. *Engineering Geology*, **2**, 363–95.

Wylde, L.J. (ed.) (1978) *Workshop on the Geology of Quarries*. Australian Road Research Board Report ARR 80.

CHAPTER 3

Sand and gravel

Sand and gravel deposits are normally considered together as sources of naturally occurring fine and coarse aggregate found mainly in alluvial environments. In former times they constituted the largest mineral commodity extracted in developed countries, but are now being overtaken by crushed rock production. Their popularity stems from being close to users, cheap to extract and easy to process (by minimal crushing, washing and screening). They are well suited to small-scale quarrying operations, providing concrete aggregates and sealing aggregates to local markets. Their present decline in many areas is partly because the best deposits have been worked out or built over, and partly a response to environmental restrictions, which tend to bear more heavily upon sand and gravel operations than on hard rock quarries.

The soundness and stable mineralogy of sand and gravel deposits result from their having been subjected to comminution and sorting by natural – mainly fluvial – processes of erosion and transportation. These selectively eliminate weak or porous lithologies and concentrate different size fractions in distinct geological locations. Alluvial sand, for example, is usually composed of more than 90% quartz grains. Gravel is more heterogeneous, but is still largely made up of siliceous metamorphic and igneous rocks. Many alluvial gravels are in fact reworked older conglomerates or terrace deposits, in which the clasts have been subjected to at least two cycles of erosion, transport and deposition. Consequently, only the hardest and most durable stone survives.

The geological origins and characteristics of some of the main sources of sand and gravel are described in case-history form later in this chapter, but can be summarized as follows:

- *Alluvial sources* include fans, terraces, channel fills (Figure 3.1) and point bars. The best deposits are usually gravel-rich remnants from pluvial stages within the Pleistocene, reworked Pleistocene glacial materials, or alluvial fans close to active faults along mountain fronts.
- *Glacial and fluvioglacial sources* include outwash sheets, eskers, kames and some moraines. The best deposits are those laid down by meltwater streams beneath the glacier or close to the ice front (Figure

Figure 3.1   Pleistocene channel fill ('prior stream') sand deposit, northern Victoria.

3.2). However, glacial drift in general, particularly the blanket moraine left by retreating ice, is too heterogeneous and clayey to be useful.

- *Aeolian (dune) sands* are composed of uniformly sized, clean and well-rounded grains that improve the workability of concrete mixes. They are often used in gap-graded mixes with crushed coarse aggregate for this purpose, or blended with crusher grit to achieve a more even fine aggregate grading.

- *Strandline deposits* of beach sand and shingle (gravel) mark higher sea and lake levels in tectonically active or formerly wetter areas. They tend to be smaller in volume and coarser in grain size than dunes, but otherwise well sorted like them. Salt and shell concentrations are sometimes a problem.

- *Marine sources* are largely drowned Pleistocene alluvial, fluvioglacial and littoral deposits (Figure 3.3) that have been variably reworked by currents. Sand and fine gravel predominate, with a considerable proportion of sand-sized shell fragments and carbonate grains. Silt and clay sizes are conspicuously deficient.

- *Residual deposits* are generally unsuitable as sources of granular material because of their unsound particles and a large excess of clay fines. Nevertheless, duricrusts (ferricrete and silcrete nodules) provide a proportion of the gravel-sized alluvium in many areas of the dry tropics. Leached sandstones and conglomerates can also be large-volume sources of relatively dirty sand and gravel.

Figure 3.2 Fluvioglacial sand and gravel deposit, southern Norway. (Photo: J.H. Whitehead.)

Figure 3.3 Bucketwheel dredger in littoral sands, Tomago, New South Wales. Note post-like 'spud' anchors and floating slurry pipeline at rear of barge.

3.1   ENVIRONMENTAL CONSTRAINTS

The environmental problems associated with sand and gravel mining arise from a combination of land-use conflicts and changing community values. Many of the best gravel deposits are located beneath prime floodplain agricultural land, or are within level and well-drained alluvial terraces, which provide flood-free residential land and transport corridors.

Off-stream extraction

Off-stream extraction of alluvial sand and gravel from abandoned channels, point bars and terraces is a heavy consumer of this land. A typical 5 m thick gravel lens produces about 8 tonnes of aggregate per square metre, while a 30 m high hard rock quarry face might yield 80 tonnes per square metre. A large off-stream operation might consist of several pits and ponds spread over many square kilometres; the Penrith Lakes scheme, west of Sydney (Chapter 21), covers 19 km², though only about a quarter of this area is being actively mined at any one time. Although these excavations can be screened by trees at ground level, they may be startlingly obvious – as white scars against a green pattern of fields – from elevated viewpoints.

On-stream extraction

The older alternative of on-stream extraction is nonetheless even more in disfavour, for two reasons:

- Dredging and on-stream washing cause *muddy plumes* to migrate downstream. This is not only visually offensive, it blankets aquatic habitats with loose silt, which furthermore is easily remobilized during subsequent floods.
- Gravel removal leaves hollows in the streambed, which the river's hydraulic processes seek to fill by *erosion upstream*, so as to maintain an even bed profile. This can cause scouring of bridge piers and abutments, undermining of weir and causeway foundations, and oversteepening of banks.

Against these environmental liabilities, which are illustrated in Figure 3.4, on-stream workings can – with care – be made less conspicuous than land pits. They can even be designed to improve river efficiency, by deepening and straightening channels. However, the removal of gravel shoals to increase stream velocity also means that oxygenation is less effective, a particular concern where the normal low-stage river flow includes a substantial proportion of sewerage effluent.

Conflict with local residents may arise where gravel mining extends

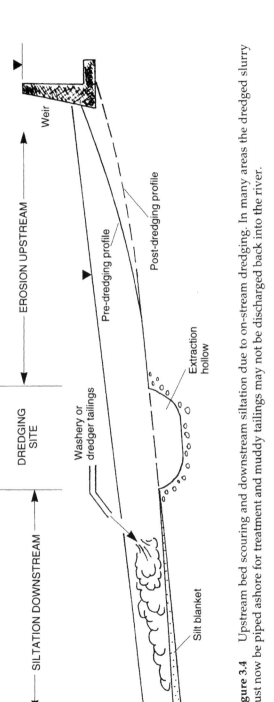

Figure 3.4 Upstream bed scouring and downstream siltation due to on-stream dredging. In many areas the dredged slurry must now be piped ashore for treatment and muddy tailings may not be discharged back into the river.

below the water table and pit dewatering is required, since existing users can find their wells drying up or their pumps clogging through sand entry. Aquifer recharge can also be inhibited by slime blankets on pond floors and walls. On the positive side, off-stream workings are by far the easiest of all quarrying operations to rehabilitate, generally as recreational and water-sports areas (Figure 3.5).

3.2   INVESTIGATION AND SAMPLING

Landforms, maps and logs

Sand and gravel deposits are usually found in distinctive *landforms* such as dunes or alluvial terraces, and consequently a knowledge of Quaternary geology and geomorphic processes is useful in prospecting for them. Some of these landforms can be interpreted from large-scale orthophotomaps with contour intervals of 1–2 m, or from airphotos; very large deposits may be visible on satellite imagery. On the other hand, many of the best sand and gravel deposits owe their formation to higher-discharge bedload streams that were active during Pleistocene time, but which are now obscured by several metres of younger silts of no engineering use (but considerable agricultural value).

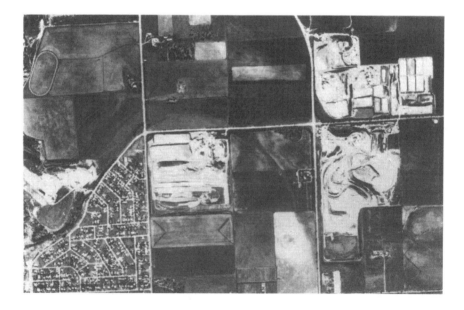

Figure 3.5   Aerial view (about 2 km across) of sand mining south of Adelaide. Note urban encroachment on left and rectangular tailings dams, which occupy half the area of some pits. The sand pit on the far left has since been rehabilitated.

Another source of information is *surficial geology maps* – where these exist. 'Drift' or Quaternary geology maps showing glacial, fluvioglacial, alluvial and other unconsolidated deposits cover much of the UK and parts of North America. Even where these surficial deposits are accurately mapped, however, they may not indicate important channel sands beneath younger floodplain silts. In poorly mapped areas, meaning most of the world, *logs* from water bores and bridge drillholes can be valuable sources of information, since the best alluvial aquifers are also sources of granular materials. Although these logs may not be geologically precise, most drillers can recognize and record coarse gravels, cobbles and clean 'running' sand. Furthermore, cable tool drilling is equally applicable to water well drilling and to gravel investigations.

Problems in investigations

The investigation of sand and gravel deposits presents a number of problems. In the first place, alluvial sediments are extremely variable, both vertically and horizontally. Representative samples need to be large, up to 100 kg or more where the particles are very coarse, and sampling locations need to be close together (25–50 m in many cases). Great care needs to be taken to ensure that both the oversize and fines fractions – the 'top and tail' of the grading curve in other words – are included. Secondly, such materials are difficult to penetrate when coarse, densely packed and even cemented. Excavation walls are prone to caving, requiring casing in boreholes and bracing in test pits. Finally, most deposits occur close to or below the water table. This magnifies the problem of excavation support and impedes sample recovery.

Trial excavations

Trial excavations by means of backhoe pits, hydraulic excavator trenches and bulldozer costeans can be very satisfactory for investigating shallow deposits, since they allow both for bulk sampling and for assessing the variability of the deposit. The depth limits for these excavators are approximately 3 m, 6 m and 10 m respectively (less in cobbles), with the cost increasing with depth. Even the coarsest gravels can be adequately sampled above the water table, and some recovery below it is also possible. Costeans can be deepened by hydraulic excavator pits in their floors, a 'trench-in-trench' layout. The main problem with such excavations, apart from their depth limitations, is the surface disturbance created. This is directly proportional to their size and a major cause of compensation claims from land-owners.

In particularly inaccessible sites, or where backhoes cannot penetrate, hand-dug shafts may still be useful. Mechanical winches, grabs and small-scale blasting will greatly assist progress. Very precise sampling of gravel and removal of large bulk samples for on-site screening are possible by this method. The drawbacks are high labour costs, the need to erect timber shoring for safety and the near-impossibility of excavating below the water table.

Sampling procedures

Sampling procedures for both trial excavations and boreholes have to take account of the likelihood that oversize blocks will be pushed aside in small-diameter drilling, and that fines will be either washed away completely or will be greatly underestimated. In gravels, 30–50 kg samples at 1 m intervals may be required, less in sands. Depth accuracy is not a serious problem in most cases, since samples are composited and related to layers rather than to specific levels. Once recovered, samples have to be carefully mixed and quartered prior to laboratory testing, and allowance made in the test reports for rejected oversize and lost fines. Because of the large samples required in gravel deposits, some on-site preparation and oversize scalping is often needed. A small grizzly, scales and even a small front-end loader may be required.

Borehole sample collection also needs some thought. Wet bailer samples should be tipped out into steel troughs to avoid loss of fines; spiral augers should be scraped rather than simply reversed to fling off the soil; and disc augers should be emptied onto canvas or plastic sheets rather than straight onto the ground.

Deleterious materials

A number of deleterious materials are commonly found in sand and gravel deposits, and these can easily be missed during drillhole sampling. The most prevalent contaminant is clay, in the form of lenses, partings, matrix and coatings on gravel clasts. Many clasts themselves, particularly in torrent bedded deposits, are unsound due to rapid erosion and short transportation distance. Gravel subject to fluctuating water tables may be weathered, stained or encrusted with iron oxides, or may accumulate carbonates and salts (mainly gypsum) in low-lying arid areas where groundwater discharges. Feldspathic and calcareous sands tend to be much weaker, dustier and less durable than quartzose sand. Other deleterious matter includes charcoal, coal, peat, wood fragments, mica, shell grit and some duricrust gravels.

3.3   DRILLING AND GEOPHYSICAL EXPLORATION

Sand and gravel deposits deeper than a few metres can usually only be investigated by means of auger or cable tool boreholes. Where these deposits are predominantly sand, are located above the water table and possess even a small percentage of clay binder and/or capillary moisture, their drilling and sampling are greatly simplified. A variety of truck-mounted augers are capable of drilling to depths of 10 m or more; usually the base of the deposit or the water table will be encountered within this depth range. Cable tool rigs can penetrate deeper, through much harder ground and below the water table, but are slower and the sample recovered is usually inferior.

Mechanical auger boreholes

Mechanical auger diameters range from about 100 mm to 1 m, and sampling accuracy is greatly improved at the larger diameters, though at the cost of much greater rig power and less vehicle mobility. Sample quality can also be improved by excluding the outer mantle of cavings. The larger rigs are capable of penetrating moderately compact fine gravels, but not densely packed or cemented cobbles or boulders. Sample recovery below the water table is close to nil, though some indication of material can be obtained from fragments jammed between cutting teeth or bound with clay to auger spirals. Three different auger designs are shown in Figure 3.6.

Bucket augers can recover large and representative samples of sand and fine gravel, but cannot handle stones larger than about 100 mm. *Disc augers* are suitable for moist or clay-bound sands or sandy gravels, and are cheaper and more mobile than bucket auger rigs. Bucket and disc augers are primarily large piling rigs, with head diameters of 0.6–1.2 m and a depth capability of about 15–20 m in favourable ground.

Continuous solid flight augers are quicker, cheaper per metre and can be mounted on lighter vehicles. They are usually only 100–200 mm in diameter, and depth capability is about 20 m. Samples are contaminated by caving and only give a rough guide to layer depths, composition and variability. Depth accuracy and sample quality can be improved by casing to the bottom of the hole and screw drilling ahead of the casing shoe. This involves augering short lengths (say 1–2 m) at a time, then withdrawing the soil core. The problem of caving is also overcome by using hollow flight augers, although much greater rig power is needed to overcome wall friction, and drilling is slower. Samples are obtained by lowering an open-drive tube sampler within the casing and pushing it into the floor of the borehole.

Suction drilling techniques

Suction drilling techniques are applicable to sand deposits only and include reverse-circulation (RC) rotary, vibrocoring and vibrosuction. These methods make use of the fluidity of saturated sand, negative pore

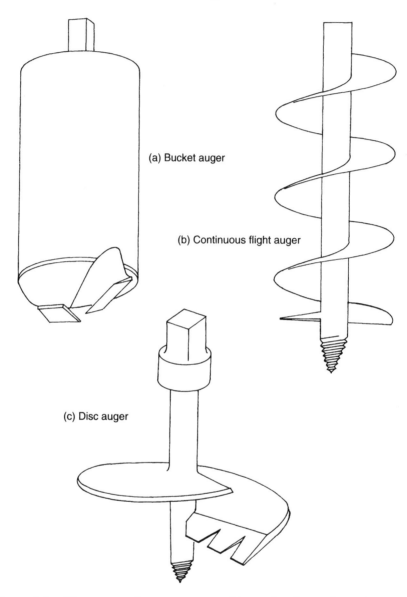

Figure 3.6   Three types of power auger used in sand and gravel sampling.

pressure in partly saturated sand and its tendency to dilate (bulk up) when disturbed. They are fast and avoid loss of fines, but even a small proportion of gravel may prevent recovery. RC uses a drill string made up of two concentric pipes, which revolve together or simply sink under their own weight. High-pressure air or water is injected down the outer annulus and returned, with the entrained sample, up the inner tube. The

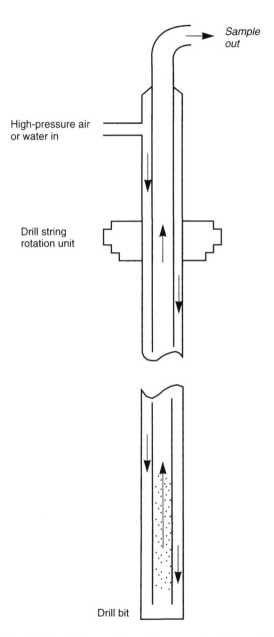

Figure 3.7 Method of operation of reverse-circulation drilling, using either air or water as the drilling fluid.

sand sample obtained is disturbed but uncontaminated, though its depth can only be estimated. Typical hole diameters are 80–100 mm, with the sample diameter about half this; hole depths can be down to about 40 m. The principles of RC drilling are illustrated in Figure 3.7.

Cable tool boreholes

Cable tool or cable percussion is the traditional method of drilling and sampling in gravels both above and below the water table. These may be large truck- or trailer-mounted rigs, or smaller folding-tripod 'shell and auger' types. The borehole is advanced by surging sediments or rock fragments at the base of the hole into a cylindrical bailer or 'shell' equipped with a non-return flap or clack valve (Figure 3.8). Hard bands, boulders and rock are broken down by a heavy chisel. Hence the chisel and bailer are used alternately at the end of different cables, moved up and down in short strokes to pulverize and then bail out the cuttings.

Cable tool drilling can penetrate the coarsest and hardest gravels in time, but the bailer samples recovered are thoroughly fragmented and washed, and hence give only a rough indication of grain size distribution. In fine gravels and sands the sample quality is better, but some fines are still lost. In dry or clayey sands the sample can be rammed into a modified bailer without a flap valve, where it is retained by a combination of friction, negative pore pressure and cohesion. Even better sample quality is ensured by hammering or pushing thick-walled tubes into the base of the borehole. It is necessary to keep the casing flush with the base of the hole, to prevent contamination due to caving and sample disturbance due to 'boiling' (quicksand conditions caused by water pressure inside the casing being less than that due to depth below the water table). In dry soils water must be added to the hole to facilitate bailing and casing driveage, but fines retention is improved where this drilling water is kept to a minimum. The largest-diameter casing in common use is about 250 mm OD, but diameters up to 1 m are desirable if coarse gravel and cobbles are to be recovered. The maximum size of particle that can be recovered unbroken is about 70% of the casing diameter.

Geophysical techniques

A number of geophysical techniques, in particular *resistivity* and *electromagnetic conductivity* (EM), have been tried for sand and gravel prospecting. These depend on the contrast between the electrical properties of fresh groundwater (high resistivity), bedrock (intermediate resistivity) and impermeable clay (low resistivity). The main purpose of electrical geophysics is to map the course of hidden channels and to select drilling and sampling sites – generally where the buried sand appears to be thickest. Granular sediments being aquifers, the presence

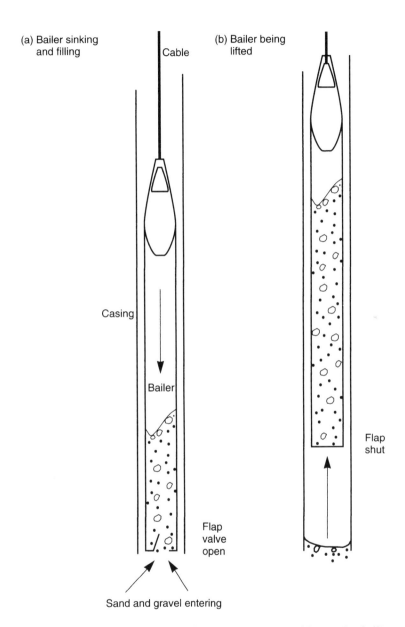

Figure 3.8 Method of sample recovery in cable tool drilling. The reciprocating motion of the bailer is imparted by alternately raising and lowering its cable.

of abundant groundwater is taken to indicate sand or gravel; the method cannot distinguish the two. Another shortcoming is that dry sands are more resistive than saturated deposits and hence appear to be a different material. EM is at present superseding resistivity because it needs no ground contact and the equipment is now portable, so measurements in the field can be greatly speeded up.

Geophysical prospecting is much more important in the search for marine aggregates. The two most important techniques employed are seismic reflection and sidescan sonar. *Seismic reflection* uses a towed vibration source ('fish') to map and probe unconsolidated shelf sediments. 'Sparker' sources generate high frequencies and offer high resolution (±100 mm) but only a few metres of penetration, while lower-frequency 'boomers' can penetrate to about 30 m with depth accuracies to reflectors of about 0.5 m. The sediment type and shape of these bodies can be estimated from the seismic profile and the character of the reflections. *Sidescan sonar* records detailed bedform information using high resolution in shallow water. This method can delineate seabed topography in some detail, and can also be used to estimate sediment type from the roughness or smoothness of the return, complementing the vertical profiling data from seismic reflection surveys.

3.4 FRIABLE SANDSTONE AND CONGLOMERATE: SYDNEY BASIN

Declining reserves of accessible alluvial sand and gravel have led to quarrying of the weak and weathered sandstones and conglomerates that outcrop extensively within the Sydney Basin. Similar deposits are worked in England, from the Bunter Pebble Beds and the Folkestone Sands. The Sydney Basin deposits are weathered, or at least poorly consolidated, to depths of 20–30 m and are generally soft enough to be dug by front-end loaders. Harder layers are ripped, but crushing is avoided due to the high abrasive quartz content of these formations. Oversize slabs may be broken down beneath bulldozer tracks or by exposure to wetting and drying (slaking) on the pit floor. The final stage of lump-breaking and disaggregation is carried out in logwashers (Figure 3.9) and screw classifiers.

The main products are sand and fine gravel, but the stronger sandstones yield a proportion of coarse (5–75 mm) fraction, which can be used as macadam roadbase for secondary roads. Because the disaggregated rock is poorly sorted (i.e. well graded), a variety of products including fine aggregate for concrete, mortar sand and filling sand can be generated. The angularity or 'sharpness' of the coarse sand particles, due to silica overgrowths, is advantageous in producing high-strength concrete for thin-walled castings and for skid-resistant

Figure 3.9   Weakly cemented sandstone being disaggregated in a logwasher (note contra-rotating paddles), Kangaloon, New South Wales.

asphalt surfacing. However, in pumped concrete the same characteristic may cause excessive pipe and pump wear.

From an environmental planning viewpoint these formations represent practically unlimited resources, in which extractive sites can be chosen so as to minimize their impact on the landscape and to relieve demand on premium aggregate sources. Present output of sand, pebbles and roadbase from friable sandstones in the Sydney region is about 5 Mtpa. Pits can be excavated so that their final shape is compatible with the adjacent topography, or they can be shaped to a specific end-use, such as residential or industrial subdivisions. Quarrying followed by landfilling is a particularly attractive proposition in the Sydney area, where sites are more valuable per cubic metre of dumping space than for the aggregate product. Sandstone landfill sites can be located in areas of low scenic and agricultural value, and can be easily screened by natural forest and topography.

The main shortcoming of friable sandstone is that it contains 10–30% of clay and silt matrix, mainly kaolinite, which must be washed out and disposed of in tailings dams. These lagoons occupy large areas of land, usually along drainage lines since the sandstone terrain is hilly. Some have been overtopped during heavy rain, resulting in downstream siltation and turbidity. Careful design of flow diversion banks and bypass spillways will reduce this risk, though off-stream tailings disposal is preferable.

The necessity for washing and sizing friable sandstone has three positive benefits:

- A number of higher-value products such as glass sand, foundry sand, granular filters and silica can be obtained in addition to fine aggregate.
- Well-crystallized kaolinite, currently discarded into tailings dams, has potential uses as a brick-making blend clay and as an industrial raw material. Brick-making requires only dewatering of the clay tailings, but higher-value end-uses necessitate further beneficiation.
- The tailings can be mixed with other discards and used for pit reshaping and as a base for topsoiling, or even as a topsoil extender where no better material is available.

3.5 MARINE AGGREGATES: NORTH SEA

Offshore sand and gravel deposits laid down on the continental shelf during Pleistocene low sea levels supply about 13% of British requirements, or about 20 Mtpa, mainly to London and southeast England. However, Japanese consumption of marine aggregates is the world's largest, at about 80 Mtpa. Marine gravels are cleaner, more rounded and more uniformly graded than their onshore equivalents; this produces concrete with improved workability and reduced water/cement ratio for the same strength. The smooth surface texture of marine aggregates means, however, that they are unsuited to high-strength concrete or wearing-course asphalt.

Although these deposits present both practical and environmental problems, and have large capital requirements in terms of sea-going dredgers and onshore terminal facilities, the availability of reserves close to urban markets is most attractive. These aggregate bodies include drowned beaches and beach ridges, buried alluvial fills, erosional lags and shelf sand bodies. The best prospecting targets appear to be adjacent to major river valleys extending out onto the continental shelf.

In the UK extraction of marine aggregates is carried out using trailing suction or – less commonly these days – forward suction (anchor) dredgers of up to 10 000 tonnes capacity (Figure 3.10). A typical British dredger has a capacity of 4500 tonnes, can fill in 2–3 hours at a rate of 9000 m^3/h (20% solids content) and discharge in 3 hours – the latter being more critical to profitability. The older ships equipped with onboard pumps work to depths of about 25 m, but newer dredgers with submerged pumps can operate down to 45 m. Sizing and washing of the sand onboard is possible, though it may be prohibited because of the resulting silt discharge. Trailing dredgers (Figure 3.10) are best suited to extensive but thin sand sheets, while anchor dredgers work over smaller areas but dig deeper – several metres instead of only 25 mm or so – and can lift gravel-sized particles.

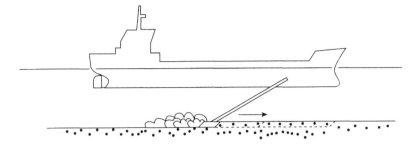

(a) Trailing suction dredger

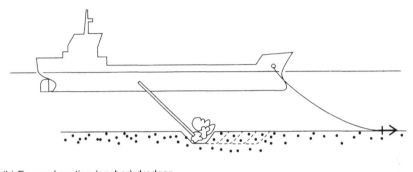

(b) Forward suction (anchor) dredger

Figure 3.10   Two types of marine sand dredger. The trailing suction head skims sand along parallel strips of the seabed, while the forward suction type excavates a pit or trench while swinging at anchor.

Practical problems

The practical problems associated with marine aggregates are mainly due to their salt content, or rather their chloride-ion content, which can corrode steel reinforcing rods and react with cement paste. Finished concrete surfaces may exhibit white efflorescence due to salt leaching out. Freshwater washing can easily reduce chloride content from 1–2% as-dredged to 0.1% or less, but 0.6% is usually considered satisfactory. Some authorities claim that in most cases the effects of salt in the aggregate and even of mixing concrete with saltwater are not significant, but care has to be taken against unusual concentrations in particular size ranges. Stockpiled sand will increase in salt content due to evaporation in hot weather, and may reach high concentrations if wet sand is repeatedly added to the pile. A high proportion of shell fragments and unsound calcareous sand grains will also reduce the value of marine sands, to the point where they can only be used as hydraulic fill for land reclamation. Moderate amounts of shell reduce concrete workability, but have little effect on strength or other properties.

Environmental problems

The environmental problems of marine aggregate extraction are related to disturbance of underwater habitats and to the erosion/deposition balance in offshore sand bodies. Dredging may induce beach erosion downcurrent by intercepting sand, or it may kill marine grass by silt deposition. Detailed studies of seafloor morphology and of wave, tide and current interaction may be required before large-scale exploitation of marine sand and gravel resources is permitted. However, dredging is now technically feasible in water depths of 50 m, and deeper extraction may be possible in the future. This means that sand beds far offshore can be worked, well beyond the likely influence of beach erosion processes.

3.6   ALLUVIAL DEPOSITS: BRISBANE, AUSTRALIA

The principal source of aggregate for the Brisbane metropolitan area over most of the past century has been gravel and sand dredged from the bed of the Brisbane River, which meanders through the centre of the city and its suburbs. During this time about 40 Mt has been extracted, from a total resource of perhaps 100 Mt (Malempre, 1990). These gravels were eroded during the Tertiary from nearby fault-uplifted blocks of igneous and metamorphic rocks, by streams that were rejuvenated by sea-level fluctuations of more than 100 m during the Pleistocene. Originally the gravel/sand ratio averaged 65/35, a high proportion for a non-glaciated terrain, in deposits up to 30 m thick. The gravel content is greatest upstream, close to the source rocks, and sand predominates in the lower reaches. In the currently worked reaches, gravel constitutes only about 10% of the dredged product.

Production fell by about half, from 3 Mtpa to 1.5 Mtpa, following a disastrous flood in 1974. The severity of the damage was partly due to riverbank collapse, which carried away many new and expensive houses. This was blamed by the public on channel overdeepening; the gravel dredgers claimed in reply that their activities had prevented many previous floods by allowing quicker passage of peak flows. Restrictions subsequently placed on depth of dredging lost the industry 30 Mt of reserves. During the 1980s the dredging companies were subjected to further criticism from riverside residents because of noise, although in this instance the operators accepted a code of self-regulation to minimize the nuisance.

Nevertheless, it had become obvious that the industry would have to move, and the North Pine and South Pine Rivers, located about 30 km north of central Brisbane, offered one suitable replacement area. Terrace, channel and point bar deposits of Pleistocene (Last Interglacial) age occur along about 10 km of the South Pine River and 7 km of the North Pine River. Deposits within this 30 km^2 area are being developed to meet strong local demand, as well as to make good declining yields from the Brisbane River. The deposits and their investigation are described by Hofmann (1980).

The project is similar, but on a smaller scale, to the Penrith Lakes scheme, west of Sydney. In both cases the geological and environmental investigations were largely carried out by third parties, to ensure that the whole resource was properly assessed and managed. This means that the deposits will be systematically and fully mined and not just 'high-graded', that mistakes made with extraction along the Brisbane River will not be repeated, and that progressive restoration will take place. Extraction remains in the hands of quarry operators working individual leases within the overall development area.

The exploration programme for the Pine Rivers project included mapping of surface geology on 1:5000 and 1:10 000 orthophotomaps, airphoto interpretation, resistivity depth probing (unsuccessful) and, most important, stratigraphic drilling. The aim of this was to delineate the total resource and its quality variations, using a total of 83 holes, which were drilled to an average depth of 10 m. Half of these were put down by 108 mm diameter continuous flight auger, 13 by a 900 mm bucket auger and the rest by reverse-circulation vibrosuction and non-core rotary drilling. From these holes 120 bulk samples were submitted for testing, mainly for particle size distributions; however, a few particle shape, sulphate soundness and alkali reactivity tests were performed.

The investigators concluded that about 70 Mt of the *in situ* resource was currently extractable in 25 deposits, within a maximum overburden/product ratio of 1/1 and maximum cover depth of 10 m. While this was regarded as a satisfactory result, some problems were highlighted and their solutions outlined:

- *Coarse aggregate* The average proportion of coarse aggregate (+5 mm)* was only 40%, where 60% would have been preferable; this deficiency will have to be made up from hard rock quarries. Oversize (+37.5 mm) averaged 10% and fines (−75 μm)* were 5–40%.
- *Off-stream pondages* Extraction by channel dredging in the past has caused erosion of nearby bridge abutments and has possibly aggravated bank collapse after floods. Future exploitation will be restricted to off-stream pondages, which can also confine turbid wash water and are suitable for restoration as ornamental lakes.

---

* Note that throughout this text we use the notation +5 mm and −75 μm, for example, to indicate material of size greater than 5 mm and material of size less than 75 μm, respectively. Also, we use micrometres (μm; 1 μm = 0.001 mm) where appropriate.

- *Dragline mining*　Most of the resource, particularly the coarsest fraction, occurs below the water table. This will necessitate dragline mining, a method that has in the past been hindered by the presence of a persistent tough clay band within the gravel. Clamshell dredges will be needed to extract below this horizon, where it is present.
- *Land ownership*　In this area land ownership is fragmented and some deposits are already being threatened by suburban encroachment. Many land-owners were unwilling to allow drilling on their property, fearing that it might be a prelude to resumption of mining. Control over exploitation is divided, in Queensland as elsewhere, between several state and local government authorities. Potential conflicts here are being overcome by adoption of a joint extraction and land rehabilitation strategy.

3.7　GLACIAL DEPOSITS: ENGLAND AND WALES

For much of the world, especially the industrialized countries of North America and northwestern Europe, sand and gravel deposits are largely synonymous with glacial and, even more so, fluvioglacial deposits. The British literature is dominated by the topic of 'drift' deposits rather than hard rock quarrying (see, for example, Smith and Collis, 1993; Merritt, 1992; Crimes *et al.*, 1994). There are several possible reasons for this interest. First, though sand and gravel now supply less than half the UK production of aggregate, they are still the preferred choice for concrete-making. Secondly, these deposits are widespread but extremely variable in size and quality, and therefore their location and evaluation are more challenging than exploration for hard rock quarries. One measure of this is the systematic evaluation carried out over 30 years by the British Geological Survey, which up to 1990 had produced about 140 reports covering perhaps one-tenth the area of Great Britain. Finally, their production costs are still relatively low, relative to marine aggregate and hard rock sources, while haulage costs are kept down by the wide distribution of the pits.

The term 'glacial deposits' includes subglacier, ice contact, fluvioglacial, glaciolacustrine and periglacial deposits, reflecting the complex history and distribution of these materials. This complexity results from multiple advances and retreats of the ice front, even within a single glacial stage, excavation and burying of older deposits, and postglacial modification by mass movement and alluvial reworking. Merritt (1992) draws a distinction between areas of northern England and Wales affected by the Last Glacial, which climaxed about 18 000 BP, and the Midlands where much older (around 450 000 BP) deposits are exposed at the surface. In the first case the glacial landforms, being younger, are more sharply defined and often easily recognizable on airphotos. They are also attractive materials sources in being less affected by post-depositional frost heaving, ice shoving and mixing. Conversely, the older deposits have been reshaped and degraded by subsequent periglacial climates and fluvial action. Though considerable resources of sand and gravel may be present, they are concealed beneath relatively featureless clay till.

The most useful sources of granular materials are eskers, sinuous ridges of relatively clean, coarse sand and gravel laid down by meltwater streams flowing beneath the ice. Kames (ice-contact deltas and terraces) also show much potential, though they include more fines. Outwash fans and terminal moraines are more extensive, but are also more variable in grain size, fining rapidly away from the ice front. Glacial debris derived from areas of sedimentary rocks is also likely to be deficient in gravel compared with that gouged from hard rock terrain.

The methods of materials investigation in this geological environment are reviewed by Crimes *et al.* (1994). They comprise:

- *Archival databases*, built up from borehole and pit logs, quarry records and test results (mainly particle size distributions). Though the volume of this information is huge, running to tens of thousands of data points for some projects, its quality is variable and the coverage very uneven. For example, most boreholes are shallow site investigation holes concentrated in urban areas. The information has to be filtered by an experienced geologist to identify the 10 or 20% that is useful.
- *Geomorphological mapping*, based largely on airphoto interpretation and walkover surveys, supported by satellite imagery. The aim is to build up a terrain model, on the assumption that the visible landform reflects the geological materials beneath. Certain areas, such as former ice sheet edges and meltwater channels, become targets for follow-up work, especially drilling and pit excavation. About $5\,km^2$ can be mapped per day for compilation at 1:50 000.
- *Shell and auger drilling*, which was the most successful exploration technique used once a potential deposit had been identified by mapping. A typical coverage is about one borehole per square kilometre, but irregularly spaced. Satisfactory samples were obtained drilling 'dry' at 200–250 mm diameter, though progress was slow in dense sands and coarse gravels.

REFERENCES AND FURTHER READING

Brown, B.V. (1990) Marine aggregates, in *Standards for Aggregates*, ed. D.C. Pike, pp. 64–91. Ellis Horwood, London.
Crimes, T.P., *et al.* (1994) Techniques used in aggregate resource analysis in four areas in the UK. *Quarterly Journal of Engineering Geology*, **27**, 165–92.

Dixon, A.J. (1988) Sampling sand and gravel deposits. *Quarry Management,* October, pp. 45–52.

Fookes, P.G. and Higginbotham, I.E. (1980) Some problems of construction aggregate in desert areas (parts I and II). *Proceedings, Institution of Civil Engineers,* **68** (i), 39–90.

Hofmann, G.W. (1980) *Aggregate Resources of the Pine Rivers Area.* Geological Survey of Queensland, Publication 375.

Malempre, J.L. (1990) Gravel dredging in the Brisbane River. *Quarry Management,* January, pp. 33–7.

Merritt, J.W. (1992) A critical review of methods used in the appraisal of onshore sand and gravel resources in Britain. *Engineering Geology,* **32,** 1–9.

Sinclair, J. (1969) *Quarrying, Opencast and Alluvial Mining.* Elsevier, Amsterdam.

Smith, M.R. and Collis, L. (1993) *Aggregates – Sand, Gravel and Crushed Rock for Construction Purposes,* 2nd edn. Geological Society, Engineering Geology Special Publication No. 9.

Thurrell, R.G. (1971) The assessment of mineral resources, with particular reference to sand and gravel. *Quarry Managers Journal,* **55** (1), 19–28.

Thurrell, R.G. (1981) The identification of bulk mineral resources: the contribution of the Institute of Geological Sciences. *Quarry Management and Products,* **8** (March), 181–93.

US Bureau of Reclamation (USBR) (1975) *Concrete Manual,* 8th edn, Ch. 2. US Bureau of Reclamation, Department of the Interior, Washington, DC.

CHAPTER 4

Natural pavement materials

These materials comprise a diverse range of naturally occurring granular soils, dirty gravels and weathered rock, which are used in relatively unprocessed form as flexible pavements for secondary roads. Roads of this type are common in rural and developing areas around the world, typically carrying 100–500 heavy vehicles per day. They are designed as two-lane, single carriageways with priority given to all-weather operation and low-cost construction over speed. The usual surface treatment is a sprayed bitumen seal topped with stone chips. Such pavement materials are variously known as natural gravel, road gravel or soil aggregate. Though called 'gravels' for convenience, they are in fact gravel–sand–silt mixtures with a small proportion of clay binder. Furthermore, 'gravel' is used in the general sense of any particles coarser than 2–3 mm, including rock fragments, concretions and water-worn pebbles.

Natural gravels are usually of marginal quality in terms of pavement material specifications for urban roads, where they might at best be acceptable as sub-base. Nonetheless they can perform satisfactorily in lightly trafficked rural highways and local roads, especially where well compacted and well drained. They are the most widely used road-making materials in the world because they are the most economical. First, they require only rudimentary processing (such as in-pit mixing), but generally no blasting, crushing or screening. Secondly, being locally available, they incur minimal haulage charges. Thirdly, they can be worked by simple equipment and are well suited to the dry tropics and to expansive clay subgrades (since they are more flexible than crushed rock pavements). In areas like southern Africa and outback Australia, production of natural road-making gravels far exceeds that of quarried basecourse materials. However, the distinction between natural gravel and prepared roadbase is fading, as more of the former is upgraded by chemical or mechanical stabilization (see Chapter 18), or by in-pit or on-road processing.

4.1   CHARACTERISTICS

Although it is difficult to generalize about such varied and widespread deposits (Figure 4.1), natural pavement materials tend to have the following characteristics:

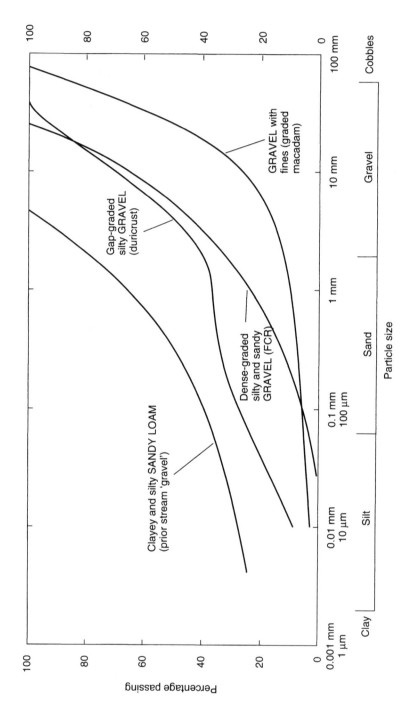

Figure 4.1 A comparison of particle size distributions for various naturally occurring road gravels with dense-graded fine crushed rock (FCR) roadbase.

- *Coarse-grained* They are often very coarse-grained (have an excess of gravel and rock fragments) and are gap-graded, due to a deficiency in medium to coarse sand (roughly 0.3–3 mm). In other words, the *soil aggregate* component is large relative to the percentage of sand–silt–clay fines.
- *Weak and unsound* The coarse particles are weak and unsound by concrete aggregate standards, often angular or ragged in shape, and may be quite porous and absorbent due to weathering. At the other end of the size range, the clay particles are often weakly cemented by soil-forming processes.
- *Moisture-sensitive* Both the soil aggregate and the fines fraction are more moisture sensitive than those in crushed rock roadbase. The coarse particles absorb water, reducing their effective strength, while the natural clay binders swell on wetting and shrink on drying.
- *Self-cementing* A self-cementing action is common and may significantly increase the strength of the pavement with time. Another strength increment comes from *soil suction*, negative pore pressure generated as the pavement material dries out.
- *Inhomogeneous deposits* These deposits tend to be very inhomogeneous, owing to their dependence on soil-forming processes of mineral solution, leaching and replacement. Clay content generally increases sharply with depth, and the proportion of gravel-sized particles may vary greatly.
- *Shallow and broad deposits* Deposits tend to be shallow and broad, typically 1–3 m deep, and pit workings are therefore laterally extensive in relation to the volume of material produced (Figure 4.2). This type of gravel pit therefore has significant environmental impact, with high restoration costs per tonne extracted.

4.2 STRENGTH AND STIFFNESS

The shear strength and resistance to deformation (stiffness) of a road pavement, and hence its load-carrying capacity, are derived from a number of factors (Figure 4.3). These include the following:

- *Tight packing*, which is achieved by having an even ('dense-graded') mixture of gravel-, sand- and silt-sized particles, and by vibratory compaction. This causes the finer grains to fill voids between coarser ones, imparting dry densities of up to 1.8–2.1 t/m^3.
- *Mechanical interlock* between coarse particles, as exemplified by macadam pavements and railway ballast. These 'open-graded' mixtures contain few fines and hence their compacted densities are relatively low, but high shearing resistance is generated by interparticle friction.

Figure 4.2 Typical road gravel pit, western New South Wales. This scrape is deeper and narrower than most, and overburden has been stockpiled at the edges for site rehabilitation.

- *Confinement* by wide shoulders or rigid kerbs, and especially by soil suction. The pavement can also be reinforced by geosynthetic membranes.
- *Cementation*, provided by natural 'petrifaction' or artificially by cement, lime or bituminous binders.
- *Effective sealing* to prevent moisture infiltration through the road surface into the pavement courses.
- *Hard foundation*, provided by a dry, well-compacted subgrade located far above the water table.

These attributes are sought in all pavements, but the particle soundness and grading deficiencies of natural gravels mean that seal integrity, subgrade strength, soil suction and self-cementation must compensate. A summary of the differences between crushed rock and natural gravel roadbases is given in Table 4.1.

The strength of gravel pavements is derived from a combination of the above factors, though the balance between these varies with the material. For example, in Figure 4.1 the graded macadam draws its strength chiefly from particle interlock, whereas suction is the significant factor in the strength of the sandy loam.

Nevertheless, *dense grading* is the primary means by which strength and stiffness are maximized in most natural gravel pavements. A high

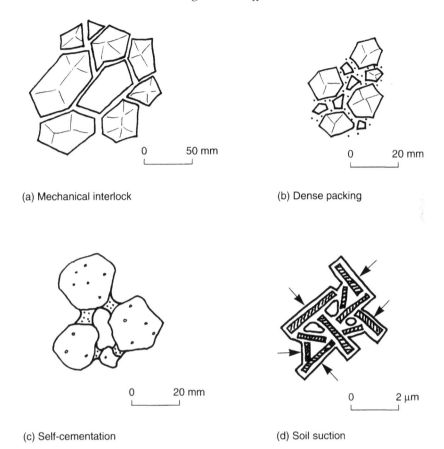

(a) Mechanical interlock

(b) Dense packing

(c) Self-cementation

(d) Soil suction

Figure 4.3　Sources of pavement strength and stiffness in natural road gravels. See text for explanation.

degree of grain-to-grain friction and a framework of coarse angular fragments spread and dissipate wheel loads. Maximum density grading and a proportion of clay binder also cause these pavement courses to be relatively impermeable, protecting moisture-sensitive subgrades from downward-percolating water.

These pavements are normally unsaturated and can therefore develop high *soil suction*, or soil tension, due to negative pore pressure. This can be thought of as a form of confinement analogous to vacuum-sealed plastic wrapping, except that in this case the membrane is a very thin film of water tightly held by capillary forces. Its magnitude (which can exceed 1 MPa) is a function of the dryness of the material and the fineness of its micropores; hence maximum strength is achieved in well-graded (but fines-rich) road gravels by heavy compaction and by 'drying back' from the optimum moisture content prior to sealing. However,

Table 4.1 Comparison of natural gravel and crushed rock pavement materials

Advantages of crushed rock
>   More uniform and consistent manufactured product
>   Grading, particle soundness, plasticity and shrinkage usually well within
>   specification (natural gravels often marginal)
>   Stiffer, higher CBR[a] pavements, hence longer traffic life (however, traffic life
>   of natural gravel may be quite adequate)
>   Negligible clay content, so relatively insensitive to water (however, secondary
>   minerals may break down to clays)
>   Less environmental cost per tonne produced due to much greater deposit
>   thickness

Advantages of natural gravels
>   About 20–30% cheaper at pit gate, but haul distances usually less than for
>   crushed rock so price delivered onto site may be much lower
>   Less permeable and more flexible than crushed rock, so may be better suited
>   to expansive subgrades
>   May be upgraded by chemical stabilization at 10–15% extra cost, though
>   on-road mixing usually less thorough than in quarry-run material

[a] CBR = California bearing ratio.

moisture content within a pavement varies with the climate, seal condition and other factors, so soil suction is looked upon as a bonus and not as part of the design strength. A fuller discussion of the compaction process is given in Chapter 10.

The usual surface treatment for natural gravel pavements is a thin *sprayed bitumen seal* protected by stone chippings (see Chapter 9). The seal generally consists of two coats of bitumen applied several months apart, so that imperfections in the roadbase that progressively become apparent can be patched prior to the final surface dressing. Periodic resealing is desirable with gravel pavements because the bitumen oxidizes and cracks, especially in hot and sunny climates, allowing moisture penetration.

To further compensate for the relatively poor quality of these pavement materials, a *dry and strong formation* is preferred. This provides a hard floor for pavement compaction and prevents loss of suction. Natural gravels are particularly appropriate in low-rainfall and deep-water-table areas, where infiltration through the road shoulders and capillary movement upwards are minimal. Sources of external moisture and some of the methods of controlling them are discussed in Chapter 11. Subgrade strength is improved by compacting in thin layers and by using the best granular material available at the top of the fill. Dry borrow areas in such environments mean that the earthworks materials can be gradually wetted up to, or slightly below, optimum moisture content.

4.3 DURICRUST ROAD GRAVELS

The most important group of natural gravels for road-making purposes are the hardened soil materials: ferricrete, calcrete and silcrete, collectively known as *duricrusts*, pedocretes or pedoderms. These occur as caprocks on flat-topped hills (mesas and buttes), as boulders and cobbles in talus, as cemented sheets along ancient drainage lines ('valley train' deposits) and as lag gravels ('gibbers') on plains. Typical geomorphic relationships between these deposits and the implications for pavement materials investigations are summarized in Figure 4.4. Their importance lies in a worldwide distribution and because – being themselves the end-products of extreme weathering and redeposition – they occur in terrains where no other hard rock materials are likely to be found. The engineering uses and problems of duricrusts are described in a classic paper by Grant and Aitchison (1970); and by also Gidigasu (1976) on ferricretes, and by Netterberg (1971, 1982) in relation to calcretes. A recent summary of the geotechnical characteristics of ferricretes (laterites) can be found in Charman (1988).

- *Ferricrete* (Figures 4.5 and 4.6) is the sesquioxide-rich ($Fe_2O_3 + Al_2O_3$) 'ironstone' horizon within a laterite profile, or transported gravel derived from this source. It generally consists of boulders and gravel-sized concretions within a silty matrix. Varieties include nodular 'ironstone gravel' and pisolitic 'pea gravel'. As a road pavement material, ferricrete compacts to a high density – partly due to the high specific gravity of the haematite concretions – and resists surface rutting when unsealed because of its gravel content. Ferricrete is probably the most extensively used road-building material in the world, but especially in central and west Africa, northern Australia and southern Asia.
- *Calcrete* (Figures 4.7 and 4.8) is calcareous duricrust, which is both more variable and perhaps even more widespread than ferricrete, owing to its relatively rapid accumulation in semi-arid soil profiles. It ranges from a light $CaCO_3$ dusting on soil fissures, through soft nodular (gravelly) horizons, and eventually to dense cemented sheets. Even in small concentrations, lime can impart useful self-cementing properties to sands deficient in clay binder. However, nodular calcretes and weaker (i.e. rippable) sheet calcretes are considered the better unprocessed pavement materials. Areas of calcrete abundance include southern and north Africa, the Middle East and southern Australia.
- *Silcrete* or siliceous duricrust is the least-used of the three main duricrust materials, although it occurs throughout the interior plains of Australia and South Africa. It normally occurs as bouldery hill cappings and talus, or as cobble-sized 'gibbers' (desert armour) on the plains, and requires crushing prior to use in road pavements.

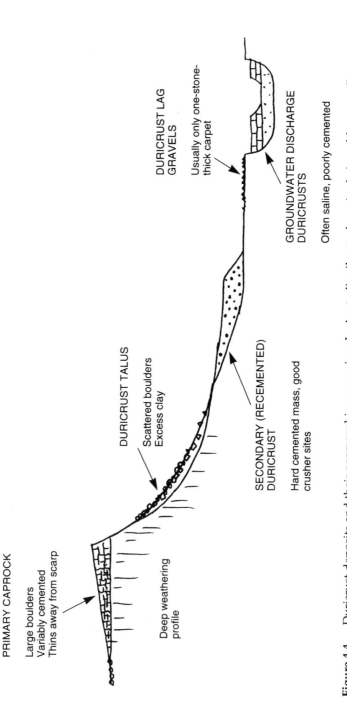

Figure 4.4 Duricrust deposits and their geomorphic expression. In Australia silcrete deposits, being oldest, usually occur as caprocks high in the landscape, while ferricretes may occupy intermediate levels. Calcretes are youngest, and tend to be developed along drainage lines and in dunes.

PRIMARY CAPROCK

Large boulders
Variably cemented
Thins away from scarp

Deep weathering
profile

DURICRUST TALUS

Scattered boulders
Excess clay

SECONDARY (RECEMENTED)
DURICRUST

Hard cemented mass, good
crusher sites

DURICRUST LAG
GRAVELS

Usually only one-stone-
thick carpet

GROUNDWATER DISCHARGE
DURICRUSTS

Often saline, poorly cemented

Figure 4.5 Laterite profile near Casino, New South Wales, showing ferricrete horizon (about 1 m thick) at top and mottled zone beneath. In extracting this material about 0.5 m of topsoil would first be stripped, and the ferricrete ripped, blended and stockpiled in the pit. The pit floor would be located at the top of the mottled zone.

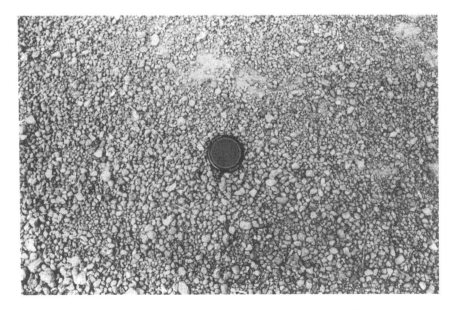

Figure 4.6 Pisolitic ferricrete ('pea gravel' or 'ironstone gravel') stockpiled for reconstruction of highway pavement near Cloncurry, Queensland.

Figure 4.7   Typical calcrete profile about 2.5 m deep in a gravel pit near Mannum, South Australia. Note the well-cemented hardpan at the top, nodular calcrete beneath and loose powder calcrete at base.

Figure 4.8   Calcrete nodules (10–25 mm diameter) agglomerated into an irregular boulder-sized mass about 1.5 m across, Broken Hill area, New South Wales. (Photo: I.R. Wilson.)

Although originally formed by weathering *in situ*, many duricrusts have been recemented, reworked and redeposited by alluvial, colluvial and groundwater processes. They tend to be gap-graded ('armchair graded', as shown in Figure 4.9) due to the absence of coarse sand, and to break down with working. The particles are porous and hence absorbent because of cavities left after the solution, redeposition and cementation phases. Hence it is believed that fines microporosity, rather than excess clay minerals, causes high liquid limits (LL) and plasticity indices (PI) in these materials.

The more indurated layers are unrippable, but paradoxically they are also difficult to blast because of their porosity. Explosive gases tend to vent (depressurize prematurely from blastholes) without fragmenting the rock, creating an airblast nuisance, or to dissipate energy in compressing the underlying soil. Well-cemented sheet and boulder duricrusts are sometimes crushed to produce dense-graded roadbase and aggregate, although the yield of coarse chips may be low and their soundness suspect.

Both calcrete and silcrete basecourses present problems of seal debonding – the bitumen can be literally rolled up in some cases. This can be due to the presence either of salt or of incompatibly charged dust particles along the basecourse/seal interface. These materials are also as highly absorbent of bitumen as they are of water, so that up to four applications – two of thinned or 'cut-back' bitumen primer followed by two high-viscosity seal coats – may be required to complete the surface dressing.

Siliceous duricrusts, both silcrete and chalcedonic calcrete, are potentially reactive with cement paste (see Chapter 8). Despite this, they have occasionally been used as coarse aggregate in low-strength concrete, through lack of any alternative source. In this situation low-alkali cement is specified and close control over the water/cement ratio (the main factor controlling hardened strength) is essential. Ideally the aggregate should be pre-soaked and mixed in a saturated, surface-dry condition, so that water intended for cement hydration is not drawn into the aggregate micropores instead.

4.4   OTHER ROAD GRAVELS

A number of other naturally occurring granular soils and weak rock materials are also used for road pavement construction. The geological origins and engineering properties of these diverse materials have never been systematically described, but an idea of their variability can be gained from the grading curves shown in Figure 4.1. The finer end of the range is represented by the prior stream 'gravel', which is in fact a sandy loam with a small proportion of granules. The graded macadam is

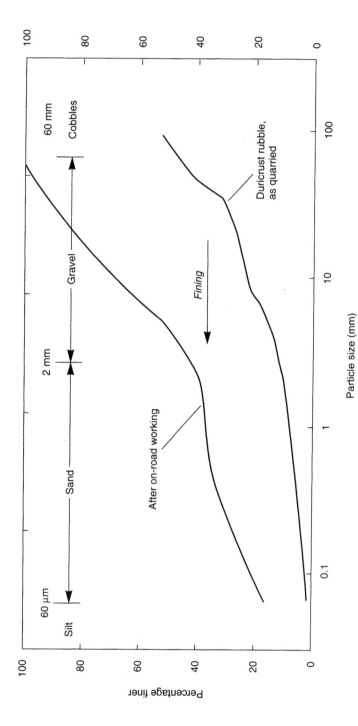

Figure 4.9 Comparison of typical duricrust gravel gradings, as-ripped and as-compacted in a road pavement. Note that the sand deficiency remains after on-road working, but the proportion of fines increases at the expense of the cobble and gravel fraction.

typical of a broken rock pavement with only about 10% by weight of sand and silt–clay fines. The grading curves for a duricrust and a dense-graded fine crushed rock basecourse are also shown for comparison.

Residual soils

Residual soils, mainly those derived from weathering of granite and related quartz-rich lithologies, were once widely used for rural highway pavements in New South Wales. Such soils tend to be well graded below the 2.36 mm sieve size and of low plasticity, but with very small quantities of fine gravel and no coarser particles. A typical road-making residual soil might be developed on top of a deeply weathered coarse granite saprolite, with a high quartz/orthoclase ratio and a low mica content. Highly feldspathic soils may be similarly well graded, but being composed of weaker particles tend to break down during working. As with the duricrusts, contamination even by small quantities of underlying clayey subsoil may make residual soil unsuitable as a basecourse.

Alluvium

Alluvium varies greatly in grain size but tends to be either too clean (that is, deficient in silt–clay binder) or, conversely, too clay-rich for use as roadbase. Clean sand and gravel are, of course, very suitable for other road-making purposes, such as concrete and sealing aggregates. Sand and gravel may also be blended with gap-graded soils to meet roadbase specifications. In southeastern Australia, abandoned channel deposits ('prior streams') of probable Late Pleistocene age contain large resources of continuous graded but silty sand (Figure 4.1), suitable for light-traffic pavement courses. Their high silt content makes them moisture-sensitive and unsuitable above shallow water tables, where they may become saturated due to capillary rise.

Older alluvial deposits (Tertiary 'Deep Leads') occurring as hill cappings, gully fills or raised river terraces are often adequately graded but weathered, and their cobble and gravel fraction is unsound as a result. This is not necessarily detrimental in a material to be used as unprocessed roadbase, since it facilitates closer packing after clast breakdown, but high plasticity can be a serious drawback. These deposits are often quite deep (5–15 m) but in the past only the leached, low-PI upper portions were worked. Lime stabilization is now commonly employed to reduce plasticity and increase bearing capacity.

Alluvial fans

Alluvial fans are significant sources of roadbase in some arid and tectonically active areas. They tend to be well graded, with excess fines carried away in suspension. Cobbles and coarse gravel may be present in quantity, and can be crushed and blended back to improve the overall grading. Many alluvial fans are, however, composed mainly of mudflow debris, and this would generally be too clayey for roadbase.

Colluvial deposits

Colluvial deposits are sometimes reasonably graded, but with an excess of clay fines and unsound coarse rock fragments, and are therefore rarely suitable for pavement materials. Some ferricrete and silcrete talus can be recemented by younger pedogenic materials (of the same or different composition), but the volume of these deposits tends to be small. Leached slopewash deposits, particularly those derived from granitic soils and fragmented duricrusts, provide minor sources of natural roadbase.

Aeolian deposits

Aeolian deposits of *carbonate sand* (aeolianite, calcarenite or dune limestone) can produce useful roadbase, since their calcrete horizons break down during working to produce angular gravel fragments. These improve the grading and hence the compacted density of the sand; facilitate mixing and uniformity of the pavement courses; and resist ravelling on unsealed road surfaces. *Coronus* or uplifted coralline sand has similar properties to aeolianite, and is the most widely used pavement material in Papua New Guinea and the islands of the western Pacific.

Quartz dune sands of the Australian interior are moderately well graded in the medium to fine sand sizes, but lack coarser material except where nodular calcrete occurs in their B-horizon. Aeolian *sand–clays* of this type are in fact sandy loams with a proportion of binder. These are used as pavement courses in remote areas of Western Australia and western Queensland, but erode rapidly where unsealed. They may also possess weak self-cementing properties because of the presence of mobile ferruginous coatings on sand grains.

Volcanic scoria

Volcanic scoria and pyroclastic debris have been used as base and sub-base on rural roads in western Victoria and South Australia, although environmental pressure to preserve the more conspicuous scoria cones is limiting further exploitation. The ash is well graded but the particles are porous and weak, and hence suitable only for very lightly trafficked roads. Scoria is a vesicular rock that breaks down under compaction, but is sometimes strong enough for use as lightweight aggregate or as sub-base. The best-quality scoria is used as a skid-resistant aggregate in wearing-course asphalt, and weaker deposits have been blended with volcanic ash and aeolianite to improve their grading.

Weak rockfill

Weak rockfill, as a macadam-type pavement material, has been used where no better source was available (see also Chapter 12). These stony materials are generally weathered and cleaved – or extremely close-jointed – metamorphic and igneous rocks or indurated shales, obtained by ripping or light blasting. Typically more than 90% of the pit-run material is coarser than 2.36 mm and composed of weak, angular, often platy rock fragments. This material is further broken down on the road by grid or vibrating drum rollers, though care has to be taken to avoid over-rolling (and hence overfilling of void space, since the strength of the pavement in this case depends on mechanical interlock rather than maximum density grading).

Sandstones and conglomerates

Sandstones and conglomerates, usually in a weathered, leached or otherwise disaggregated state, are used as sources of ripped rock and aggregate in some areas. The end-product is usually a conventional maximum-density compacted roadbase with a proportion of coarse harder rock fragments, rather than an open-graded macadam pavement.

4.5   LOCATION AND SAMPLING

Terrain evaluation

The location and evaluation of duricrust, residual, alluvial, aeolian and pyroclastic road-making materials depends to a great extent on recognition of the landforms associated with these deposits. Hence image interpretation, or terrain evaluation in its more systematic form, is a much more useful exploration technique for natural gravels than it is for hard rock quarry sites. This is particularly so when major development projects, such as trunk roads and mine sites, are being planned for remote and semi-arid areas. In these situations there is generally a close relationship between geology, soils and vegetation, and good-quality airphotos or satellite imagery are available. Land ownership and environmental constraints play a smaller part in the selection of quarry sites and borrow pits than they do in settled areas, so materials searching can be more extensive.

Landsat MSS imagery has in the past not lived up to engineering expectations, owing to poor resolution (80 m wide pixels), lack of spectral contrast between soil units and user-unfriendly processing techniques. The latest interactive programs allow the geologist who does the fieldwork to complete the interpretation on his/her own PC. Promising results are being obtained using Thematic Mapper (TM) imagery with 30 m resolution, and even finer detail is visible on 10 m resolution SPOT images. Airborne remote sensing, using infra-red linescan, thermal imagery and side-looking radar (SLAR), has in the past been used experimentally for locating construction materials over large areas, and is now becoming cost-effective for routine applications. The latest systems can be mounted in a light plane and offer pixel resolutions in the range 1–10 m, depending on the swath width required, and up to 20 data channels in the visible and infra-red regions.

Airphoto interpretation

Airphoto interpretation, nevertheless, remains the basis of terrain evaluation for engineering purposes. It is best performed as a three-stage process, beginning with a preliminary appraisal of available airphotos, maps and the existing geological information in the office. This should highlight possible deposits – or at least landforms likely to be associated with such deposits – while at the same time allowing large non-prospective areas to be discarded. The second stage, the site survey or 'ground truth' stage, involves sampling and reconnaissance mapping in the field. The geologist should not attempt to map the entire area (a corridor 5–10 km wide on either side of a proposed road, for example), but should select typical samples of the landscape that coincide with distinctive photographic patterns and are most conveniently located for visiting. Once the geology underlying these patterns is established, the interpretation can be extended throughout the surveyed area in a final compilation stage.

Sampling

Sampling during broad-scale preliminary investigations of this type is best carried out using mobile excavation plant such as backhoes or truck-mounted augers. With these machines it may be possible to sample and log 10–20 holes per day along 5–20 km of traverse line. Backhoes can excavate to depths of 3–4 m, which is quite adequate for most superficial deposits. They can also penetrate loose nodular (but not bouldery or sheet) duricrusts, and can give a good picture of layering and horizontal variation in material properties. The most useful augers are 0.3–0.6 m diameter disc or short-spiral types, which can quickly penetrate to depths of 5–10 m. Both types can rapidly penetrate slightly cohesive granular soils above the water table, and the more powerful ones can penetrate weak rock.

Before undertaking detailed sampling of naturally occurring construction material deposits, it is necessary to establish (by scout drilling, pilot sampling and index testing) whether the soils present are within or close to specification, and likely to be present in sufficient quantities to justify further investigation. If this is so, the next step is to determine the lateral and vertical variability present within the deposit by grid drilling (30–50 m centres is a typical interval) and systematic sampling. Individual samples should be taken from the principal layers present in each hole, so that the properties of each horizon can be established. Not all samples will be tested, but it is far better to have taken too many than too few! This is also important because, with careful in-pit blending, two substandard soils can be combined to generate a product that meets specification. Conversely, poor mixing – for example, including clay bands or clay floor in the stockpiled product – can degrade a material otherwise just within specification limits.

Because many of these natural materials contain a high proportion of coarse particles (weathered rock fragments, duricrust boulders, alluvial cobbles and volcanic bombs), large volumes are required for representative samples. These can be reduced to manageable proportions by removing and weighing oversize particles – those thicker than the pavement layer (say 100 mm) – and carefully quartering the remainder. In some cases the size of the bulk sample may be so large that a backhoe or front-end loader are needed to assist with quartering. Bulldozer trenches (costeans) are another means of exposing gravel faces for logging and bulk sampling.

4.6   TESTING AND EVALUATION

Natural roadbase materials are subjected to the same acceptance tests – including particle size distribution (grading), consistency (Atterberg) limits and compacted (Proctor) density – as crushed rock. The main differences lie in the specification limits set and the tolerances permitted. These are determined by climate, expected traffic volume and material availability. Dry climates and low traffic usually permit higher values of plasticity index and liquid limit, in particular, to be accepted.

The most important characteristics of many natural pavement materials, to be kept in mind when evaluating their suitability, are as follows:

• Both their coarse fraction and their fines should be composed of cemented and porous particles, which will break down when excessively compacted. Hence it is desirable that the laboratory pretreatment of samples be no more severe than that which will be applied on the road. Surprisingly, this is not generally the case –

laboratory samples are subjected to intense pounding during Proctor compaction.
- Furthermore, as a consequence of oversize removal during sample pretreatment, laboratory gradings may bear little relation to the material as compacted on the road. In particular, on-road gradings are likely to be much coarser and the proportion of fines less.

Conventional test *sample preparation* methods include removal of +20 mm stones after pounding, oven drying and disaggregation of cemented fines. This not only increases the proportion of particles finer than 425 μm, but also raises the PI of these fines relative to the on-road material. In some halloysitic clay soils, oven drying at 105°C causes irreversible dehydration and granulation. The use of mechanical agitation and dispersants in settling velocity tests to estimate the clay content of the −75 μm fraction likewise tends to overstate this component, since no such deflocculation occurs in the road pavement. Instead, repeated compaction cycles under standard laboratory conditions should be used to simulate, as far as practicable, the coarse particle breakdown due to handling and working on the road.

4.7   UNSEALED PAVEMENT MATERIALS

Material characteristics

About two-thirds of Australia's public roads system, as in other developing countries, remains unsealed because of small traffic volumes, generally less than 200 vpd (vehicles per day). These are mostly rural secondary roads, for which the cost of sealing – which means not just the cost of the bitumen and chippings, but also the expense of meeting higher geometrical design standards, using better-quality pavement gravel, and ongoing seal maintenance – cannot yet be justified. Although their pavement materials are more varied than those in sealed roads, it does not follow that these materials are unspecified and untested. Broadly speaking, the characteristics sought in wearing-course materials for unsealed roads, relative to those used in sealed construction, are the following:

- *Low-cost*   The material must be low-cost, which in practical terms means that it must be unprocessed and taken from a pit close to the job, even if this means that quality is somewhat compromised.
- *Wear-resistant*   Good wear resistance to surface abrasion by tyres, runoff and wind erosion is needed. This dictates a high proportion of gravel, preferably 20–60%, which also improves on-road mixing. Angular and rough-textured, but weak, rock fragments are preferred to rounded river gravels.
- *Ease of grading*   Adequate ease of grading and ride quality are

facilitated by a maximum particle size of 26 mm (preferably) or 53 mm, although this is at odds with wear resistance. A related requirement is to minimize tyre wear by avoiding hard and sharp-edged siliceous rock fragments.

- *Plastic clay binder* A proportion of plastic clay binder must be present to resist wear, to increase dry strength and to make the compacted surface as impermeable as possible. However, too much clay will cause the material to soften and become slippery on wetting. Hence it has to be diluted with sand and silt fines, which also fill the voids between the aggregate particles. Too much sand and the road surface will 'ravel' or disintegrate under traffic; too much silt and the surface becomes very dusty in dry weather, and sticky when wet.
- *Deformability* Moderate deformability (loss of pavement shape) and lack of shear strength are also tolerable in unsealed gravels. Pavement shape is in any case difficult to maintain, even in good-quality unsealed roadbase, and can be restored by periodic regrading (of the road surface, that is, not of its particle size distribution!).

The life of wearing-course material is relatively short, as the name implies, owing to erosion and grader trimming losses, hence it is usual to resheet with a new gravel layer every few years (where a sealed pavement would be designed for about 30 years). This is another reason why sub-specification unsealed materials are acceptable, since anything better would eventually be wasted.

Because of these requirements, most of the better-performing unsealed pavement materials have a higher gravel content than those in sealed roads, and this fraction is also coarser. Their fines have somewhat higher liquid limits, plasticity indices and linear shrinkages. The *dust ratio* (which is defined as the ratio of particles $-75\,\mu m$ to particles $-2.36\,mm$, roughly the fines to sand ratio) should ideally be 0.2–0.6. In many unsealed basecourses, however, the main difference lies in the plasticity allowed rather than in the overall particle size distribution. This is one reason why good unsealed gravels – such as coronus (coral) or ferricrete – can sometimes be later upgraded for sealing by lime stabilization.

Staged construction

Unsealed roads built to a high geometrical standard are often the first step in staged construction. By this means a cheap pavement on a low embankment is provided initially, where the main requirement is for an all-weather road rather than for high vehicle speeds. When the growth of traffic justifies the expense, perhaps 10–20 years later, a superior-quality base is laid and sealed, with the former unsealed pavement acting as a sub-base. This also allows the underlying subgrade to compact slowly under traffic and to reach its equilibrium moisture content.

Haul road pavement

A somewhat different approach is used in heavy-duty haul road pavement design. These roads are unsealed because the maintenance of a 'blacktop' would be too expensive, and adequate strength can be obtained from a thick waterbound macadam pavement made up of coarse, angular but free-draining particles (Figure 4.1). The trafficability of these roads under heavy wheel loading at low speeds is maintained by continual watering and grading, and by frequent resheeting. Compaction is provided by the truck wheels themselves and some care is needed in selection of wearing-course stone, since tyre wear can be a significant cost where sharp fragments are abundant. In open-cut coal-mines the base is often coarse washery reject, unweathered stone fragments with a top size of about 75 mm and very little –2.56 mm fines, but any source of relatively sound rock will suffice.

4.8   BENEFICIATION

Beneficiation or upgrading of *in situ* marginal roadbase materials is becoming more common than not in Australia, as better-quality deposits of natural gravel diminish and pavement design standards improve. Upgrading can take place in the pit or on the road, and strategies for achieving this, roughly in order of increasing cost, are now outlined:

- *In-pit blending* or on-road mixing (granular stabilization) of substandard soils to create one meeting specification grading limits.
- *Scalping off oversize* in the pit using a grizzly (a robust bar screen) and either discarding or crushing it.
- *Breaking up oversize* on the road, using a grid roller, a cleated vibrating roller or a 'rockbuster' (a small tractor-drawn impact crusher).
- *Single-stage crushing* run-of-pit material, or only oversize – sometimes a proportion of the fines is removed as undersize (note that this is only feasible where the soil is very dry and loose).
- *Stabilization* of run-of-pit or processed material using lime and/or cement.

Some of these procedures are discussed further in Chapters 5 and 18.

4.9   CASE HISTORY: CALCRETES, SOUTHERN AFRICA

The geological occurrence and engineering properties of calcareous duricrusts have been described comprehensively in a series of papers from South Africa (in particular, Netterberg, 1971, 1982). These deposits are widespread in areas of low rainfall throughout southern Africa, where they are the predominant sources of unprocessed roadbase. They form by cementation and replacement of pre-existing soil particles, the carbonate content ranging from 10% to 90%. It occurs as silty matrix, crack infillings, nodules, concretionary boulders and sheet-like hardpans. There are two basic types: pedogenic and groundwater calcretes. *Pedogenic calcretes* form by capillary rise and downward infiltration in the unsaturated zone of a soil profile. They are predominantly nodular, variably cemented and often weakly agglomerated into irregular masses, and are widespread in sand-dune terrains. *Groundwater calcretes* develop where lime-bearing phreatic waters discharge and evaporate, usually along ephemeral watercourses. These are limited in areal distribution but more cemented and larger in volume than pedogenic deposits, and are therefore better suited to crushing and screening operations.

Compared to ferruginous and siliceous duricrusts, most calcretes are geologically young, sometimes only a few thousand years old, though the older varieties (being more cemented) make the better gravel sources. One consequence of their recent age is their occurrence relatively low in the terrain, in landforms and beneath vegetation that are distinctive on airphotos. Other prospecting aids include probing rods and backhoe pits, since useful deposits tend to be most dense at their top and to be present at depths of 1–2 m. This top crust may be smooth-surfaced and difficult to rip, yet also difficult to blast because of abundant cavities. Conversely, the calcrete may be bouldery and generate oversize that cannot easily be broken down by tracking. Deposits are generally inhomogeneous and shallow, becoming finer and looser with depth; selecting a cutoff depth takes some skill on the part of the bulldozer operator.

Particle size distributions for calcrete materials are not especially diagnostic of their properties, since these depend on the degree of in-pit and on-road breakage to which they have been subjected. However, they are almost invariably gap-graded with a deficiency in the coarse sand and fine gravel sizes. Atterberg limits are similarly misleading because of particle microporosity and water absorption; high liquid limits indicate this characteristic rather than clay content. The clay itself may be palygorskite or illite. Perhaps surprisingly, many calcretes respond well to lime stabilization due to their clay or diatom (i.e. silica) content. Natural self-cementation can also increase the strength of some calcretes threefold in laboratory wetting and drying tests.

However, a number of problems have been reported with calcrete pavements:

- Because of their deviation from maximum-density grading, sensitivity to sample preparation techniques (especially oversize removal and oven drying) and ambiguous Atterberg results, it can be difficult to distinguish between adequate and inferior-quality materials. Self-cementing characteristics, in particular, are difficult to predict.

- Most calcretes contain some clay and salt, and in large amounts these can cause seal debonding.
- Their coarse particles are relatively weak, making most unprocessed calcretes unsuitable for even moderately trafficked (say, 500–1000 vpd) trunk roads.

Hence the harder calcretes are being increasingly crushed and screened for prepared roadbase and aggregate. This can result in a high proportion of dust and fine sand-sized waste, and very extensive pits requiring rehabilitation. The best crushed calcrete is comparable with other limestone aggregates, and may exhibit high polishing resistance and low flakiness, but a high proportion of particles are usually unsound to some degree. Siliceous calcrete may be subject to alkali–silica reaction, and all calcrete aggregates may present problems in concrete mixes due to their microporosity and clay content. In short, calcrete is a useful source of marginal-quality aggregate and roadbase in remote areas, but deposits require considerable care in selecting, evaluating and working.

REFERENCES AND FURTHER READING

Bullen, F. (1990) Materials problems associated with providing a highway infrastructure for remote areas. *Proceedings, 15th Australian Road Research Board Conference*, **15** (2), 217–31.
Carseldine, H.B. and Parkinson, B.D. (1986) Experience with silcrete materials in road construction in northwest Queensland. *Proceedings, 13th Australian Road Research Board Conference*, **13** (5), 146–61.
Charman, J.H. (1988) *Laterite in Road Pavements*. CIRIA/TRRL Special Report 47. Construction Industry Research and Information Association, and Transport and Road Research Laboratory, London.
Cocks, G.C. and Hamory, G. (1988) Road construction using lateritic gravel in Western Australia. *Proceedings, 2nd International Conference on Geomechanics in Tropical Soils*, Singapore, December, pp. 369–84.
Gidigasu, M.D. (1976) *Laterite Soil Engineering*. Elsevier, Amsterdam.
Grant, K. and Aitchison, G.D. (1970) The engineering significance of silcretes and ferricretes in Australia. *Engineering Geology*, **4**, 93–120.
Hight, D.W., Toll, D.G. and Grace, H. (1988) Naturally occurring gravels for road construction. *Proceedings, 2nd International Conference on Geomechanics in Tropical Soils*, Singapore, December, pp. 405–12.
Kilvington, D. and Hamory, G. (1986) The performance of natural gravels as basecourse in a sealed pavement. *Proceedings, 13th Australian Road Research Board Conference*, **13** (5), 86–101.
Lawrance, C.J. and Toole, T. (1984) *The Location, Selection and Use of Calcrete for Bituminous Road Construction in Botswana*. TRRL, Crowthorne, Berks, LR 1122.
Metcalf, J.B. (1991) Use of naturally-occurring but non-standard materials in low-cost road construction. *Geotechnical and Geological Engineering*, **9**, 155–65.

Netterberg, F. (1971) *Calcrete in Road Construction.* National Institute for Road Research, CSIR South Africa, Bulletin 10.

Netterberg, F. (1982) Geotechnical properties and behaviour of calcretes in South and South West Africa, in *Geotechnical Properties, Behaviour and Performance of Calcareous Soils*, eds K.R. Demars and R.C. Chaney, ASTM Special Technical Publication No. 777, pp. 296–309.

Netterberg, F. and Paige-Green, P. (1988) *Pavement Materials for Low Volume Roads in Southern Africa: A Review.* National Institute for Transport and Road Research, CSIR South Africa, Report RR 630.

Toll, D.G. (1991) Towards understanding the behaviour of naturally-occurring road construction materials. *Geotechnical and Geological Engineering*, **9**, 197–217.

Weinert, H.H. (1981) *The Natural Road Construction Materials of Southern Africa.* Academia, Pretoria and Cape Town.

CHAPTER 5

Blasting and crushing

Rock size reduction in quarries and mines is referred to as *comminution* and the processes include blasting, crushing and grinding. Blasting produces rock spalls with a maximum size typically 300–800 mm, which can be fed into a medium-sized primary crusher for reduction down to 100–200 mm. This material is then passed on to secondary and tertiary crushers for further size reduction and particle shaping. Typical nominal product sizes are 20 mm, 14 mm, 10 mm and 7 mm.

Most comminution techniques used in quarries are scaled-down versions of those developed for overburden removal and mineral processing. One important difference, however, is that quarry stone is generally stronger and tougher than overburden (and many ores), yet it must be blasted finer. On the other hand, the finest quarry product size is about 7mm, so further comminution by milling and grinding are not usually required.

Further information on blasting can be found in Sen (1995) and in manuals published by the explosives manufacturers. The standard reference book on comminution is that by Lowrison (1974), but much useful material can be found in textbooks on mineral processing, such as those by Kelly and Spottiswood (1982) and Wills (1988). Crushing in quarries is addressed specifically by Mellor (1990), and Australian quarrying practice is well summarized in Armstrong and Dutton (1993).

5.1 QUARRY BLASTING AND EXPLOSIVES

Elements of blast design

The basic elements of blast design can be thought of as the following:

- *Choice of explosive*, its distribution within the blasthole and the measures used to 'partition' the ratio of heave (gas pressure) to strain (shattering) energy.
- *Blasthole geometry*, its diameter, depth, angle of inclination, stemming length and backfill depth.

- *Blasthole pattern,* including spacing within rows, row spacing (burden distance) and delay configuration.

All of these parameters are influenced by the geomechanical properties of the rock, but even more so by the distribution of fractures within the *in situ* rockmass. The main considerations in quarry blasting are to generate minimum oversize and to satisfy environmental restrictions on dust and ground vibration. Good fragmentation with acceptable vibration is achieved by limiting the weight of explosive fired at any instant, and by carefully distributing this charge within blastholes (see Chapter 20). Oversize blocks are not always unwelcome, however, because they may sell at a premium for use as breakwater armourstone.

Quarry blastholes tend to be narrow, heavily charged (high *powder factor*) and closely spaced relative to mine overburden blasts. Quarry blasts are relatively small (Figure 5.1) – a blasting round might fragment $6000 \, m^3$ (say 13 000 t) of stone with one or two rows of holes. A large coal overburden round, by comparison, might blast $100 000 \, m^3$ and be 10 rows deep. Many quarries blast only once a week, usually at a predictable time like midday, while mine blasts occur daily. Drilling and blasting costs are less important in quarrying than in mining, since money spent here can reduce downstream processing costs and prevent complaints from neighbours (who also tend to live closer in the case of quarries!).

Explosives

The main types of explosives used in quarrying are compared in Table 5.1. Traditional nitroglycerine-based high explosives, such as AN gelignite 60, are little used today. The most commonly used explosive, actually a blasting agent since its constituents are non-explosive, is ANFO (a mixture of 94% ammonium nitrate and 6% fuel oil). This is cheap, energetic and safe to transport, store and load.

ANFO was the first of the bulk explosives that could be poured straight into blastholes, thus simplifying loading and improving charge-to-rock energy coupling. It is well suited to closely jointed hard rock, since it generates a higher proportion of its energy as gas pressure (heave energy) than any other explosive. It is less satisfactory in strong and massive rocks where more strain energy is required, and may not detonate at all in small-diameter blastholes. However, the main drawback of ANFO is its water-sensitivity; it will not initiate in wet holes and may only partially detonate in damp ones.

Increased shattering power (brisance) for ANFO was obtained initially by denser packing – note the increased bulk (volumetric) strength of pressure-loaded ANFO in Table 5.1 – and later by the addition of aluminium dust (hence ALANFO). From the 1960s *watergels* or 'slurries'

Blasting and crushing

Figure 5.1 Bench blast, Riverview Quarry, Adelaide. Although this is a major quarry producing aggregates from dolomite and quartzite, a typical blast might include only 20 holes. (Photo: M. Drechsler.)

began to replace ANFO in wet holes. Watergels are saturated aqueous solutions with gelling agents and ammonium nitrate (AN) prills (spherical grains) in suspension. By varying the ingredients, their strength can be tailored to the rock, though at a cost penalty relative to ANFO.

The latest generation of explosives are *emulsions* of AN-saturated droplets about 1 µm (0.001 mm) in diameter, dispersed in an oil–wax

Table 5.1 Comparison of quarrying explosives

Explosive	Weight strength (%)	Bulk strength (%)	Density (g/cm³)	VODᵃ (m/s)	Energy outputᵇ (s/h %)
ANFO, loose-poured	100	100	0.80	3500	13/87H 25/75S
ANFO, pressure-loaded	100	119	0.95	–	–
ANFO with 5% aluminium dust	115	121	0.84	–	–
ANFO with 10% aluminium dust	128	142	0.89	–	–
ANFO/emulsion mix ('heavy' ANFO)	90	120	1.10	3700	–
Watergel, low-density	80	112	1.15	4300	21/79H 33/67S
Watergel, high-density	115	182	1.30	4500	–
Watergel, cartridged	–	–	1.23	4300	–
Emulsion, low-density	65	100	1.15	4500	29/71H 48/52S
Emulsion, high-density	70	110	1.30	5000	–
AN gelignite 60	130	230	1.41	3000	18/82H 30/70S

ᵃ VOD = velocity of detonation.
ᵇ Energy output is the percentage ratio of strain energy to heave energy (s/h); H indicates proportion of each in hard rock, S in soft rock.

mixture. Emulsions are waterproof, offer the highest brisance of any of the bulk explosives, and hence are well suited to hard rock quarrying. Because of their very rapid reaction rate, indicated by a velocity of detonation (VOD) up to 5000 m/s, their strain energy component is about three times that of ANFO. Like the watergels, emulsion mixtures can be adjusted to suit the rock, and can be pumped into blastholes. Blends of ANFO with emulsion, or 'heavy ANFO', may be more economical in some rockmasses.

5.2   BLAST DESIGN

Blast pattern design

A typical ANFO-loaded quarry blasthole design is illustrated in Figure 5.2. Holes are drilled using tracked rotary-percussion rigs, generally at diameters of 90–150 mm. Blastholes may be vertical or, more commonly these days, inclined at about 70° from the horizontal. Inclined holes are more difficult to drill straight, but result in finer and more uniform fragmentation. After charging, blastholes are stemmed with crushed stone to depths of at least 2 m. Shorter stemming depths can result in flyrock, airblast and cutoffs (breaks in the detonating cord, which isolate later-firing parts of the pattern and prevent their ignition). Long stemming columns, on the other hand, inhibit airblast but may increase

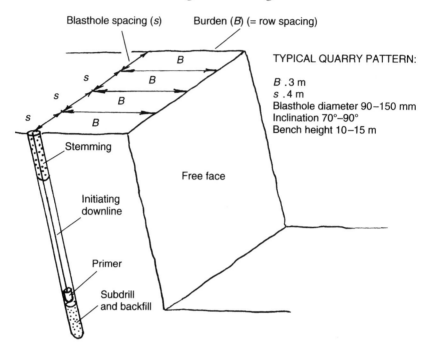

Figure 5.2   Typical ANFO-loaded blasthole design and explanation of some blasting terms.

ground vibration by over-confining explosive energy; they also cause poor fragmentation around the collar of the blasthole.

Some of the terms used in blast pattern design are illustrated in Figures 5.2 and 5.3. In quarry blasting the inter-row distance or *burden* is typically about 3 m, blasthole *spacing* within rows is 3–5 m and bench height is 10–15 m. Drilling and blasting costs diminish as these dimensions increase, but so do the risks of generating oversize and excessive ground vibrations. A *free face* is any plane that can reflect strain energy, which includes bench faces and open joints. Blasthole patterns are most commonly rectangular, with spacing around 1.3 times burden, but may also be square or even oblique.

Initiation sequence and delays

The blasthole initiation sequence by delayed firing is even more important in quarrying than in mining, since it largely determines the degree of rockmass fracturing and the ground vibration intensity. *Delays* are slow fuses inserted into detonating circuits; their purpose is to limit the weight of explosive firing at any instant. Blasts may therefore be

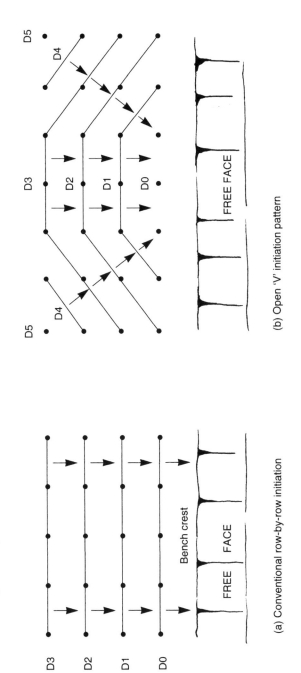

D = Delay number (D0 = zero time, D1 = 25 ms, D2 = 50 ms, etc.)

(a) Conventional row-by-row initiation

(b) Open 'V' initiation pattern

Figure 5.3 Blasthole patterns and initiation sequences. Row-by-row initiation is more common in quarrying, though the 'V' pattern offers better fragmentation.

initiated hole-by-hole, row-by-row or in the 'V' pattern shown in Figure 5.3, which improves fragmentation by throwing rock towards the centre of the 'V' and thereby maximizes block-to-block impacts. Note that the drilling pattern is identical in both blast designs; only the firing sequence changes.

Inter-row delays are usually set at 17–35 ms (milliseconds). Longer delays impair fragmentation and cause cutoffs, while shorter delays may cause 'choking' (overconfinement) of rear rows, hence backbreak and flyrock. *Backbreak* is fracturing that extends into the rockmass beyond the rear row, resulting in bench face instability when the broken rock is removed. Short delays are becoming more common as their timing accuracy improves, although previously they could overlap and conspicuously increase ground vibration amplitude. Another innovation with relevance to quarry blasting is the use of non-electric (Nonel) initiation, which, among other virtues, eliminates the high-frequency airblast associated with detonating cord firing on the ground surface.

5.3   FRAGMENTATION MECHANISMS

The fragmentation mechanisms produced by blast detonation depend largely on whether the rock is closely jointed or massive. In the first case gas pressure or *heave energy* is the dominant factor, and most fragments will be bedding- or joint-bounded. Hence the size distribution of the blasted rock fragments will be largely determined by the original jointing intensity (joint area per cubic metre). In massive rock *strain wave energy* will be more important, and most fractures will result from radial cracking and gas-pressure-driven joint extension. Some additional comminution is likely to be caused by flexure during rockmass dilation and rock-to-rock collisions.

5.4   GEOLOGICAL ASPECTS OF BLAST DESIGN

Intact rock properties

Most lithological or intact rock properties (Table 5.2) have a greater effect on crushing than on blasting, but tensile strength and elastic modulus are important in both cases, with one proviso. In blasting we are concerned with *dynamic* properties and the effects of almost instantaneous loading; rock is both stronger and stiffer under these conditions, so even soft rock can behave in a brittle manner. Crushing involves much slower loading, and hence *static* geomechanical properties, obtained from laboratory tests, are more relevant.

Table 5.2 Geological factors in blast and crusher design

Lithological/intact rock properties

Rock type, mineralogy (especially quartz content, clay content) and fabric

Strength: point load strength index (PLSI), uniaxial compressive strength (UCS) and indirect tensile strength (ITS)

Elastic modulus (static, E_s), modulus ratio (E/UCS), toughness/brittleness characteristics

Bulk density, porosity or water absorption, moisture content

Ultrasonic velocities (V_p and V_s), dynamic modulus (E_d), E_s/E_d ratio

Structural/rockmass properties

Joint sets present: dip and dip direction, prevalence of each set

Joint persistence, spacing (and whether evenly spaced, clustered or randomly distributed)

Joint aperture (tightness), dampness and infill/cement

Layer thickness, location of soft/hard bands and beds, strike and dip

Rockmass properties

Conversely, structural or rockmass properties, especially jointing, are much more significant in blasting than they are in crushing. 'Joint' is used here in the engineering sense, meaning any structural discontinuity within a rockmass. This includes true (geological) joints but also bedding, foliation, cleavage and flow layering, as well as fractures induced by previous blasts.

Joint aperture (width or openness) is important in that slightly open joints (say +1 mm) can act as free faces, confining the strain energy within single joint-bounded blocks. The charged blocks become excessively fragmented, leaving the surrounding ones unbroken (and hence generating oversize). Open joints can also waste blast energy by premature venting of explosive gases, which also causes airblast. Finally, open joints act as groundwater conduits, thereby precluding the use of ANFO below the water table. However, closely spaced tight joints greatly improve fragmentation throughout the rockmass. They transmit most of the strain energy (especially if damp), while at the same time providing a network of fractures for gas pressure to act on.

The *orientation and attitude* (dip) of the principal joint set can affect fragmentation and face stability. Blast comminution is improved when the face is aligned parallel to the main joint direction, and stability will be maintained so long as the joint dip is steeper than the face (that is, these joints do not 'daylight'). Methods of quarry and pit slope stability investigation, and open-pit blast design, are extensively described in Hoek and Bray (1977). Bench slope stability appears to be much less of a problem in quarries than it is in open-cut mines, since the former are almost always shallower and work stronger rock. Some instability can, however, be expected in bedded and dipping deposits, such as some limestones and quartzites.

The effects of *groundwater* in tight rockmasses may be negligble, or even beneficial (owing to improved coupling across tight joint faces), but in general its presence will require that ANFO be replaced by watergels at about three times the cost. Furthermore, damp fines may choke crushers, accelerate abrasion of crusher plates, clog screens and overload conveyor belting. In some cases, such as cavernous limestones, fractured basalt flows and jointed quartzites, the quarry rock itself may be an aquifer, and its dewatering may be a major operating expense.

5.5 PRIMARY CRUSHING

Crusher classification

It is generally not possible to comminute blasted rock satisfactorily in one operation, so crushers are described as primary, secondary or tertiary depending on the size of the feed. The principal types are shown in Figures 5.4 to 5.9. Most are available in either primary or secondary versions, but jaws and gyratories are usually primary crushers, while impactors and cones perform secondary duty. The maximum particle size usually processed is about 800 mm for primaries and 200 mm for secondaries.

One means of classifying crushers is in terms of *reduction ratio*, the ratio of maximum feed size to maximum product size. Alternatively, this can be defined as the ratio of 80% of entry width (gape) to 80% of exit width (closed side setting). Compression crushers (jaws and gyratories) have maximum reduction ratios of 8:1 to 10:1 but normally operate at about half this, since high ratios result in excessive fines due to interparticle rubbing and generate platy particles through shearing. Impact crushers perform satisfactorily at much higher reduction ratios, in the range 10:1 to 20:1.

Most primary crushers have a feeder, which evens out the flow of broken rock and separates (scalps) undersize using an inclined bar screen. Spalls retained on the screen trickle into the primary crusher, while the undersize bypasses this stage and is directed either straight into the secondary crusher, or to waste if it is weathered. This arrangement increases the throughput or *capacity* of the primary machine and prevents fines packing in its crushing chamber. Coarse oversize is separated at the blast face prior to loading, or fractured by a hydraulic-arm rock breaker when it jams in the primary crusher entry slot.

Jaw crushers

Jaw crushers (Figure 5.4) break rock by slow compression–release cycles between ribbed plates, one fixed and one moving, on opposite sides of a wedge-shaped chamber. This narrows downwards, so that after blocks are split on the compression stroke the pieces slip further down on the release stroke. At some point they become jammed again and the cycle is repeated, until eventually the fragments fall out of the base of the crusher.

Jaw crushers can be classified into those with a single or double toggle action. Double-toggle machines are able to exert larger compressive loadings, and are therefore used for handling the strongest (UCS up to 500 MPa), most blocky (up to 3 m) and most-abrasive feed. However, in most quarrying applications, single-toggle machines perform satisfactorily and are less costly. They are well suited to small to medium-sized operations, including mobile plant, where intermittent production is normal. A number of jaw arrangements involving different pivot points, curved and planar plates are also offered. Jaws are cheaper

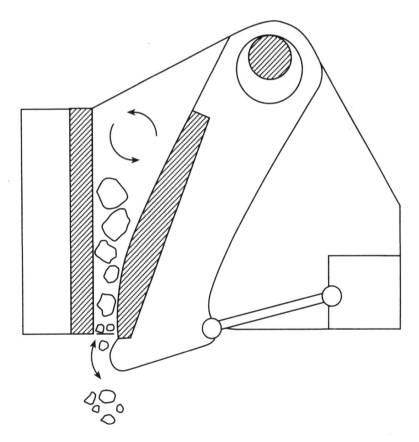

Figure 5.4   Single-toggle jaw crusher, schematic. The left side plate (shaded) is fixed, with the right-hand jaw moving in an elliptical path.

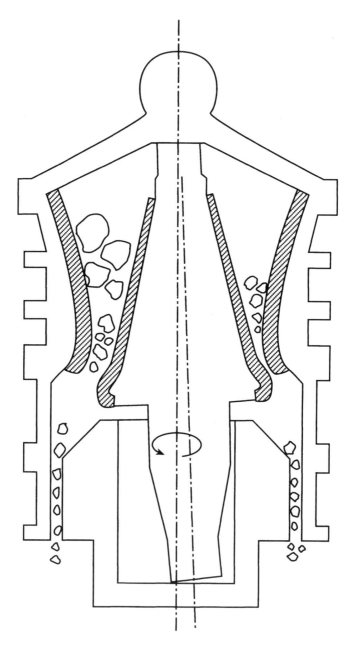

Figure 5.5   Gyratory crusher, schematic. The inner cone moves eccentrically, but the outer plate (mantle) is fixed.

in both capital and maintenance costs than gyratories handling the same feed size, but their production rate is much less. They are economic for up to 500–1000 tph (tonnes per hour), above which gyratories are favoured.

Gyratory crushers

A gyratory crusher resembles a pestle in a narrow open-base mortar, or two cones, one inverted within the other (Figures 5.5 and 5.6). The inner pestle-like solid cone, known as the *head*, moves eccentrically around the fixed outer bowl, alternately closing and opening gaps around the lower rim. It functions as a multitude of jaw crushers, and can therefore be much smaller for the same capacity. Gyratory crushers tend to be chosen where high throughput (up to 5000 tph) is required and where a relatively fine and uniformly blasted feedstock is assured, since they are less tolerant of oversize than jaws. They are more compact than jaw crushers of the same capacity and more efficient in terms of energy consumption per unit of output. Because their construction is lighter, they require less elaborate foundations (but more headroom) than jaw crushers and their feeding arrangements are simpler. They are 'choke fed', meaning that the machine works best when buried in rock

Figure 5.6   Mortar-and-pestle arrangement of a gyratory crusher at Prospect quarry, near Sydney, New South Wales. When operating, this would be buried in blasted rock spalls ('choke feeding').

fragments, as opposed to the 'trickle feed' arrangements needed for jaw crushers.

Both types of compression crushers turn out a fairly uniformly sized product, though the gyratory output has somewhat more fines. In hard rock the gyratory action is said to produce more cubic chips, although both types tend to generate a proportion of sheared and unsound particles. Product grading curves for both of these compression machines are similar: the particle size range is narrower than from impact crushing, but wider than from a rolls crusher.

Rolls crushers

A third primary type, the rolls crusher (Figure 5.7), is limited to weaker (UCS less than 100 MPa) and non-abrasive rock, such as limestone and brick shale. The machine may be equipped with single, double or multiple rollers, and these can be fitted with teeth, cleated or ribbed. Double rolls are most common, with one drum axis fixed and the other spring-loaded to accommodate oversize slabs.

Rolls crushers are relatively cheap in relation to their high capacity, economical in power consumption, easily transportable and low in

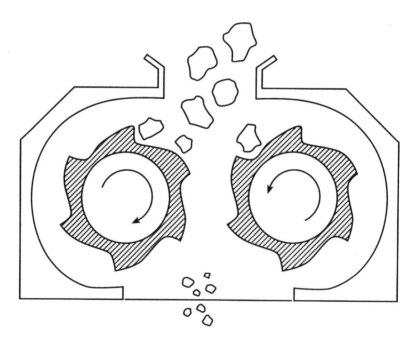

Figure 5.7   Double rolls crusher with interlocking teeth. Both rolls rotate towards the centre, but one is fixed and the other sprung to allow oversize blocks to pass through.

profile when erected (which simplifies feeding). These characteristics make them a good choice for crushing demolition rubble, especially concrete slabs and brickwork, or weak excavated rock for select fill on construction sites. Rolls crushers are particularly suited to clay-rich feed, and to processing plastic and slabby bedded rocks, which would tend to block jaw or gyratory crushers.

5.6   SECONDARY AND TERTIARY CRUSHING

Secondary or reduction crushers have two functions in quarrying. They further comminute primary crusher output to marketable sizes, and may at the same time correct particle shape and soundness (that is, make platy particles more cubic and disintegrate porous or microfractured ones). In some materials primary comminution may be unnecessary, since single-stage crushing is possible with large reduction ratio (10:1 or more) impactors. Tertiary crushers are employed to shape flaky aggregates and to increase the percentage of fines generated. They generally are similar in design to secondaries, but have a closer set and hence very small reduction ratios (only about 2:1).

A wide variety of machines are used for secondary and tertiary comminution, including scaled-down jaw crushers (granulators) and short-head gyratories, but the most common are cone and impact crushers. Cones are the most popular secondary machines in Australia, while impact crushers dominate the tertiary market. The main operational problems experienced are as follows:

- *Capacity mismatches* between primary and secondary crushers, the first producing too much or too little product for the efficient operation of the second stage.
- *Wear in the primary crusher* enlarging outlet settings and producing oversized feed for the later stages.
- *Excessive fines* passing through the secondary crushers, causing increased wear and reducing their capacity.

Cone crushers

Cone crushers are really small gyratories in which the oscillating inner cone and fixed outer cone are roughly parallel, and the inner cone is pivoted from below rather than suspended from above. Cone crushers have large capacities in relation to their small size, and can handle hard and abrasive feed. Compared to impact crushers they tend to produce a narrower range of particle sizes, less fines and more flaky particles. The two crushing surfaces are held together by springs, allowing tough oversize particles to pass through without damaging the machine.

Impact crushers

Impact crushers (Figures 5.8 and 5.9) break rock by the action of rapidly rotating hammers or beaters attached to a central shaft, which may be horizontally or vertically mounted. The feed particles cascade into the crushing chamber and shatter on impact with the whirling beaters, or are deflected by them to strike hardened breaker plates lining the chamber. The beaters may be hinged swing hammers or, more commonly, fixed

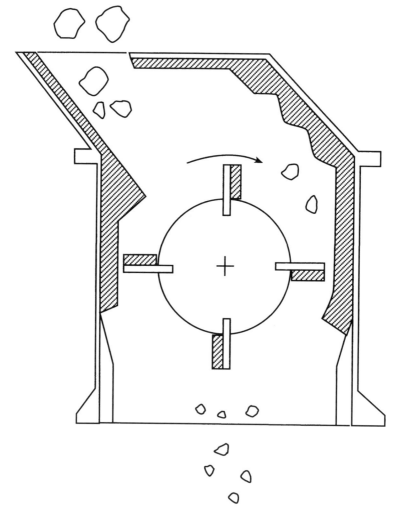

Figure 5.8 Impact crusher, schematic. Rock pieces falling from above are struck by fast-moving blow bars or swing hammers, and either shatter immediately or are flung against the armoured lining (breaker plates).

Figure 5.9   Small portable impact crusher being used for comminution of demolition rubble. Note hopper and elevating feeder conveyor, and product stockpile at rear. The right-hand conveyor feeds to a screening unit.

blow bars rotating at 250–3000 rpm. Impact crushers are relatively light and cheap for their capacity, and do not require elaborate foundations. Their main disadvantage is the cost of frictional and chipping wear to beaters and breaker plates, and the downtime required for their replacement.

In general terms, impact crushers are limited to rock with UCS values below about 150 MPa and granular quartz ('free silica') contents below 5–7%, though some machines can handle stronger feed. The beating action is advantageous with wet and sticky clay-rich river gravels, though clean gravels are too hard and abrasive for this type of comminution. Product particle shape is particularly good, since high-speed impact shatters unsound fragments and tends to generate near-cubic chips and abundant fines.

Impact crushers as a group are divided between large single-stage and smaller reduction machines. The larger *impact breakers* are characterized by very high reduction ratios (up to 40:1, but 10:1 to 20:1 is more usual) and relatively low velocities (250–500 rpm). They can accept large blocks as feed and some impactors offer capacities up to 1500 tph – comparable with primary gyratories in both respects, though not in the quality of rock that can be crushed. Smaller secondary impactors operate at higher rotational speeds (500–2500 rpm) but with finer feedstock (–200 mm) and generate more fines.

Vertical shaft impact (VSI) crushers have no hammers at all, but depend on high-speed collisions between rock fragments (rock-on-rock breakage). A ring of breaker plates is protected from wear by a cushion of fallen stone, against which other particles are thrown horizontally at high speed. These are dedicated tertiary machines with reduction ratios of only 2:1, used for shaping hard but flaky particles with a maximum size of about 40 mm. They are also unlike other impact crushers in producing few fines and being unaffected by abrasive feed.

At some sites it is necessary to manufacture fine aggregate (sand) from crushed rock, either because the natural material is insufficient or because it fails to satisfy grading, soundness or roundness requirements. *Sand plants* for this purpose are in effect quaternary crushers, rod or ball mills that grind –25 mm feed from tertiary machines. Alternatively, fine crushed rock can be recirculated through tertiary cones until sufficient sand is generated (closed-circuit crushing).

5.7   GEOLOGICAL INPUT FOR CRUSHER SPECIFICATION

Crushing mechanisms

There are four basic ways to crush rock – by compression, impact, attrition and shear (Figure 5.10). Most crushers employ a combination of these mechanisms:

- *Indirect tensile failure*, due to slow compression, predominates in jaw and gyratory crushers where the feed is strong and brittle, as is generally the case with aggregate-quality rock. Tensile breakage tends to produce a small quantity of similarly sized particles with few fines, which is one reason why secondary crushing is required.
- *Dynamic (impulsive) loading*, due to high-speed rock-on-steel or interparticle collision, is the basis of impact crushing. In these conditions even soft rock behaves in a brittle manner, clean tensile fractures can form, and the resulting particles are equidimensional and unstrained. Disintegration of unsound particles adds a proportion of fines and hence the product is well graded.
- *Attrition*, or interparticle rubbing, is a much more important reduction mechanism in milling than it is in crushing. It is associated with particle rounding in choked crushers, but is otherwise insignificant.
- *Shearing* is also a minor crushing mechanism, associated mainly with plastically deforming shales and with the tearing action of toothed rolls crushers. It generates platy, strained particles.

The general requirements for aggregate are that the rock chips should be clean, tough, durable and roughly cubical. The requirements for roadbase are similar but less restrictive, since a proportion of fines is not

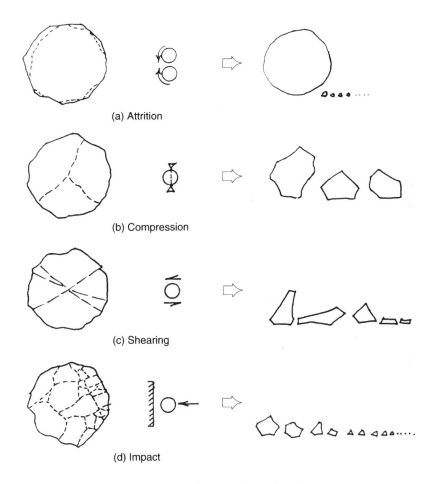

Figure 5.10 Crushing mechanisms. See text for explanation.

only tolerated but necessary. These specifications can be met by most unweathered and unaltered igneous and metamorphic rocks, and by a few well-cemented sedimentary rocks. Such lithologies typically have uniaxial compressive strength (UCS) values in the range 50–500 MPa, densities around 2.7 t/m^3, porosities of less than 1% and elastic moduli around 500 times their UCS.

Measures of rock crushability

One method of classifying rock suitable for aggregate is in terms of *modulus ratio* (E/UCS), as shown in Figure 5.11. In this diagram compressive strength is plotted against elastic modulus, and the group of

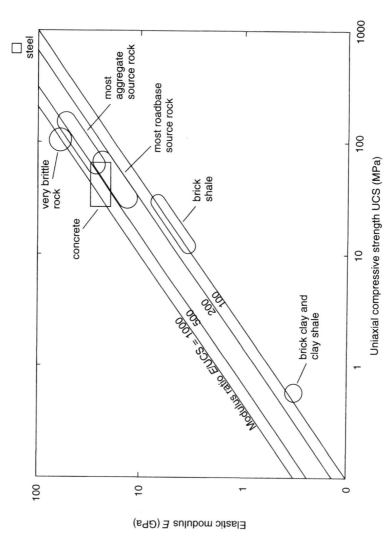

Figure 5.11 Classification of rock materials in terms of modulus ratios (E/UCS), with steel and high-strength concrete for comparison.

aggregate-quality rocks are shown to be stronger and stiffer than the roadbase sources, although there is some overlap. Some very brittle cherts and limestones, with modulus ratios around 1000, are also indicated; these can be blasted and crushed easily, but break down under repeated impacts in road surfacing.

In crusher specifications, rock strength is usually quoted in terms of its *uniaxial compressive strength* (UCS), even though the main breakage mechanism is believed to be indirect tensile failure caused by compressional or impact loading. Tensile strength averages about 7% of UCS, but varies from 2% (for very brittle rock) to 20% (for ductile shales). The usual rule-of-thumb adopted for primary crusher design is to assume that 99% of the feed particles will have a strength less than their mean UCS plus three standard deviations (SD). Given that the SD for UCS values is commonly about 30% of the mean, this can mean designing for twice the average rock strength measured.

A more useful measure of rock 'crushability' than strength is the work expended in breaking a unit weight of the material, otherwise known as the specific energy of fracture. This is most widely expressed in terms of the *Bond work index* (BWI), in kilowatt hours per tonne (kWh/t). The procedure for determining BWI, known as the 'double pendulum' test, measures the impact energy required to fracture a rock lump struck on opposite sides by swinging hammers.

Toughness can be thought of as the rock's resistance to crack penetration, or simply as the opposite of brittleness. Brittle rocks are easily crushed, even though their surface can be hard, and fail at very low strains (typically less than 0.5%). Tough aggregates require more energy in comminution and considerably reduce crusher capacity, but are necessary for road surfacing, high-strength concrete and railway ballast. Toughness is related to strength and elasticity, so that a *toughness index* (resilient modulus) equal to $UCS^2/2E$ has been defined. One measure of toughness is the area under the stress–strain curve recorded during the UCS test, which is proportional to the strain energy stored in the specimen prior to failure.

The other parameter important in crusher selection and design is the *rock abrasivity*. It is quantified in terms of milligrams of metal wear per kilogram of rock fragments in paddle abrasion (stirring) tests. Abrasivity is a function of compressive strength (the main cause of chipping or impact wear) and granular quartz content (the source of most frictional wear on crusher plates, roller teeth and gyratory mantles). These two types of wear require different countermeasures. Impact wear or chipping resistance dictates hard-surfaced but brittle tungsten alloys, while abrasive wear is best minimized by the use of tough, work-hardening manganese steel components. The influence of abrasive minerals – primarily granular quartz, but including all framework silicates – is most apparent when comparing wear rates between limestone crushers (low) and those handling quartzose rocks (high).

A number of other geological factors influence crusher design, but are less easily quantified. These include: maximum expected particle size and shape (estimated from joint and bedding spacing); proportion of clay fines liberated during comminution or present in the feed material; moisture content of the feed; and whether or not hard inclusions (such as siderite nodules) are present within the rockmass.

REFERENCES AND FURTHER READING

Armstrong, A.T. and Dutton, A.H. (1993) *Handbook on Quarrying*, 5th edn. South Australia Department of Mines and Energy, Adelaide.

Ball, M.J. (1988) A review of blast design considerations in quarrying and opencast mining, Parts I and II. *Quarry Management*, June, pp. 35–9, and July, pp. 23–7.

Ball, M.J. (1990) Emulsion explosives. *Mine and Quarry*, June, pp. 14–16.

Dalgleish, I. (1989) Cost-effective drilling and blasting. *Quarry Management*, January, pp. 19–27.

Hagan, T. (1980) Effects of some structural properties of rock on the design and results of blasting. *Proceedings, 3rd Australian–New Zealand Geomechanics Conference*, Wellington, NZ.

Hoek, E. and Bray, J.W. (1977) *Rock Slope Engineering*, 2nd edn. Institution of Mining and Metallurgy, London.

Kelly, E.G. and Spottiswood, D.J. (1982) *Introduction to Mineral Processing*. John Wiley, New York.

Lowrison, G.C. (1974) *Crushing and Grinding*. Butterworths, London.

McKenzie, C. (1993) Quarry blast diagnosis – How can it help? *Quarry Management*, July, pp. 17–23.

Mellor, S.H. (1990) *An Introduction to Crushing and Screening*. Institute of Quarrying, UK.

Rorke, A.J. (1988) A scientific approach to blast design – the influence of rock characteristics. *Quarry Management*, October, pp. 33–43.

Sen, G.C. (1995) *Blasting Technology*. University of New South Wales Press, Sydney.

Wang, H., Latham, J.P. and Poole, A.B. (1991) Prediction of block size distribution for quarrying. *Quarterly Journal of Engineering Geology*, **24**, 91–9.

Wills, B.A. (1988) *Mineral Processing Technology*, 4th edn. Pergamon, Oxford.

Wylde, L.J. (ed.) (1978) *Workshop on Aggregate and Road Base Production with Mobile Equipment*. Australian Road Research Board Report ARR 98.

CHAPTER 6

Sizing and processing

The downstream treatment of crusher products includes separation into size fractions, cleaning and blending to meet differing specifications. In this chapter emphasis is placed on processing of alluvial sand and gravel, but the screening requirements for crushed hard rock are similar. Gravel differs from crushed stone chiefly in having smoother and more rounded particle faces, and in being derived from harder and more durable rock types.

Sand and gravel production processes can be grouped into *extraction* (digging, loading and transporting), *preparation* (clay matrix disaggregation, washing, scrubbing and cobble crushing), *sizing* (gravel screening and sand classification) and *beneficiation* (removal of deleterious particles, desliming and dewatering). The present tendency is towards more elaborate plants, with more diverse products obtained from less-promising feedstocks. Other recent developments are the use of friable, weakly cemented sandstone and conglomerate as sources, and the wider use of dredged marine sands.

Sand and gravel processing are well covered in the British technical literature, reflecting a former dependence (which is declining) on fluvioglacial outwash deposits rather than hard rock quarrying for aggregates. Three recent and comprehensive references are those by Mellor (1990), Littler (1990) and Smith and Collis (1993). An older, but still relevant, summary of aggregate processing techniques based on American practice is that by Rockwood (1948).

6.1 EXTRACTION

The first consideration in selecting mining equipment for alluvial pits is whether they are to be worked wet or dry, i.e. above or below the water table. In most cases it is not economic to dewater the excavation, and hence different types of excavators must be used in dry and wet situations.

Dry pits

Excavation in dry pits is generally carried out using conventional earth-moving plant such as bulldozers, scrapers, front-end loaders (FELs) and trucks. Face shovels and hydraulic excavators, smaller versions of the overburden removal equipment at opencast mines, are used in some alluvial quarries. Cemented or densely packed gravels, particularly those with a stiff clay matrix, can be loosened by bulldozer ripping or – much less commonly – by blasting. The main geotechnical problems in dry pits are face collapse due to oversteepening in gravels and piping in sands, and the presence of occasional large boulders and of layers cemented by ferricrete, calcrete or silcrete. Selective working is much more feasible in dry pits than in wet ones.

Wet pits

In wet pits sand is generally mined by suction-cutter dredging, using a barge-mounted pump (Figure 6.1). Usually the pond floor is simply agitated to cause the sand to go into suspension, but where the material is bound by clay matrix a more vigorous action is required and this is provided by renewable knives or water jets mounted on the cutter head. The sand slurry, typically 10–20% solids by weight, is pumped through a floating pipeline to shore for processing and stockpiling. The pumps used are centrifugal (impeller) types for the most part, but jet pumps (operating on the siphon principle) and air-lift (reverse-circulation) pumps are used in special circumstances. Centrifugal pumps can draw sand from depths down to about 25 m, while jet and air-lift pumps actually improve in efficiency with depth.

Gravel extraction underwater is more difficult, with correspondingly lower productivity. Fine gravel dredging can be carried out by suction-cutter barges equipped with powerful pumps, but operational penalties include increased impeller and pipeline wear relative to sand. Much more power is consumed because of the higher fluid velocities required to keep the gravel in suspension, and because a greater volume of water has to be moved for a given amount of solids.

Small draglines are generally preferred for gravel extraction in wet pits, despite their working depth limitation (5–10 m) and inability to dig tough clayey materials. For coarser or more compact gravels, a variety of bucketwheel and ladder dredges, floating grabs and clamshells are used.

The geotechnical information needed for assessing whether a material can be economically dredged or not includes the following:

- *Particle size distribution*, with particular emphasis on the percentage of clay fines and of oversize cobbles. Both size extremes can easily be underestimated from drilling samples. The quartz content and angularity of the sand fraction should also be noted, since coarse 'sharp' sand is especially abrasive in pumps and pipes.
- *Consistency* (stiffness) of the matrix, its silt/clay ratio and its dispersivity. This is important for estimating cutter resistance,

Figure 6.1 Suction-cutter barge working an offstream mineral sands deposit, Tomago, New South Wales. Note the lifting frame for manoeuvring the submerged suction pipe, the floating pipeline and the sand delta created by discharge from the treatment plant.

washing characteristics, settling time in tailings ponds, and turbidity if released into downstream river waters.

- *Obstructions* such as cemented or densely packed gravel bands, tree trunks, boulders and bedrock pinnacles; and the presence of deleterious layers such as clay lenses or peat bodies.
- *Aquifer characteristics* of the alluvium (such as static water level, transmissivity, storage coefficient and water quality) if pool lowering is planned.

6.2 WASHING, SCRUBBING AND CRUSHING

Alluvial sands and gravels are prepared for size separation by oversize removal and fines reduction, in other words 'topping and tailing' the grading curve. More consolidated materials require lump breaking, matrix disaggregation and scrubbing.

Washing

Washing, as the term is used here, means stripping off relatively small quantities of silt weakly adhering to gravel-sized particles. The quantity of water and the degree of agitation needed are relatively small. At its simplest, washing is performed by water jets directed onto the material as it passes across a vibrating deck or through a rotating cylindrical screen. The water sprays not only clean the aggregate, but also increase the screen capacity and reduce its rate of wear.

Scrubbing

Scrubbing is a more energetic process, required when a gravel deposit is so fines-rich that it must be broken down by agitation and thoroughly flushed. These fines are not only more abundant, they are also more clay-rich, creating a considerable disposal problem. As a result, scrubbing is more expensive than washing and would only be used when less-clayey deposits were unavailable.

In mechanical terms the main distinction between scrubbing and washing lies in the energy involved. Drum washers operate at about half the rotational speed of drum scrubbers, but are otherwise similar in design. They are inclined revolving steel cylinders lined with projecting shelves to lift, turn and drop clayey gravel in a continuous water shower. The disaggregating action is due to stone-on-stone impact, complemented by jostling of particles immersed in the lower half of the drum. The operating principles of drum washers and scrubbers are illustrated in Figure 6.2.

In another type of scrubber, known as a logwasher, the agitation is performed by rotating paddles in a sloping trough similar to a heavy-duty spiral classifier (Figure 6.3). The paddles are attached to two thick, parallel, contra-rotating axles or 'logs'. Logwashers are well suited to disaggregating dense sands and weak sandstones, and helical screws can be substituted for the logs.

Crushing

Crushing is considerably less significant in gravel pits than it is in hard rock quarries, since oversize here usually means coarser than about 100 mm and may constitute only 5–10% of the material. In some pits these cobbles are simply discarded, but they may also be sold as landscaping or beaching stone. The usual arrangement is a small single-stage gyratory or cone crusher reducing from 100 mm down to 20 mm. Impact crushers, though cheaper, are not widely used because of the severe wear caused by hard siliceous cobbles.

6.3   COARSE AGGREGATE SCREENING

Purpose of screening

Screening provides the means by which run-of-crusher materials down to a size of about 3 mm are separated into saleable products. In addition

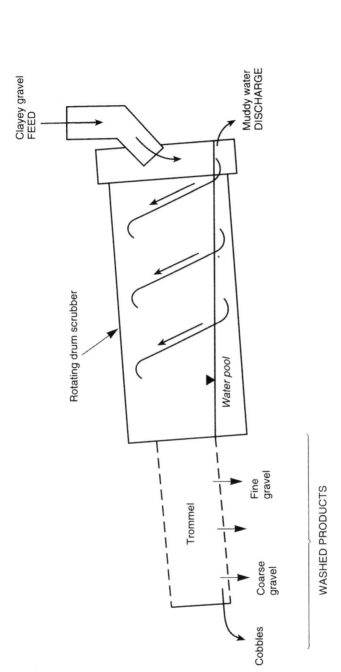

Figure 6.2 Drum scrubber and coarse-aperture cylindrical screen (trommel), mainly used for disaggregating and cleaning clayey gravels.

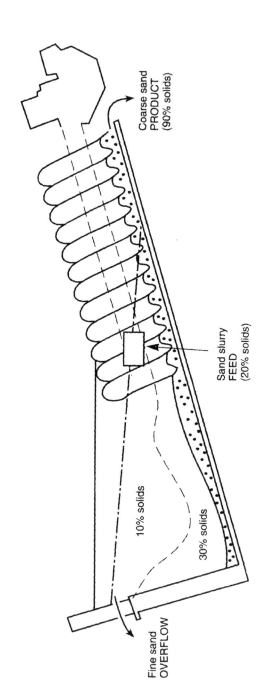

Figure 6.3 Spiral sand classifier, schematic. The sand slurry is agitated by the helical screw, which also elevates and drains the coarse sand product. Note the different solids contents of the feed, product and waste (overflow).

Coarse sand
PRODUCT
(90% solids)

Sand slurry
FEED
(20% solids)

10% solids

30% solids

Fine sand
OVERFLOW

to this primary sizing task, screens are used for scalping oversize, waste removal and dewatering.

- *Scalping* of blasted rock is carried out over bar screens in order to separate coarse spalls, which will be passed through the primary crusher, from undersize, which can be routed straight into the secondaries. The bar spacing (100–200 mm) equals the exit width of the primary, thereby effectively increasing its capacity, since only a third or so of the blasted rock need pass through. Scalping also reduces the risk of blockage due to fines packing, and may improve product shape.
- *Waste removal* by screening usually means the separation of topsoil and weathered rock as undersize. Sounder spalls are then passed on for primary and secondary crushing (this is sometimes referred to as pre-screening). Oversize trash such as timber, reinforcing bars and clay balls may also be skimmed off in this way.
- *Dewatering* of sand (clean gravels have negligible moisture adhesion) may be carried out over fine vibrating screens, which reduce the moisture content from over 20% to less than 10%. Since sand is sold by weight, this is still a considerable proportion, and further moisture reduction by cycloning or air drying is often required to avoid contract penalties, particularly where long road hauls to markets are involved. Nonetheless, sand drying below about 5% moisture content is usually impracticable.

A wide variety of screens are used in aggregate processing, but the most common type is the vibrating multi-deck set (Figure 6.4). The vibratory motion is induced mechanically, by electromagnetic induction or by resonance. Both frequency (10–50 Hz) and amplitude can be varied. Deck inclination is generally in the range 12–20°, but can alter along the screen length. Decks usually consist of two to four screens, typically 40 mm down to 3 mm aperture.

The screens themselves may be of round, square, hexagonal or slotted punched steel plate; woven wire or welded steel rods; and hard rubber or polyurethane mesh. The latter materials are rapidly gaining in market share, despite higher cost, owing to their low noise output and – surprisingly – excellent wear resistance.

Particles are sized according to their intermediate dimension, since the long axis can align itself perpendicular to the screen opening. The vibratory motion and inclination of the screen causes the crushed stone to segregate and stratify, with finer particles sinking to the bottom and either passing through the holes or gradually migrating down the screen. Screen performance, meaning throughput rate plus completeness of size separation, is affected both by mechanical factors and by the characteristics of the feed material.

Sizing and processing

Figure 6.4 Screen decks and product bins in a limestone quarry, Kuala Lumpur. Normally these are hidden behind cladding to prevent dust and noise escaping. The two multi-deck vibrating screens enable six products to be generated.

Screen capacity

Screen capacity, or rate of throughput, is directly proportional to the hole width (about 10% larger than the nominal product diameter), which means that coarser screens have the greater capacity. Other factors affecting capacity include the following:

- Percentage of *open area* on the screen (the ratio of aperture space to total screen area). This is proportional to aperture width, but is also dependent on the screen material. Punched plate screen, for example, has less open area for the same aperture than welded mesh.
- *Screen width* (width of the deck, not aperture of mesh), with wider screens having greater capacity.
- *Screen inclination*, which for most screens has an optimum around 15°. Below this, the screen's capacity is reduced; steeper, and feed material moves too quickly down the screen and undersize is retained.

Furthermore, capacity also varies with the position on the screen, reaching a peak about one-third along its length, as shown in Figure 6.5. By this point about 75% of the undersize has passed through the screen, and the remaining length merely improves the precision of the size separation or 'cut'. In other words, capacity is largely unaffected by screen length beyond this critical distance, but efficiency is improved;

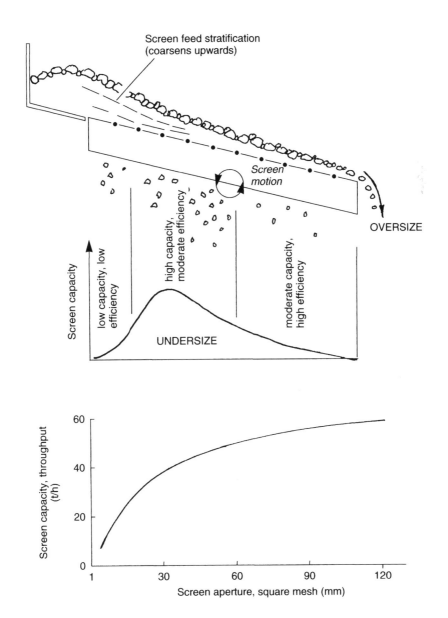

Figure 6.5 Single-deck vibratory screening, illustrating the difference between screen capacity (rate of throughput) and efficiency (degree of size separation). Note, in the lower diagram, how screen capacity increases with aperture.

long vibrating screens present more opportunities for undersize particles to slip through.

Screen efficiency

Screen efficiency is a measure of how complete size separation is achieved between oversize (particles retained) and undersize (material passing) for a given screen. Coarse particles are most efficiently sorted by low-frequency, high-amplitude vibrations, while passage of finer sizes is facilitated by high frequencies and short strokes. Screen efficiency must generally be traded off against capacity, as it is uneconomic to allow the low feed-rate and long screen residence time necessary for complete size separation.

The nature of the feed material is probably the most important factor determining screen performance. Screen efficiency will be reduced by the following factors:

- A high proportion of particles close to the nominal aperture size (say 70–110% of its width), which tend to jam in the slots and reduce capacity. This is known as *pegging*, and prismatic fragments ('carrots') are particularly prone to jam in this way. Furthermore, misshapen fragments adversely affect grading, since flaky particles tend to be retained one deck higher than their volume justifies, while prismatic particles slip through their nominal aperture size. In other words, flaky aggregates appear to be coarser, while prismatic aggregates appear to be finer than their true sizing.
- Accumulations of damp clay and silt grains around mesh openings, which eventually block off or *blind* the screen. These fines will also adhere to the larger particles, producing a dirty aggregate. Where very fine sizes are prevalent in the feed, it should be screened either perfectly dry or wet. Wet screening is technically superior, but incurs drying costs plus clean water supply and mud disposal problems.

Apart from the vibratory multi-deck inclined screens described above, two other types are important. *Grizzlies* are robust inclined frames with widely spaced parallel bars, used for scalping. Rotating cylindrical screens called *trommels* (see Figure 6.2) are used for small-scale wet screening of gravels. By varying the screen aperture along the length of the drum, several sizes can be delivered in one operation, with oversize tumbling off the lower end.

6.4   FINE AGGREGATE CLASSIFICATION

Separation via classification

As screen capacity and efficiency diminish in the fine aggregate sizes, and mesh wear accelerates, screening is replaced by size separation through particle settling velocity, or classification. This is based on Stokes' law, whereby settling velocity is proportional to volume for spherical particles of the same specific gravity (SG) – generally assumed to be quartz, SG 2.65. Complications arise because non-spherical, and particularly rough-textured and flaky, particles settle more slowly. Their movement is further retarded by the presence of suspended clay platelets in the water, making it effectively a dense medium.

Hindered settling of this kind, rather than the free settling assumed by Stokes' law, is therefore the normal situation with sand slurries. Furthermore, clay platelets attach themselves to the early-sinking coarse sand particles, resulting in a dirty product. This can be counteracted by gentle upwelling or elutriating currents, which keep the clay in suspension but allow clean sand to fall to the base of the settling column.

Sizing by classification is cruder than the cut obtained by screening; separation efficiencies are only 30–80%, compared with 90–95% when screening coarse aggregates. Fortunately, only two products are normally obtained from classifiers: a relatively clean, coarse sand for concrete mixes (the high-value product), and a silty fine sand suitable for mortar or filling and other low-value applications. Much greater sizing precision is required for glass and foundry sands.

Classifiers come in even more varied designs than screens, but have two features in common – some means of agitating a sand slurry, and of separating a suspended *overflow* of fine sand and silt from a settling *underflow* or bottom load of coarser grains. Usually these are the only products, but several size fractions in the range 60 µm–2 mm can be obtained where the devices are arranged in series, such that the overflow from the first classifier becomes the feed for the second and so on.

Hydraulic classifiers

Hydraulic classifiers include fluidized-bed classifiers, elutriators and classification tanks. *Fluidized-bed classifiers* (hydrosizers) make use of hindered settlement, by which upward-flowing water in a sorting column opposes particles sinking under gravity (Figure 6.6). The upward velocity is adjusted until only sand coarser than a certain diameter will settle, with the rest being carried away in the overflow. The accumulated coarse sand is periodically drawn out at the conical or pyramidal base. Possibly the most common type of hydraulic classifier is the *elutriator* (Figure 6.7). This also depends on hindered settlement, but has no moving parts and no introduced clean water. Instead, the descending sand slurry is deflected upwards such that it splits into two streams, a coarser underflow that is drawn off at the base of the vessel and a fine overflow that spills over at the top. Sand can also be sized by horizontal

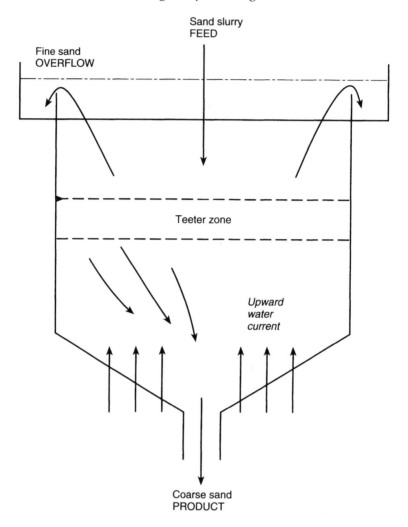

Figure 6.6 The principle of hindered settlement, as demonstrated in a fluidized-bed classifier. The 'teeter zone' is where fine particles are prevented from sinking further under gravity by upwelling water currents.

fractionation in a *classification tank* (Figure 6.8). Here the sand slurry enters a long, narrow trough with some velocity and particles follow the current in a ballistic trajectory, with the coarsest grains settling at the feed end and the finest close to the muddy water discharge lip. A series of spigots, typically 8–10 along the base, allow closely graded fractions to be drawn off; these may subsequently be reblended if the specification demands.

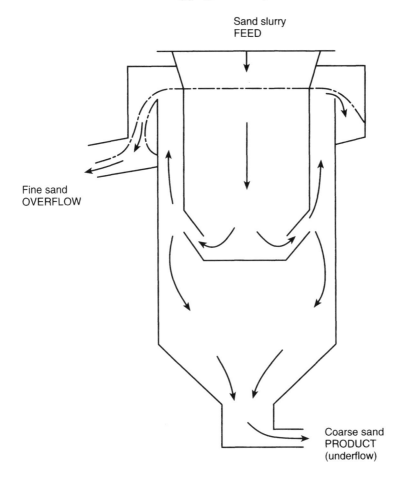

Figure 6.7 Hydraulic classification by vertical separation in an elutriator. In this case the water currents are provided by the downward motion of the slurry and directed by baffles in the cylindrical vessel.

Mechanical classifiers

Mechanical classifiers work on the principle of free settling in a box-like tank. Sand slurry is poured into a pool at the lower end of the trough and the coarsest fraction quickly settles. Coarse sand is removed from the tank by lifting, pushing or scraping it up an inclined trough, while muddy water is discharged over a weir at the opposite end. The most common type of mechanical classifier uses a helical spiral to agitate and lift the coarse product (see Figure 6.3), and the upward movement allows it to be partially dewatered and deslimed. Figure 6.9 shows a small mechanical classification plant.

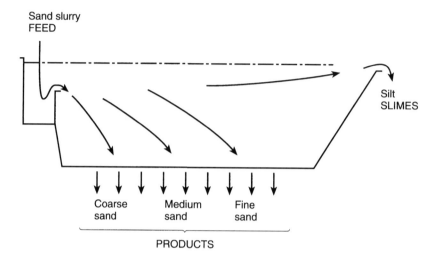

Sand slurry
FEED

Silt
SLIMES

Coarse
sand

Medium
sand

Fine
sand

PRODUCTS

Figure 6.8 Horizontal size separation in a classification tank. Here a variety of sand products can be drawn off at different points along the length of a narrow, deep, 'V'-section trough.

Figure 6.9 Portable sand classification plant, Glenlee, New South Wales. Coarse lumps are shed on the bar screen (at right) and finer clay balls by the vibrating screen (top left). The dry scalped feed is then dropped into a long, narrow, water-filled trough, where it is agitated by a helical screw. Clean coarse sand is lifted out of the trough by the bucketwheel and the fine sand residue is drained away.

6.5 DEWATERING, DESLIMING AND BENEFICIATION

Dewatering

Dewatering is necessary to improve the saleability of washed sand and to allow washwater to be recirculated. Classification, washing and scrubbing all generate heavy suspensions of clay- and silt-sized grains (1–60 µm) in the circulating water. These must be precipitated before the water can be recirculated within the washing plant or discharged into rivers.

Hydrocyclones (Figure 6.10) are widely used for dewatering because they have no moving parts, and hence are simple to operate and cheap to build. Their main shortcoming is that internal scouring by sharp sand limits their working life. Waste water from classifiers is introduced under pressure into the conical hydrocyclone chamber, where its centrifugal movement creates a vortex in the centre. Heavier grains are forced outwards and downwards towards the apex of the cone, while the suspended fines are drawn upwards and out of the chamber. Hydrocyclones can be used for sizing as well as for dewatering, but since the former depends on fluid velocity – and this is fairly unpredictable in

Figure 6.10 Dewatering hydrocyclone (conical object at the top of the tower), Calga, New South Wales. The lower (high pressure) of the inclined twin pipes carries the feed slurry and the upper pipe drains off silt-laden water to tailings dams. Two product stockpiles are visible.

vortices – the grain size of the underflow product tends to be inconsistent. Although hydrocyclones are less efficient than hydraulic classifiers in fractionating within the sand size (but superior to mechanical classifiers), they are very effective in removing silt and clay in the muddy overflow.

Desliming

The process just described is known, rather inelegantly, as desliming. The muddy water is discharged into tailings dams, which may be either old pits or shallow lagoons constructed for the purpose. The settling time for silt is only a few hours, but finer clay may require weeks or months. The simplest solution is to decant the upper water as soon as it clears, then allow the rest to evaporate. This can require large areas of level land and a dry climate, and in any case wastes water. Drying can be accelerated by scooping up and turning over the tailings once the surface water has drained off or evaporated. Relatively coarse non-cycloned tailings can sometimes be sold after drying as bricklaying sand, or may be reserved for pit restoration.

Where land is scarce or the climate too wet for evaporation, finer clay suspensions may have to be chemically flocculated in thickening tanks. These products have been experimentally used as brick-making and ceramic clays. However, it has been found that tailings pond deposits are inhomogeneous and may have only 30–40% solids content, despite a firm-looking and dry surface crust, even after years in place. Further dewatering, possibly by electro-osmosis or filter drainage, is necessary before the tailings can be put to commercial use.

Beneficiation

Low-grade sand and gravels can be upgraded, or beneficiated, by removing deleterious matter using gravity separation techniques. These are generally of low specific gravity relative to quality aggregates, which are mostly SG 2.65–2.70. They include wood fragments, shells, coal chips, lignite and weak or unsound rock (SG 2.00–2.35) such as chalk, porous sandstone, pumice and shale. Beneficiation techniques have not been much used by the sand and gravel industry up to the present, but could become more common in the future as prime deposits are further depleted. Their main potential appears to lie in removing shell fragments from marine sands, unsound particles from crusher grit, and clay from ground limestone.

Like much of the comminution and sizing technology used in quarrying, beneficiation procedures have been adapted and scaled down from those used in extractive metallurgy. The two principal methods of gravity separation are heavy media separation and jigging.

REFERENCES AND FURTHER READING

Kelly, E.G. and Spottiswood, D.J. (1982) *Introduction to Mineral Processing*, Chs 9 and 10. John Wiley, New York.

Littler, A. (1990) *Sand and Gravel Production*, Chs 1–3. Institute of Quarrying, UK.

Mellor, S.H. (1990) *An Introduction to Crushing and Screening*, Ch. 2. Institute of Quarrying, UK.

Raspass, F.W. (1980) Developments in fine aggregate processing. *Quarry Management and Products*, **7** (August), 217–28.

Rockwood, N.C. (1948) Production and manufacture of fine and coarse aggregates, in *ASTM Symposium on Mineral Aggregates*, Detroit, June. ASTM Special Technical Publication No. 83, pp. 88–116.

Sand and Gravel Association of Great Britain (SAGA) (1967) *Pit and Quarry Textbook*. MacDonald, London.

Smith, M.R. and Collis, L. (1993) *Aggregates: Sand, Gravel and Crushed Rock Aggregates for Construction Purposes*, 2nd edn, Chs 4 and 5. Geological Society of London, Engineering Group Special Publication No. 9.

Trahair, J.W. (1981) Sand extraction and processing in South Australia. *Quarry Management and Products*, **8** (September), 603–13.

Wills, B.A. (1988) *Mineral Processing Technology*, 4th edn. Pergamon, Oxford.

CHAPTER 7

Aggregate and roadbase testing

This chapter discusses the methods of testing and specification applied to crushed rock aggregates, including roadbase. It is worth emphasizing that not all end-uses require the same toughness, durability and wear resistance. The most demanding applications are as road surfacing aggregate and railway ballast; somewhat lesser-quality stone is acceptable for most concrete mixes, lower-course asphalt and unbound basecourse. Relatively poor aggregate may be adequate for road sub-base, backfill concrete and sub-ballast courses.

In this book 'aggregate' refers to any crushed rock product, plus some naturally fragmented materials such as river gravel and sand. *Coarse aggregate* is that which is larger than about 5 mm, and *fine aggregate* comprises particles from 5 mm down to 75 μm (0.075 mm). Particles finer than 75 μm are referred to as filler, binder or dust. *Roadbase*, as the term is used here, is any unbound pavement material; it can be either a natural granular mixture (see Chapter 4) or processed fine crushed rock. The better sort of roadbase is known as *basecourse* or simply as base, and the lesser-quality one (lower course) is *sub-base*.

The purpose of aggregate testing, as with all engineering materials, is, first, to provide a basis for comparison with satisfactory or failed materials used in the past. In practice the highest recommendation that can be given to a quarry product is that it has consistently performed well over a long period. Specification limits, especially for roadbase, are set up with these successful materials in mind rather than from any theoretical considerations. The other main purpose of testing is to provide a basis for accepting or rejecting a material, or in other words to ensure compliance with a specification.

The tests prescribed for compliance vary from country to country, state to state, and even between construction organizations – but generally involve at least:

- Particle size distribution and shape
- Crushing resistance
- Durability (susceptibility to in-service weathering)

More specialized tests to determine such things as chemical reactivity, skid resistance and bitumen adhesion are sometimes also required. The tests stipulated, and the limits imposed, vary somewhat because of local practice and material availability. The desirable characteristics of an aggregate testing procedure might therefore be the following:

- It is *widely used and accepted* as a valid test, and carried out under more or less standard conditions with similar equipment.
- It is a *reasonable simulation* of in-service conditions (for example, accelerated weathering by wetting and drying cycles).
- The procedures give *repeatable results* (that is, the same operator can achieve nearly the same result each time on identical samples), and these are *reproducible* (meaning that a different operator in a different laboratory can produce a similar result).
- The test is *economical* to perform (that is, it can be performed by an operator of average skill, using simple apparatus and easily obtained samples).

In this chapter the aim is to compare aggregate tests grouped into broad procedural categories (Table 7.1 gives a listing), emphasizing their applications and shortcomings. To describe these tests in detail would be tedious and detract from the main purpose, and in any case is best done in the relevant national standards. These include those of the British Standards Institution (BSI), the American Society for Testing and Materials (ASTM) and the Standards Association of Australia (SAA). Note that these are really suggested methods for good practice rather than true standards, and that it is quite common for major users of rock products – such as state highways departments – to set their own test procedures and specification limits. Good summaries of the most common test procedures and their limitations are given in Lay (1990) and Smith and Collis (1993).

7.1   AGGREGATE SAMPLING

Sampling categories

Aggregate sampling is nowadays much involved with statistical analysis, but the remarks here will be confined to the practicalities of sample-taking. Interested readers are referred to the chapter by Harris and Sym in Pike (1990) for a discussion on the application of statistical methods to the sampling and testing of aggregates.

Aggregate samples can be thought of as falling into three broad categories:

- *Continuous samples* such as rock spalls or drillcore are taken across the full width or depth of a deposit, usually during the exploration stage. Core drilling at the largest possible diameter (generally HQ, 61 mm, in

Table 7.1 Principal aggregate tests

Intact rock properties

Performed on solid drill core and rock specimens subcored from block samples
   Uniaxial compressive strength (UCS)
   Elastic modulus (E)
   Tensile strength: indirect (ITS), flexural, point load (PLSI)
   Particle density, porosity, water absorption
   Petrographic description
   Moh's hardness, Schmidt hardness

Particle size and shape

Performed on crushed samples up to 50 kg for grading and on selected chips for shape
   Particle size distribution (PSD) by sieving (or grading)
   Flakiness index, flakiness ratio, elongation index
   Average least dimension (ALD)

Surface properties

Performed mostly on individual stones, groups of stones and bitumen-coated plates with embedded chips
   Surface texture description
   Bitumen adhesion (stripping)
   Polished stone value (PSV)
   Aggregate abrasion value (AAV) (Dorry test)

Durability/soundness

Performed on graded coarse aggregate samples, measuring fines generated under standard test conditions
   Sulphate soundness test
   Freeze/thaw soundness test
   Slake durability test
   Wet/dry strength variation in TPF test (see below)
   Secondary minerals content (on thin section, by point count)
   Methylene blue absorption (MBA) test or X-ray diffraction (for clay content
   and activity)

Abradability (wear resistance)

Performed on graded coarse aggregate samples (up to 40 mm particle size and 5 kg weight); measures fines generated under standard test conditions
   Los Angeles abrasion (LAA) test
   Texas ball mill (TBM) test
   Washington degradation test (WDT)

Particulate properties

Primarily crushing resistance and compacted stiffness; mainly performed on 1–2 kg samples of 10–14 mm aggregate under confined conditions
   Bulk density, voids ratio, permeability
   Aggregate crushing value (ACV)
   Ten per cent fines (TPF) test
   Aggregate impact value (AIV)
   Texas triaxial confined compression test
   Repeated load (cyclic) triaxial test

Fines properties

Performed on $-425\,\mu m$ fraction
   Particle size distribution (PSD) by settling velocity in water
   Atterberg consistency limits: liquid limit (LL), plastic limit (PL), plasticity index (PI)
   Sand equivalent test (and modified versions)
   X-ray diffraction (for clay mineralogy)

Australia) is used to recover the maximum weight of sample per metre intersected. Sometimes drillholes are duplicated to obtain sufficient material for a variety of tests, since two adjacent slim holes are cheaper than one large-diameter hole.

- *Type samples* are meant to represent the full range of variability within a product size fraction, or to represent a batch being supplied for a major contract. To encompass this variability, samples have to be very large, say 100–400 kg, and must be reduced by quartering at the sampling site. This should produce manageable, but still representative, quantities of sample – generally 50–100 kg, the weight increasing with nominal particle diameter.
- *Typical samples* are much smaller and less rigorously collected, but more common. They are intended to demonstrate the material properties within a particular bed, layer, rock type or stockpile in the quarry. Their reliability is very much dependent on the expertise of the sampler and the geological complexity of the deposit.

Sampling from stockpiles

Sampling from stockpiles requires particular care because, although the procedures are straightforward, the results may be used to penalize a contractor or even to reject aggregate batches outright. Since the contractor will already have processed the material and probably delivered it on site, he may well dispute the sampling method or the test procedures. For a start, therefore, stockpiles need to be laid out systematically for sampling and volume measurement. They are typically long parallel ridges of trapezoidal cross-section, about 100–200 m long by 3–4 m high, as illustrated in Figure 7.1.

Sample increments are taken at the base, midslope and top, and at regular intervals along the length of the stockpile, as shown in Figure 7.1. Conical stockpiles should be sampled in a spiral fashion, allowing for *segregation* whereby the coarser particles tend to roll to the base of the pile. Sample increments should also be taken at least 200 mm below the surface, since this is where fines washed downwards by rain accumulate and where the moisture content is representative of the bulk of the stockpile (Figure 7.2). In sand stockpiles the moisture content at this depth may be 5–10% higher than at the surface, a significant difference when calculating water/cement ratios for concrete mixes.

7.2   INTACT ROCK PROPERTIES

The geomechanical properties of intact parent rock are less important in aggregate evaluation than might at first be supposed. This is because aggregate is a particulate material, so its behaviour as a mass is more

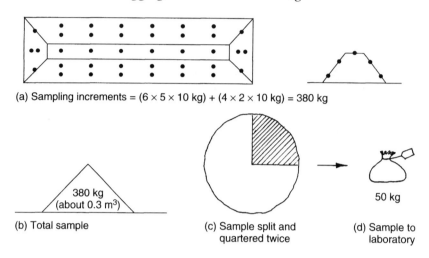

(a) Sampling increments = (6 × 5 × 10 kg) + (4 × 2 × 10 kg) = 380 kg

380 kg
(about 0.3 m³)

(b) Total sample

(c) Sample split and
quartered twice

50 kg

(d) Sample to
laboratory

Figure 7.1 Typical aggregate stockpile and steps required to obtain a representative 50 kg specimen.

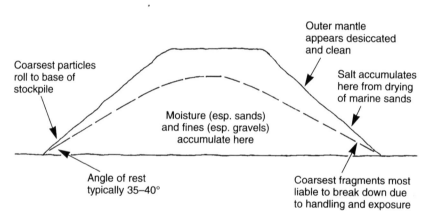

Outer mantle
appears desiccated
and clean

Coarsest particles
roll to base of
stockpile

Salt accumulates
here from drying
of marine sands

Moisture (esp. sands)
and fines (esp. gravels)
accumulate here

Angle of rest
typically 35–40°

Coarsest fragments most
liable to break down due
to handling and exposure

Figure 7.2 Problems of segregation, inhomogeneous moisture and fines distribution in aggregate stockpiles.

relevant than the properties of individual fragments. Nonetheless, rock mechanics index tests are useful at the exploration stage because they provide a good basis for comparison between proposed quarry sites. They are also well suited to small drillcore samples, whereas standard aggregate tests require larger quantities of rock.

Intact strength

The intact strength of aggregate source rock is conventionally measured by the *uniaxial compressive strength* (UCS) test, performed on cylindrical samples about 50 mm in diameter by 120 mm long. This is called the 'crushing strength' in some of the literature, but should not be confused with the aggregate crushing value (ACV), a test carried out on stone chips. In general, rock with UCS values less than 50 MPa is useless as aggregate and may be suitable only as low-grade roadbase. Values of 50–100 MPa indicate more indurated sedimentary rock or weathered igneous lithologies; it may be suitable for crushed roadbase or aggregate in low-strength concrete. UCS values of 100–200 MPa include most sources of good-quality crushed roadbase and concrete aggregate, but even higher values are desirable for very-high-strength concrete and best-quality ballast. Extremely strong rocks (above 200 MPa) are usually difficult to crush, generating few fines and resulting in a harsh or 'boney' product.

The *point load strength index* (PLSI) test is a convenient and widely used method for estimating the strength variation in drillcore. As a means of predicting UCS it has limitations, since the multiplier – usually quoted as 24 – may range from about 12 for weak rock to 30 for very strong but brittle material.

Elastic modulus

The elastic modulus (E) of intact rock proposed as a source of aggregate is most useful as a factor in two derived parameters, *modulus ratio* (E/UCS) and *toughness* or resilience ($UCS^2/2E$). Normally it is obtained as part of a UCS test, by attaching strain gauges to the specimen. High values of both UCS and E indicate that the rock resists crack penetration and is difficult to crush (i.e. stores strain energy and therefore reduces crusher throughput).

Hardness

Strictly speaking, hardness, as defined by Moh's scale of scratch resistance (M), is a property of minerals rather than of intact rock, which is usually a composite of many minerals. It is, however, related to *abradability*, the resistance of an aggregate to tyre wear and polishing, and explains the quite different wear characteristics of limestone (mainly calcite, $M = 3$) and granite (mainly quartz, $M = 7$, and feldspar, $M = 6$).

Density, porosity and absorption

Particle density, porosity and water absorption are interrelated properties, since the specific gravities (SGs) of most rock-forming silicates are in the range 2.6–2.7. Dry particle densities below 2.5 t/m^3 indicate significant porosity in silicate rocks. This in turn implies low strength, reduced durability, high binder absorption and high abrasion wear rates in aggregates from such sources. Note that particle density, often called particle SG, should not to be confused with the *bulk density* (from about 1.4 t/m³ loose, up to 2.4 t/m³ densely compacted) of crushed aggregate mixtures. It is also necessary to note whether the particle density determinations relate to oven-dry, air-dried (AD) or saturated/surface-dry (SSD) conditions.

Water absorption (WA) of aggregate particles after 24 hours soaking is usually measured in preference to porosity, which requires more elaborate equipment. WA is generally less than porosity, often only half, since it depends not only on the pore space available but also on the diameter of the pores and their dryness. Nevertheless it is a useful indicator of non-durability in some rock types.

7.3   SIZING AND GRADING

Sizing of aggregates

Aggregate sizing is performed dry, using a series of standard sieves whose apertures generally range from about 40 mm down to 75 μm (0.075 mm), each being roughly half the mesh width of the preceding one. The nominal size of each fraction is the smallest aperture at which 100% passes. The proportion retained on each sieve is weighed and plotted as a cumulative percentage passing versus particle diameter diagram such as Figure 7.3. It is assumed that the particles are equidimensional – hence sieve aperture equals grain diameter – and that their specific gravity (SG) is uniform. Elongate or prismatic particles cause the coarser fractions to be underestimated, while platy or flaky ones cause it to be overestimated. In other words, prismatic particles are retained one deck lower than their volume requires, while platy particles are held one deck too high.

Particle sizing below 75 μm is impracticable using sieves, and instead is based on settling velocity in a water column dosed with dispersant. However, although the percentage of –75 μm fines in aggregate samples is of interest, the silt/clay proportion can more easily be estimated from Atterberg limits. The dust of crushing, predominantly non-plastic and silt-sized, adheres to damp aggregate chips – hence it is frequently underestimated by dry sieving – and must be released by washing.

Grading of aggregates

Grading, as the term is used in this book, is synonymous with *particle size distribution* (PSD), the end-result of the sieving process. The PSD of an aggregate, roadbase or subgrade soil has important implications for its geomechanical properties such as bearing capacity, shearing resistance, voids ratio (hence permeability) and compaction behaviour. Three types of PSD are important for such applications as concrete and asphaltic mix design, pavement material evaluation and road surfacing: dense-graded, open-graded and gap-graded aggregate mixes (Figure 7.3).

- *Dense-graded* mixes (also termed well graded, continuous graded, straight line graded) are characterized by an even blend of size fractions from coarse gravel to silt, such that the finer grains can – with the help of vibration, watering and compaction – fit between the coarser ones. This closer packing is evident in a higher mix density and reduced void space following compaction, which in turn means increased shear strength and layer stiffness.

  Dense-graded aggregate mixtures are used as unbound roadbase and in asphaltic concrete. In roadbase they may be composed of either fine crushed rock or natural gravel–sand–fines mixtures. Their strength and stiffness derive partly from interparticle friction and partly from the confinement offered by a low-voids compacted mass.
- *Open-graded* aggregate blends (also termed no fines, harshly graded) contain an even mixture of coarse particle sizes, but little or no void-filling fines. They are characterized by a steep PSD curve, and depend on friction between interlocking angular, coarse and rough-textured

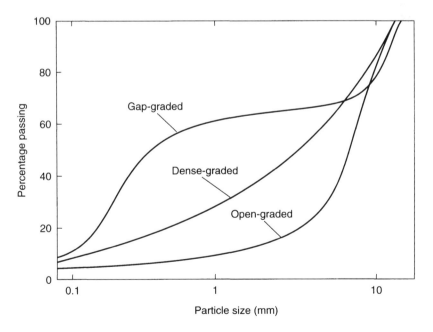

Figure 7.3 Particle size distribution curves for different aggregate mixtures used for concrete, unbound roadbase and asphalt.

fragments for their strength and layer stiffness. This internal friction is greatest where the number of point-to-point and face-to-face contacts is maximized. Because of this, stronger and more abrasion-resistant aggregate particles are preferred than is the case with dense-graded roadbase.

The other significant property of open-graded mixes is that they are more or less free-draining and not subject to capillary rise – an important consideration in shallow-water-table or frost-prone sites. The main disadvantage is that they are 'harsh' materials, meaning that they are difficult to spread and compact to their highest density – which is low relative to dense-graded mixes anyway. The classic open-graded aggregate mixture is railway ballast, but the same principle is used in the design of permeable asphalts, no-fines concretes, free-draining sub-bases and macadam pavements.

A special case of open grading is *uniform grading*, in which only a single size fraction is represented. Two examples of this are the monosized cover aggregate used on sprayed seals (chip seals), and dune sand, used as fine aggregate in some concrete mixes. Uniform grading is indicated by near-vertical PSD curves, with negligible coarse tops and fine tails.

- *Gap-graded* mixtures (also termed skip graded or 'armchair' graded) have an intermediate size fraction missing, generally coarse sand or fine gravel (say, 1–5 mm). This reduces their compacted density, leaves a proportion of voids unfilled, and makes them prone to segregation. Gap grading is common in natural road gravels, particularly in those derived from duricrusts, and is sometimes corrected by granular stabilization. Some concrete mixes are gap-graded, either by necessity – because only coarse aggregate and uniform sand are available – or because the use of rounded dune sand improves their workability.

Grading curves

Particle size distributions in aggregate mixtures are represented pictorially by grading curves, also known as Fuller or Talbot curves, such as those shown in Figure 7.3. Because of the wide particle size range represented on these curves, sometimes 1 μm (0.001 mm) to 100 mm, the horizontal scale is logarithmic. Usually the coarsest sieve sizes are plotted on the right, but this may be reversed. Cumulative percentages finer than a particular grain size are shown in linear scale on the vertical axis. The PSD may be drawn as straight-line segments between sieve aperture sizes, or as smooth curves, or as *grading envelopes* enclosing the range of sizes acceptable in a particular product. These curves are mathematically described by the equation:

$$P = 100(d/D)^n$$

where P is the percentage of particles finer than the sieve aperture d, D is the maximum particle size and n is Talbot's factor.

The shape of a PSD curve is defined by the Talbot exponent (n), with values in the range 0.4–0.5 for dense-graded roadbases, 1.5 or more for open-graded mixes, and around 0.3 for well-graded sands with an excess of fines. Dense-graded mixtures plot in Figure 7.3 as slightly concave curves spanning a wide range of particle sizes. Open gradings are shown by steep sigmoidal curves at the coarse (right-hand) end, while uniform (single-sized) sands, though not shown here, are even steeper ($n = 3$–5). Gap-graded mixtures show up as stepped curves, with the flat segment indicating the deficient size fraction.

'*Poorly graded*' mixtures are any that depart significantly from the maximum density curve; hence they may be gap-graded, open-graded or uniformly graded. This is therefore a term that should be either discarded or only used when qualified.

7.4 PARTICLE SHAPE

In an ideal aggregate mixture all particles should be equidimensional and angular, but not spherical (unless good workability has a higher priority than shear strength). In practice, a proportion of misshapen particles – flaky or elongate – is always present, but their percentage is severely limited in most specifications. The reasons for this stipulation, summarized in Figure 7.4, are as follows:

- Some flaky particles are more easily stripped from bitumen seals, because they are held by only a narrow face in contact with the binder.
- Elongated particles are more prone to flexural breakage, causing mixtures to become finer than their nominal grading.
- Concrete mixes rich in misshapen fragments are difficult to pump and to compact.
- These flakes may bridge across open voids in the mix, weakening it, or, if filled, will increase the cement demand.

The causes of flakiness may be inherent (foliation or flow banding in the parent rock) or induced (by single-stage crushing, by reduction ratios above 3:1 or 4:1, and by excessive fines in the feedstock). Induced flakiness appears to be the more common, since foliated rocks are rarely crushed for aggregate, and flaky particles are common even where the feed stone has no obvious planar fabric. Hard, brittle rocks also appear to be more likely to generate misshapen fragments by splintering than tougher lithologies.

The degree of departure from cubic shape is measured by the *flakiness index*, the weight percentage of particles in a sample that have a

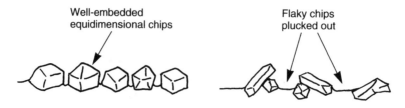

(a) Stripping of cover aggregate from sprayed bituminous seals

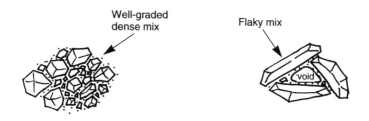

(b) Porous, low-strength concrete or high cement demand

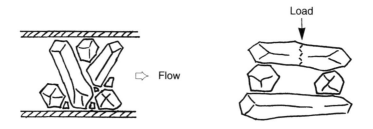

(c) Blockages in pump lines or tensile fracture of bridging particles

Figure 7.4 Consequences of flaky and elongate particles on aggregate behaviour in chip seals, concrete and asphalt.

minimum dimension less than 0.6 times the mean, or by the *elongation index*, the proportion of prismatic particles longer than 1.8 times the mean. The main problem with these tests lies in selecting a truly representative group of stones to perform the required 200 measurements, and the fact that flaky or prismatic particles may be more prevalent in some size fractions than in others.

There is some dispute as to whether jaws, gyratory or cone crushers give rise to the more flaky products, but it is accepted that impact crushers generate the most-equant particles. Hence poorly shaped products may be improved by tertiary crushing in an impactor, which shatters particles by high-velocity blows rather than by squeezing (see Chapter 5).

7.5   SURFACE PROPERTIES

'Surface properties' of aggregates include all those factors influencing the bond between chips and bituminous binder in asphalt, and that between cement paste and aggregate in concrete. The skid resistance and tenacity of bituminous road surfacing materials is also dependent on these characteristics of sealing aggregates. The relevant aggregate properties comprise particle roughness, dustiness, moisture content and surface chemistry.

Aggregate surface texture

Aggregate surface texture is a function of mineral grain size and rock fabric. Fine-grained igneous rocks tend to have smooth or glassy fracture surfaces, medium- and coarse-grained ones are rougher textured, while some are visibly porous or even vesicular. The rougher the texture, the better the bond strength; one indication of this is the extent to which concrete test cylinder failure surfaces pass through aggregate particles (good bonding) or around them (weak bonding). Another piece of evidence is the fact that some relatively weak (UCS < 100 MPa) but porous aggregates like slag and scoria can produce strong concrete. This is especially true of lightweight aggregates, which can yield up to 40 MPa compressive strengths.

Bond strength in bituminous mixtures, on the other hand, is more dependent on the absence of clay coatings and moisture films from the particle faces. Some aggregate lithologies, such as granite and quartzite, attract water films because of their surface charge and then repel bitumen coatings. This tendency is countered by precoating the chips, or by adding adhesion agents to the hot bitumen.

The likelihood of aggregate debonding from bituminous surfaces is assessed by *stripping tests* using variously precoated, dry, dusty and moist stones set into bitumen-coated plates. These are then plucked out with pliers and their relative tenacity and the amount of bitumen adhering noted. Basic igneous rocks, other than glassy basalts, tend to adhere well. The bonding of limestone and most metamorphic rocks is also good, but untreated acid igneous rocks strip readily.

Skid resistance

Skid resistance of surfacing aggregates is largely dependent on their tendency to polish under traffic (i.e. to become rounder and smoother in service). This is assessed primarily by service experience; hence certain quarries and rock types are favoured on the basis of past performance, since the quantities required are small in relation to the aggregate market as a whole. An accelerated polishing test is available, though the results do not always agree with field experience. A rubber-tyred model wheel, coated with abrasive paste, is run across a series of aggregate-surfaced plates several thousand times to simulate years of traffic. At the end of the procedure, the frictional coefficient of the aggregate mats is measured by a pendulum tester and compared with its original condition, and a *polished stone value* (PSV) calculated.

Highly polish-resistant aggregates have PSV > 45 and their parent rocks include some finely vesicular basalts, harder scorias and slags, and some coarse but well-cemented orthoquartzites ('gritstones'). Surprisingly, a few altered igneous rock types perform well, the secondary minerals being soft and the primary minerals harder. Erosion of the soft matrix minerals under traffic exposes rough crystals of the harder species, maintaining the rough microtexture while the chip diminishes in size.

7.6   SOUNDNESS, DURABILITY AND ABRADABILITY

These are three properties that describe the tendency for aggregate to break down within its usual service life, say 10–100 years. This breakdown generally involves a reduction in chip size and an increase in the proportion of clay fines, plus weakening of the remaining coarse particles. The processes involved are similar to those of physical weathering. Chemical weathering (decomposition) is believed to contribute little, since the mineral species liberated are the same as those present in the parent rock prior to quarrying.

What has changed as result of crushing, however, is the surface area of the source rock now exposed to the effects of atmospheric oxygen, stress relief, fluctuating moisture content and thermal cycling. Consider a typical blasted rock spall of volume $1\,m^3$ in a quarry muckpile: crushing this to nominal 10 mm cubes increases its area a millionfold. Clays and fibrous secondary minerals previously locked up within the intact rock can now draw in airborne moisture, swell and further split the particles. The combination of cracking and water infusion can cause slaking, whereby moisture is sucked into newly created microfractures by capillary tension and compresses the air within, to the point where the rock chips literally explode.

These are the reasons why some altered igneous rocks, which appear fresh and hard when first exposed, can deteriorate in stockpiles or road pavements within months. The effects of diminishing particle size on the rate of breakdown are illustrated in Figure 7.5. Note that the effects of degradation by salt crystallization increase dramatically at sizes finer than 20 mm, and that the surface area per unit weight curve almost exactly parallels the degradation curve.

- *Soundness* is the vaguest of the three 'weatherability' parameters used in engineering. Originally it may have referred to the fact that rock that is liable to break down in service is porous; when struck by a hammer it emits a dull thud, whereas better-quality rock will 'ring'. As the term is presently understood, it means the extent to which aggregate chips or rock cubes will disintegrate in salt crystallization ('sulphate soundness') tests.
- *Abradability* is more specific: it is the tendency of aggregate particles to wear away by abrasion (tyre-on-stone wear) or attrition (stone-on-stone rubbing).
- *Durability* is another general term that can cover any form of in-service degradation, although it is best restricted to moisture-dependent, non-wearing processes.

Abradability tests

Abradability tests simulate on-road wear by ball milling, interparticle attrition and dropped weight impacts, in either the wet or dry state. The

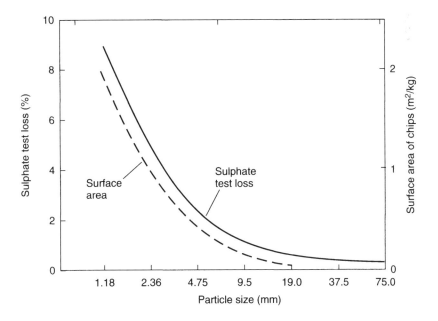

Figure 7.5 Relationship between soundness (as measured by sulphate test loss) and particle diameter for single-sized chips from the same source. Note the close connection between surface area and weatherability, assuming all chips to be cubic.

samples used are closely sized coarse aggregate fractions, and the weight percentage of 'fines' (−2.36 mm or −1.18 mm) produced under standard conditions is taken as the measure of wear potential. The best known and oldest of the abradability tests is the *Los Angeles abrasion* (LAA) test, a dry milling procedure using steel balls. Its values are widely quoted and just as widely criticized, mainly on the grounds that steel ball impacts are unduly severe in simulating road conditions and that the test fails to allow for the weakening effects of water. Typical allowable losses as fines from good-quality aggregates are 25–40% in the LAA test. The *Washington degradation test* (WDT) more closely mimics on-road conditions, since water is added but there are no steel balls and the attriting action – 20 minutes agitation in a sieve shaker – is more gentle. The WDT scale is also a more complex one, which gives a measure of both the amount of fines generated and their activity (i.e. clay content plus clay mineralogy).

Durability tests

Durability tests, or accelerated weathering tests, examine the wetting and drying behaviour of aggregates, with or without the addition of chemicals to hasten breakdown. The best known of these is the *sulphate soundness test* (SST), a method that combines salt wedging – by crystallization of Na_2SO_4 or $MgSO_4$ – with alternate soaking and desiccation. Like the LAA test, it has been criticized for being unrealistically severe.

Another way of assessing durability is to compare the saturated and oven-dry strengths, assuming that porous rocks are prone to disintegrate as moisture moves into and out of the interparticle voids. In Australia the most favoured aggregate durability test is the wet/dry strength variation in the *ten per cent fines* (TPF) test. The TPF test itself is primarily a means of assessing the crushing resistance of roadbase mixtures, and is described later in this chapter. The wet/dry strength variation is determined by performing the test on identical saturated surface-dry (SSD) and oven-dry specimens. Large differences between the ram loads required to generate 10% of fines indicate water-sensitive and therefore non-durable aggregate; similar low wet/dry strength ratios are characteristic of UCS results from weak and porous rock like sandstone.

7.7 STRENGTH AND STIFFNESS OF PARTICULATE MIXES

As pointed out earlier, the properties of rock in a particulate mass – such as an aggregate mixture – differ greatly from those of the intact parent rock. Particle mixtures are weaker in compression, more deformable and have negligible tensile strength. This is not a problem in concrete because the imposed load is carried by the hardened cement paste, but it may become so with unbound roadbase, since in this case the coarse aggregate framework supports and redistributes most of the stresses due to moving vehicles.

Strength

Strength in road pavement courses is defined by the point at which compressive strains become very large, due to some aggregate chips fracturing and rearranging themselves into a denser packing. However, general failure of a pavement section from a single loading is rare, because the stresses imposed by traffic are less than 10% of the ultimate (failure) stress for most granular materials. The problem therefore lies not in assessing the effects of one grossly overloaded truck, but from the passage of (say) 10 million heavy vehicles over 20 years. This is because each loading cycle imposes both recoverable (elastic or resilient) strains and irrecoverable (permanent or residual) strains on the compacted roadbase, as illustrated in Chapter 11. These permanent strains can accumulate to the point where general *fatigue failure* of the pavement occurs, as indicated by wheelpath rutting or other severe surface deformations.

Stiffness

Strength and deformability are thus closely related in roadbase materials, the stiffness modulus (or simply 'stiffness') being 50–100 times the peak strength. The term 'stiffness' is generally used in preference to 'elastic modulus' for particulate mixtures because their stress–strain behaviour is non-linear (i.e. not truly elastic). At the low stresses typical of wheel loadings the stress–strain curve is flatter (i.e. stiffness is less) than it is when approaching peak stress. Hence roadbase becomes both stiffer and more nearly elastic in its behaviour due to closer packing at high confining pressures. It also responds more stiffly to dynamic loading than to static or slowly applied loads. This is the reason why slow-moving trucks cause more furrowing in climbing lanes than in fast lanes, and why airport taxiways should be thicker than runways.

Consider, for example, a hard limestone crushed to produce a well-graded highway basecourse. The parent rock in the intact state might have a UCS of 100 MPa, an elastic modulus of 40 GPa and a density of $2.7 \, t/m^3$ at 4% porosity. As a crushed and recompacted roadbase mixture its dry compressive strength might be only 5 MPa, its stiffness 300 MPa, its bulk density $2.1 \, t/m^3$ and its porosity 25%. Although the strength and stiffness of the aggregate mixture is much less than that of the intact rock, it is nevertheless quite adequate to handle the low level of traffic-imposed stress, typically about 0.8 MPa (static) per truck wheel. In addition, both the strength and stiffness of this roadbase are increased by confinement within a pavement.

7.8   TESTING OF PARTICULATE MIXES

Triaxial testing

Triaxial testing differs from the other methods used with particulate materials in that it can simulate confinement, repetitive traffic loadings, and variable moisture and compacted density. The test results – the shape and steepness of the Mohr failure envelope, stiffness and Poisson's ratio – are directly usable as input for numerical models.

Tests can be performed using a *Texas triaxial cell* or equivalent apparatus. The procedure is multistage, carried out on a single 100 mm × 150 mm cylindrical specimen at five successive lateral pressures in the range 0–200 kPa. The resulting failure envelope is compared with a set of standard curves, illustrated in Figure 7.6. Classes 1 and 2 are suitable as basecourses, class 3 as sub-base, and the flatter curves represent clay-rich materials unsuitable for pavement courses but of varying quality as subgrades.

The *repeated load triaxial test* is a development of the standard Texas method, which simulates the effects of moving traffic by cyclic loading of the test cylinder. Ten minutes of vertical load pulsing at 30 Hz can represent 10–20 years of truck traffic. The specimen is first 'conditioned' by about 1000 pulse repetitions, to mimic the bedding-down effects of traffic before the *resilient modulus* is measured. This is the material stiffness as indicated by the unloading portion of the stress–strain curve, which is thought to be more consistent than the loading half-cycle.

Crushing resistance tests

Routine methods of aggregate and roadbase assessment include several crushing resistance tests (not to be confused with UCS). These measure the weight percentage of fines generated by a standard compactive effort. A monosized coarse aggregate specimen contained within a cylindrical steel mould is loaded by the ram of a compression testing machine or struck repeatedly by a sliding weight (Figure 7.7). The proportion of –2.36 mm fines produced is considered to be inversely proportional to the sample's strength and durability.

In the *aggregate crushing value* (ACV) test, the compressive force is gradually raised to 400 kN and the weight percentage of fines liberated is measured. A high value indicates a weak, potentially non-durable material. This test is a useful alternative to the LAA test, because the crushing action is non-violent and hence better replicates on-road

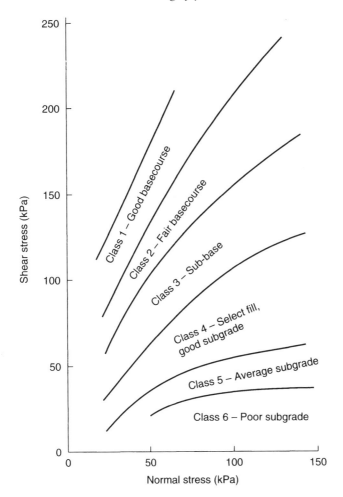

Figure 7.6 Texas triaxial test results (Mohr envelopes) for granular materials. The steep curves indicate high internal friction and rapid strength increase with confinement, hence suitability as unbound basecourse in flexible pavements.

conditions. However, with some softer rocks used for crushed roadbase the voids tend to clog and prevent further ram penetration, putting an upper limit on the ACV. Another problem with the test is that flaky aggregates can increase the ACV by up to 60%, so particle shape must therefore be allowed for in the assessment.

The *aggregate impact value* (AIV) test produces results that are numerically similar to ACV, but the loading is applied by a 14 kg dropped weight or sliding hammer. Hence the apparatus is portable and more suited to field laboratory use, though AIV is less widely used than either the ACV or TPF tests.

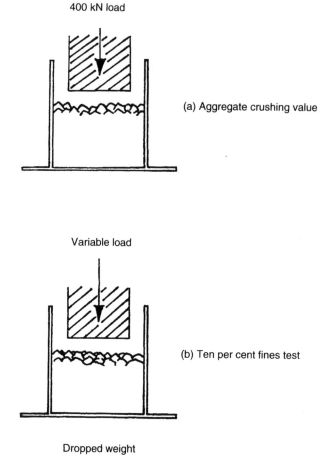

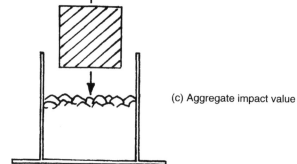

Figure 7.7   Test methods for comparing aggregate strength, based on the proportion of fines generated under standard loading conditions.

The *ten per cent fines* (TPF) test was devised to remedy the problem of void clogging in the ACV test, by limiting the amount of fines produced and recording the ram load (kN) to achieve this instead. Clearly, some judgement is required to generate exactly 10% fines; in practice, two tests are usually performed, with the aim of liberating about 7% and 13% fines. The ram load required for 10% fines is then estimated by proportioning between these values. The TPF test is also carried out on wet and dry aggregate samples, the wet/dry strength ratio being taken as a measure of durability (high wet/dry ratios indicating a sound material).

The TPF is distantly related to the *California bearing ratio* (CBR) test, which was originally devised as a penetration resistance test for fine crushed rock roadbase, although it is now almost exclusively used for subgrade strength estimation. However, the CBR ram is much smaller than that in a compression machine, the apparatus is much less robust, and penetration is only 2.5 mm (compared to 15–25 mm for the TPF ram, the depth increasing as the aggregate becomes stronger).

7.9 PROPERTIES OF FINES

The presence of 'fines' in aggregate mixtures – meaning, in this context, particles finer than 75 μm – may be necessary, tolerated, or severely limited, depending on the proposed use. Fines are necessary as binders in otherwise cohesionless roadbase or as void fillers in asphalt; they are tolerable up to a few per cent in concrete aggregate; but they can be a nuisance in sealing aggregate. Dense-graded roadbase specifications include up to 10% of low-plasticity fines, predominantly silt. Most concrete mixes stipulate a maximum of 2–4% fines, though much larger percentages of fly ash (PFA) filler are often acceptable.

Asphaltic mixes include up to 12% of –75 μm filler as a percentage of total aggregate weight, although 3–7% is more common. Its function here is not only to fill the finer voids, but also to stiffen the bitumen binder. Sealing aggregates, on the other hand, are required to be 'dry and dust-free'. They are normally precoated to improve the bitumen-to-stone bond and thereby reduce any tendency to strip.

The type of fines and their occurrence, as well as their amount, are important as well. Evenly distributed coarse inert silt 'rock flour' or crusher dust is much more acceptable than active clay minerals. Limonite coatings, fine mica flakes, organic matter and adhering mud are also objectionable to varying degrees, depending on the end-use.

The desirable properties of mineral *filler*, as opposed to deleterious clay, are not stated in specifications, but appear to be the following:

- There should be a predominance of coarse silt-sized particles, say 20–75 μm, which should be dry, and made up of quartz or carbonate grains (but not mica flakes).

- Particles should preferably be rounded, like fly ash, cement or ball-milled limestone dust. Surface charge should be minimal.
- The filler grains should not be attached to aggregate particles, but added dry in weighted quantities during batching to ensure even distribution through the mixture.

In assessing the significance of fines in aggregate mixes, it is usually sufficient to know the weight proportion of –425 µm material (the fraction used for consistency limit tests), that of –75 µm material, and the consistency limits themselves. The latter are better known as *Atterberg limits* (liquid limit, LL; plastic limit, PL; and plasticity index, PI = LL – PL). Silt-dominant mixtures are indicated by LL < 20 and low plasticity, and this is typical of the fines produced by crushing. Clay dominance is indicated by higher LL and PI values, and by a relatively high proportion of the –5 µm to 10 µm fraction, which is usually the finest size recorded. This is typical of fines produced by weathering in natural gravel roadbase.

A rough idea of the sand/silt/clay ratio can be obtained from a modified *sand equivalent test*. In this procedure the fines fraction is vigorously shaken with dispersant and water, and then allowed to settle in a graduated cylinder. The sand settles out within 1 min and most of the silt within 10 min; hence the height of the sediment column at 1 min (relative to its original saturated height) is the sand content, the extra height after 10 min is silt, and the remainder still in suspension is assumed to be clay.

7.10   AGGREGATE AND CONCRETE PETROGRAPHY

Aggregate petrography

Petrographic examination of aggregate thin sections is not yet a routine test procedure, although it is becoming more common. It is usually carried out when a new hard rock quarry site is being investigated, where rapid deterioration of aggregate has been observed in stockpiles or in service, or where anomalous test results require an explanation. It is also useful where the amount of sample is insufficient for routine testing, for example when drillcore is being evaluated. Concrete and slag, being artificial rocks, are also amenable to petrographic study.

Much has been written on the petrographic aspects of aggregate evaluation, mainly for predicting mineral reactivity in concrete, and the most comprehensive treatment is that by Dolar Mantuani (1983). The petrology of sealing aggregates is dealt with by Hartley (1974), and the current state of concrete petrographic techniques by French (1991). The use and testing of marginal-quality aggregates as roadbase in Australia is reviewed by Wylde (1979).

Although aggregate petrography is at a disadvantage relative to conventional physical testing because it cannot give quantitative answers, the areas where it has shown the greatest benefit include the following:

- *Durability*   Predicting the durability of aggregates, especially where this is suspect because weathering or alteration minerals are present in the stone. These secondary minerals may not be apparent in hand specimens and their effects may not be detected in some durability tests, such as LAA or ACV.
- *Reactivity*   Predicting potential reactivity between free alkalis in the cement paste and certain siliceous and dolomitic aggregates (note, however, that this may not translate into actual swelling and cracking). There are two main types: alkali–silica reaction (ASR) and alkali–carbonate reaction (ACR).
- *Deleterious minerals*   Detecting other deleterious minerals such as gypsum (a source of sulphate attack in concrete), sulphides (which oxidize, generate acid and stain finished concrete), fine micas (in sand) and clays (swelling and otherwise).
- *Abrasive wear*   Estimating the abrasive wear potential of siliceous rocks on crusher plates and screens, in terms of their granular quartz ('free silica') content. This is most important where the quartz content of the rock exceeds 50%, or where the crystals are coarser than about 2 mm, angular and tightly bound (i.e. in quartzite and granite).
- *Bond strength*   Investigating the aggregate–paste bond strength in concrete, particularly where reaction rims form due to pozzolanic action. This can be quite significant with slag and some limestone aggregates, where large flexural strengths can be achieved with stone of only moderate UCS values.
- *Surface microtexture*   Investigating the surface microtexture of aggregate chips, mainly with regard to their skid resistance. Fracture surface asperities, mixtures of hard and soft minerals, medium grain size and fine porosity are all indicators of a non-polishing aggregate.
- *Flakiness*   Investigating the lithological, as distinct from the crushing-induced, causes of flakiness. These include metamorphic foliation and flow banding in acid lavas.

However, the issue of non-durability tends to loom largest in the aggregate literature. One reason for this prominence over the past 30 years has been the move away from highly durable, mainly siliceous, crushed river gravels – some of which had been reworked several times by alluvial processes – to hard rock aggregate sources. These are variably weathered and altered, and a large quarry usually includes some proportion of inferior rock types. This could be a kaolinized zone in a granite, a weakly cemented bed in a quartzite sequence, or an olivine-rich layer at the base of a dolerite sill. Bear in mind also that even slightly weathered igneous rocks, though visibly sound, may have UCS and E values only 60% of those obtained from unweathered rock. At the same time, there has also been a great increase in the amount of roadbase crushed from low-grade igneous sources such as volcanic breccia.

Alteration

The favoured lithologies for hard rock quarries – mainly basaltic, doleritic and acid volcanic – are subject to alteration, especially in the intrusive bodies that are otherwise very suitable as quarry sites. In the basic rocks this is mainly *deuteric alteration* of plagioclase, olivine and pyroxenes to chlorite, zeolites and smectitic clay minerals; while in the acid-to-intermediate lithologies it takes the form of *hydrothermal alteration* to kaolinite. Alteration products are less obvious than those from weathering because they may be the same colour and strength as the unaltered rock when first exposed, or be so pervasive that the whole rockmass appears homogeneous and therefore fresh. Also, unlike weathering, the effects of alteration may become more, not less, severe with depth in an igneous body.

The engineering problem with secondary minerals in aggregates is twofold. First, many secondary minerals can absorb small quantities of water, even if they are not highly expansive. In a rock whose *in situ* porosity is less than 1%, even this small amount of swelling can cause extensive microcracking and a dramatic reduction in elastic modulus and durability. Furthermore, crushing releases much dusty, fibrous and clay-size secondary mineral, which clings to the sounder particles and inhibits cement or bituminous bonding.

Concrete petrography

Concrete petrography is largely concerned with investigating the properties of hardened concrete and the reasons for its inferior performance in service. As well as demonstrating the tenacity of the cement paste–aggregate bonding, thin sections or polished sections can provide information on the homogeneity of mixing, the size and distribution of air voids, and the quality and condition of aggregate. In addition, the effects of additives (such as air-entraining agents) and fillers (such as fly ash) can be evaluated. Finally, petrographic examination provides a check on the durability of old concrete and clues for design improvements in future mixes.

REFERENCES AND FURTHER READING

British Standards Institution (BSI) (1975–1989) *Methods for Sampling and Testing of Mineral Aggregates, Sands and Fillers*. British Standards Institution, BS 812.

Brown, E.T. (ed.) (1981) *Rock Characterisation, Testing and Monitoring.* ISRM/Pergamon, London.

Dolar Mantuani, L. (1983) *Handbook of Concrete Aggregates, A Petrographic and Technological Evaluation.* Noyes, New Jersey.

French, W.J. (1991) Concrete petrography: a review. *Quarterly Journal of Engineering Geology,* **24**, 17–48.

Hartley, A. (1974) A review of the geological factors influencing the mechanical properties of road surface aggregates. *Quarterly Journal of Engineering Geology,* **7**, 69–100.

Hobbs, D.W. (1988) *Alkali–Silica Reaction in Concrete.* Thomas Telford, London.

International Association of Engineering Geology (IAEG) (1984) International Symposium On Aggregates, Nice, May. *IAEG Bulletin,* Nos 29 and 30.

Lay, M.G. (1990) *Handbook of Road Technology,* vol. I, 2nd edn, Chs 8, 11. Gordon and Breach, New York.

Orchard, D.F. (1976) *Concrete Technology,* vol. 3, *Properties and Testing of Aggregates,* 3rd edn. Applied Science, London.

Pike, D.C. (ed.) (1990) *Standards for Aggregate.* Ellis Horwood, London.

Smith, M.R. and Collis, L. (1993) *Aggregates,* 2nd edn, Chs 6, 7. Geological Society Engineering Geology Special Publication No. 9.

Standards Association of Australia (SAA) (1974a) *Test Methods for Aggregate.* Standards Association of Australia, AS1141.

Standards Association of Australia (SAA) (1974b) *Dense Natural Aggregates for Concrete.* Standards Association of Australia, AS1465.

Unisearch/ARRB (1976) *The Production, Properties and Testing of Aggregates. Conference Proceedings,* Sydney, September.

Wylde, L.J. (1979) *Marginal Quality Aggregates Used in Australia.* Australian Road Research Board, Report ARR 97.

CHAPTER 8

Concrete materials and mix design

Concrete is an artificial rock made by blending coarse and fine aggregates, cement powder, water and – increasingly – chemical additives. In the wet state it can be poured into complex shapes within formwork, where the cement and water react together by hydration. This forms a gel, referred to here as the 'cement paste', which sets (stiffens) within a few hours and then hardens at a diminishing rate over the following days, weeks and even years. About 30% of the ultimate strength is typically achieved within 24 hours, about 50% within one week, and upwards of 80% after four weeks (Figure 8.1).

Water is required both for the hydration process and to make the wet concrete *workable* so that it can be pumped, poured and compacted. Concrete mixes of low workability, usually because of insufficient fine aggregate or water, are said to be 'harsh'. However, excess water in the mix greatly reduces its hardened strength and increases its shrinkage on drying. Hence the *water/cement ratio* (WCR) by weight is an important factor in mix design. A 'low' WCR is in the range 0.3–0.4, but mixtures of this consistency are very stiff and require chemical additives to make them workable. On the other hand, compressive strengths at a WCR of 0.7 may be only half those at 0.4, as shown in Figure 8.1.

Many of the properties of hardened concrete are similar to those of moderately strong sandstone, though concrete is much more homogeneous, stiffer and more brittle, particularly at low strengths. The modulus ratio (E/UCS) of concrete is considerably higher, 500–1300 (compared with about 200 for most sandstones). Normal concrete has compressive strengths in the range 7–40 MPa, while 'high' strengths are 40–100 MPa. Like rock, concrete is strong in compression and weak in tension, but unlike rock it can have its flexural (bending tensile) strength increased by reinforcing bars or even fibreglass strands. Hence the two main classes of concrete are plain, unreinforced or *mass concrete*, which is used for backfilling or where only compressive loads are expected, and *reinforced concrete*, which can withstand bending and twisting forces as well.

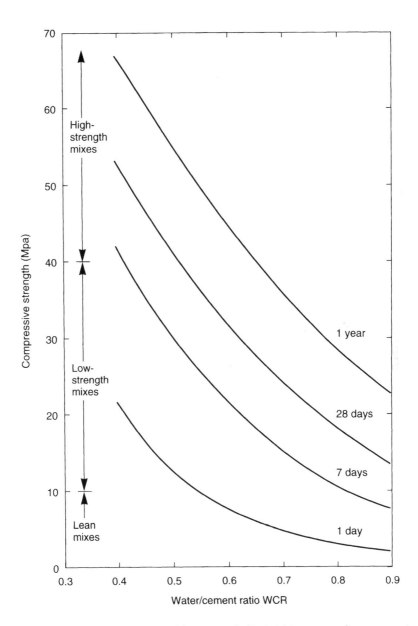

Figure 8.1 Strength increases with age and diminishing water/cement ratio for typical UK concrete mixes. Approximate 28-day strength classes on left. (After Neville, 1981.)

- *Mortar* is a mixture of cement, sand and water mainly used for bonding masonry, external rendering and plastering. The fine aggregate includes a significant proportion of –75 µm filler, to impart cohesiveness and plasticity. Mortar WCRs are large compared with normal concrete, since high strength is not required and some allowance has to be made for moisture absorption by the bricks.
- *Shotcrete* is a pneumatically applied form of high-strength mortar, which on drying shrinks and hardens to a tight, tough skin – making it useful for protection of weak and friable rock exposed in excavations.
- *Grout* is a thin cement–water suspension, often containing a large proportion of fly ash filler. It is used for filling narrow voids in fractured rockmasses, either to reduce their permeability or to increase their stiffness.

A wide variety of other characteristics such as low density (for high-rise building floors), high density (for breakwater blocks) and thermal insulating properties can be designed into concrete mixes. Mixes can also be made 'lean' and therefore cheaper, with a low cement content; porous, by reducing the fines content or by foaming; frost-resistant, by air-entrainment; and highly impervious, to resist surface attack by chemicals and salt water.

This chapter presents an introduction to concrete technology, emphasizing the effects of aggregate properties on those of the fresh (wet) and hardened product. Cement is dealt with in Chapter 15. Much more comprehensive discussions of both cement properties and concrete technology in all its aspects are provided in the standard texts by Neville (1981), Neville and Brooks (1987) and Day (1995), in manuals such as that of the US Bureau of Reclamation (1975), and in trade publications.

8.1 CONCRETE AGGREGATES

Mineral aggregates make up 60–80% of concrete by volume because they are cheaper than cement, reduce drying shrinkage and improve the properties of the finished product. Normally two size fractions are combined in the mix, *coarse aggregate* (gravel) and *fine aggregate* (sand plus fine gravel up to about 5 mm), as illustrated in Figure 8.2. A wide variety of medium- to fine-grained igneous and metamorphic rocks, and even a few well-indurated sedimentary ones, are used for making coarse aggregate. River gravels (Chapter 3) were formerly the preferred sources of concrete aggregate – because both coarse and fine sizes occur together, no blasting and minimal crushing are required, and particles are likely to be tough and durable. Quarried rock (Chapter 2) is now being used more extensively, following depletion of the alluvial deposits.

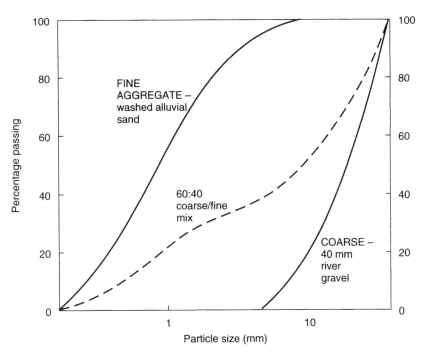

Figure 8.2 A typical concrete mix design based on clean medium-to-coarse river gravel (60% by weight) and well-graded sand (40%). Note that this mix is slightly gap-graded, due to a deficiency in the 2–5 mm range.

Concrete aggregates are required to comply with standard specifications, the main provisions of which are the following:

- There should be a mixture of coarse and fine sizes, so that the amount of cement paste needed to fill the remaining void space is minimized.
- The particles must be sufficiently dense, hard and well shaped to meet strength, workability and durability requirements.
- No components are present that might react adversely with the cement and either reduce its target strength, affect its rate of strength development, or impair its long-term durability.

In addition, most aggregate specifications concede that, if an aggregate fails to pass a particular test but has been shown to be satisfactory in service, it may be used in concrete of the same strength.

Chemically deleterious substances

Chemically deleterious substances in aggregates include salts, natural set retarders and accelerators, reactive minerals and coatings.

Figure 8.3   Concrete deterioration, Marree, South Australia. The primary cause of breakdown was probably dirty and gypsiferous aggregate, which was dug from the adjacent river bed in an arid area. However, silcrete pebbles predominate in the coarse aggregate and may have caused alkali–silica reaction (ASR).

- *Soluble salts* may cause efflorescence (white surface encrustations) or reinforcement corrosion. A maximum chloride-ion content of 0.4–0.6% is generally specified, but this may be reduced to only 0.1% in prestressed concrete. Salt content is much more a problem with sands than with gravels, and with fine sands especially, owing to their large surface area per unit weight (hence high moisture content). Gypsum is the most common natural salt after halite (sodium chloride); it can cause premature setting in wet concrete and sulphate attack in hardened cement paste (Figure 8.3).

- *Organic materials* (peat, humus, wood fragments and vegetable matter) act as setting retarders and may also discolour the concrete. They are sometimes detected by means of a colorimetric test, but this has serious limitations (for example, sugar, a strong retarder, causes no reaction).
- *Reactive aggregates* are those which chemically interact with excess alkalis in the cement paste, forming expansive secondary minerals. Though rare, the best known of these destructive phenomena are alkali–silica reaction (ASR) and alkali–carbonate reaction (ACR). Aggregates prone to ASR include some siliceous rocks, such as silcrete, argillite and chert; some glassy volcanic rocks; and some dolomitic shales.
- *Sulphide minerals,* especially chalcopyrite and marcasite, may oxidize to sulphuric acid, which in turn reacts with the cement to generate swelling gels. Pyrite, however, is sometimes considered to be stable in the concrete environment, although like the other sulphides it may give rise to rust-coloured surface staining.
- *Other deleterious substances* can include chemical coatings (such as limonite, calcite and evaporites); dust, mica and clay adhering to aggregate particles; and water-sensitive minerals (mainly alteration or weathering products). These may prevent a good bond between aggregate and paste, increase water demand, or even react with the paste. Nonetheless, a small proportion can be tolerated in most cases, especially for low-strength concrete.

Physical properties

The physical properties of aggregates were formerly thought to be closely related to the quality of concrete produced, but it is now thought that aggregate–paste bonding and the physicochemical properties of the paste itself are more important. Aggregate properties that require consideration in mix design, particularly for high-strength or high-durability concrete, are listed below.

- *Compressive strength* Concrete aggregate can have compressive strengths from about 70 to 350 MPa or more (i.e. 2–10 times that of the concrete in which they are used). Within the normal range of aggregates and mixes, however, concrete strength is mainly dependent on that of the hardened cement paste.
- *Elastic modulus* Aggregates of high elastic modulus reduce concrete deformability, drying shrinkage and creep. On the other hand, low-modulus aggregates are sometimes favoured because they absorb thermal and swelling strains, and thereby prevent cracking.
- *Porosity* In general, low-porosity rocks (i.e. low water absorption) are also strong, stiff and durable – all desirable properties in an aggregate. Low-porosity aggregate also enhances resistance to freezing and chemical attack. Conversely, porous lithologies tend to be weak and

unsound, though particles may have a rough surface texture, which improves bonding.

- *'Unsound' particles* These are of three types: those which break down with handling and mixing, those which are finely porous, and those which weather rapidly in service. Shale, mudrocks, sandstones, schist, slate and phyllite – in other words, most sedimentary and low-grade metamorphic rocks – are generally regarded as sources of unsound aggregates. In addition, these rock types may be variously weak, swell-prone, reactive and flaky. Some altered basic rocks containing olivines, chlorite and zeolites ('green' basalts and picrite) may deteriorate rapidly in exposed concrete.

- *Thermal compatibility* Aggregate–matrix thermal incompatibility can give rise to problems in very large pours, where much heat is generated by hydration, and in concrete used for low-temperature facilities. The problem seems to be most acute where the coefficient of thermal expansion of the aggregate is lower than that of the cement, as is the case with some granites and quartzites.

- *Bond strength* This is increased by an interlocking framework of coarse, well-graded, angular aggregate particles; by clean, rough-textured fracture surfaces on these particles; and by a dense, strong cement matrix. Rounded and smooth-textured water-worn gravels reduce bond strength, so that some specifications require at least one fractured face per particle. Many limestones and slags produce high bond strengths because of reaction rims between cement paste and aggregate particles, making these attractive source rocks despite their low strength.

- *Volume changes* In aggregates, volume changes caused by with-drawal of porewater during cement hydration may lead to excessive shrinkage of the concrete. Where this is restrained by reinforcing, cracking occurs (Figure 8.4). Such cracks are aesthetically undesirable and may lead to reinforcement corrosion. This problem is associated with aggregates containing expansive clays or water-sensitive secondary minerals, which are generally unsuitable for other reasons as well.

- *Misshapen particles* Prismatic and platy particles distort gradings by appearing to be finer or coarser respectively than their volume allows. They also decrease workability, increase void space – and hence reduce both strength and durability – by bridging, and increase cement and water demand when these voids are filled.

8.2   FINE AGGREGATES

The preceding section was mainly concerned with the properties of coarse aggregates, though many of the remarks are equally applicable to sands. Nonetheless, fine aggregate presents some specific problems as a

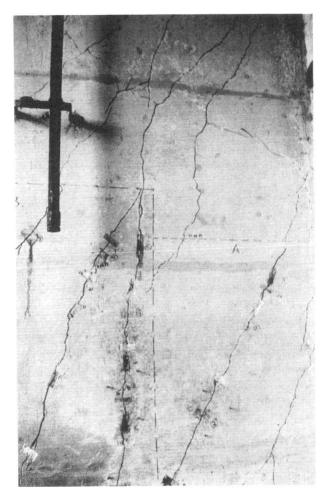

Figure 8.4 Concrete foundation deterioration, Sydney. In this case the aggregate was an olivine dolerite with a high proportion of water-sensitive secondary minerals.

concrete material. These can become more acute where a high proportion of sand is required in the mix design – either for lack of suitable coarse aggregate, or in thin-walled castings such as culvert pipes.

Mineralogy

The mineralogy of concrete sand essentially comprises quartz, carbonates and feldspars. By far the most common species is crystalline quartz, which fortunately performs best. However, amorphous quartz (opaline silica) in the form of silcrete sand is potentially reactive and hence much less satisfactory. Carbonate sands are common offshore, in raised reefs and in variably cemented dunes (aeolianites). Their particles tend to be weak, platy and unsound due to the presence of shell fragments; a significant proportion of carbonate silt is often present. These characteristics increase water and cement demand. Feldspathic sands are much less common, except in glacial outwash deposits, but particles are often unsound, debond with cement due to dissimilar thermal properties, and generate fines on working. Where a quartz sand is contaminated by a substantial proportion of unsound carbonate or feldspathic granules, heavy–medium separation may be needed before it can be used in high-quality concrete.

Particle size distributions

Particle size distributions in natural sands are generally poorly graded (i.e. well sorted), the most common deficiency being in coarse grains. Well-graded sands are often sedimentologically immature, with the coarse fraction made up of unstable rock fragments and weathered feldspar grains. Mature alluvial and marine sands, on the other hand, tend to be finer and uniformly graded. The *roundness* and sphericity of sand grains also increases with maturity and size reduction, which improves concrete workability and reduces void space between coarser stones. Hence aeolian sands, which are especially well rounded, make useful fine aggregate despite their uniform grading. Coarse manufactured sand (crusher grit) is, by contrast, distinctly angular or 'sharp', and particles are often misshapen unless artificially rounded by ball milling. These characteristics increase water demand and reduce workability, though they may contribute to bond strength.

Bulking

One problem peculiar to fine aggregates is that of bulking, whereby damp sands increase in volume by 20–30% relative to their dry or fully saturated states (Figure 8.5). This is due to the repulsive forces generated by water films surrounding the sand grains. Fine sand is particularly troublesome in this respect, since its surface-damp moisture content may be as high as 10% and this can cause bulking of up to 40%. This complicates concrete batching by weight, since wet sands are heavier than dry sand of the same mineral volume. Furthermore, the outside of a sand stockpile may appear quite dry while the interior remains moist.

8.3   FILLERS

The shortage of very fine (–0.1 mm) but non-clay material needed to fill the narrowest spaces between aggregate particles in concrete mixes can be

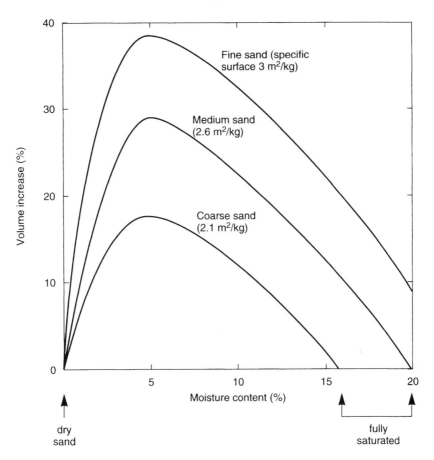

Figure 8.5 Volumetric increase in partly saturated sand due to bulking. Note that this is directly related to grain surface area and diminishes as the moisture films around particles coalesce.

compensated by an increase in cement content beyond that required for binding or, alternatively, by the use of a fine-grained filler. The latter not only reduces the cost of the concrete, but can improve its fluidity and other properties. The filler may even react to form part of the cementitious phase. Mineral fillers may therefore be classified into three groups:

- *Chemically inert materials* such as ground limestone (stone dust), hydrated lime and powdered crystalline quartz.
- *Cementitious materials* including natural cement, granulated blast furnace slag, slag–cement blends and hydraulic lime.
- *Pozzolans* (natural and artificial) such as fly ash, silica fume, diatomaceous earth, volcanic glass and some expanded shales or clays.

However, the non-reactive fillers also increase water demand and with it the water/cement ratio; hence they are only applicable in low-strength concrete. Cementitious materials and pozzolans, particularly ground granulated slag (GGS) and fly ash (PFA), are much more commonly used. All have silt-sized particles of similar diameter to those of Portland cement. Some, such as fly ash and silica fume (a very fine flue dust), influence the concrete mix properties by their spherical particle shape, glassiness and fineness, while others merely occupy the finer voids in place of cement.

They improve wet concrete by decreasing the amount of bleeding and segregation, by increasing workability without sacrificing slump, and by reducing viscosity (hence pumping resistance). In hardened concrete they can reduce heat generation, suppress sulphate- and chloride-ion attack and – sometimes – increase ultimate strength and durability.

8.4   CHEMICAL ADMIXTURES

A great variety of chemical additives for concrete manufacture are now available on the market. Most either speed up or slow down the setting and hardening of fresh concrete, or improve its pumping and compaction behaviour. They can also reduce the tendency of the mixture to *bleed* (i.e. for the cement paste to exude free water) or to *segregate* (i.e. for the coarse aggregate to separate from the cement matrix). However, the risk is that an improvement in one property may only be gained at the expense of another, and some additives have produced undesirable long-term effects such as cracking. The common admixture types are:

- *Set accelerators*, which speed up early strength development, particularly at low temperatures. These are necessary where formwork has to be re-used in later pours, or where it is necessary to allow traffic onto fresh concrete with least delay.
- *Set retarders* have the opposite effect, and are used where there is likely to be some delay between mixing and placing. The effects are usually temporary, so that concrete retarded four hours will have similar 24 hour strength to unmodified concrete.
- *Water reducing agents* allow concrete to be made with less water (i.e. lower WCR), and hence stronger, while still meeting fluidity requirements. The practice of using these as 'cement savers' – that is, reducing both water and cement content of a mix while maintaining a constant WCR – is not encouraged by concrete authorities.
- *Plasticizers* are also water reduction agents, but are used primarily to improve the fluidity of concrete. They may be so effective that the poured concrete is self-levelling and no compaction is necessary. They also make high-strength concrete workable, despite WCRs as low as 0.28.

- *Air-entraining agents* improve workability by reducing friction between aggregate particles, particularly in 'lean' (low-strength) mixes. However, the main application appears to be in frost-resistant concrete.
- *Thickeners* are flocculating agents that are used as pumping aids, to inhibit bleeding, to increase cohesiveness in lean under-sanded mixes, and to improve the penetration and adhesion of cement grouts.

8.5   CONCRETE MIX DESIGN

The usual aim of mix design is to achieve the required strength, durability and workability in the finished concrete as economically as possible. This generally means using the lowest proportion of the most expensive ingredient – cement – and making best use of the locally available aggregates, which may be less than ideal. Mix proportioning inevitably calls for compromises, since some requirements can only be achieved at the expense of others. The main factors involved are as follows:

- *Maximum particle size*   A proportion of the coarsest available aggregate, up to about 40 mm, can reduce cement and water requirements for a specified strength (Figure 8.6). However, coarse mixes are difficult to compact, especially in heavily reinforced or narrow sections, and large particles are more likely to segregate or break down during handling.
- *Particle size distribution*   Continuous-graded aggregate blends in theory reduce cement demand, but are difficult to work because of interparticle friction. Sometimes this is overcome by air entrainment or by the use of plasticizers. Such mixes are also less prone to segregate than gap-graded ones, which usually lack coarse sand and fine gravel (2–10 mm) sizes. The idea behind gap-graded mixes is that the uniform sand will be easily vibrated in to fill the spaces between the gravel chips; unfortunately this sand–cement slurry can sometimes drain away, leaving voids.
- *Surface area*   Fine aggregates have much larger surface area per unit weight (i.e. more 'specific surface') than coarse aggregates and hence require more cement to coat the particles fully. To achieve the same strength, therefore, mixes with a relatively large proportion of fine sand need more cement, and hence more water, than do coarser mixes. Where the water/cement ratio is held constant, strength decreases with increasing specific surface.

Usually two or more aggregate size fractions are combined to make up a concrete mix, as demonstrated in Figure 8.2. The proportions of a typical mix might be nominal 38 mm and 19 mm sized crushed basalt blended

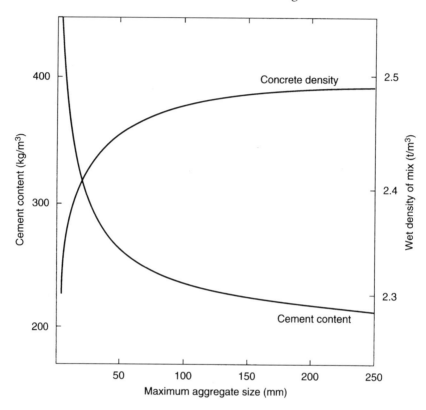

Figure 8.6   Increasing mix density and decreased cement demand as maximum particle size increases (hence specific surface decreases). However, these data are from large masonry dams – maximum particle sizes above 50 mm are unusual in reinforced concrete. (After US Bureau of Reclamation, 1975.)

with washed river sand in the ratio 1/1/2.5 by weight. Before such a design could be written into a specification, however, a series of *trial mixes* would be made up at different water/cement ratios and strength tested after 28 days of curing at constant temperature and humidity.

8.6   SETTING AND HARDENING

For some time after mixing, concrete remains plastic, and may readily be moulded and vibrated into steel shutters or timber formwork. After a period that depends on the temperature and the particular cement used, but is typically 1–10 h, the mix stiffens (*sets*) and subsequently begins to *harden* or gain strength. The strength gain is very rapid at first ('early strength'), but it will continue, although increasingly slowly, for several years. Evaporation from the hardening concrete should be limited, since just enough water is present for hydration and workability. This is known as *curing*, and generally it involves simply keeping the surface moist, but steam curing and wax coatings are also used.

From a purely chemical point of view, Portland cement powder requires between 20 and 25% of its own weight of water for complete hydration, but in practice not all this water is available for chemical combination. Cement can fully hydrate only if the water/cement ratio is greater than 1.2 by volume, or 0.38 by weight. For WCRs below this value the paste will always contain unhydrated cement, while above this it will always contain some space not filled by hardened cement. Concrete mixes in general use for structural work have WCRs in the range 0.35–0.50, rising to 1.00 in lower-strength mixes. Excess mixing water also adversely affects concrete durability, permeability, abrasion resistance and drying shrinkage.

Hydrated cement consists predominantly of a calcium aluminosilicate gel made up of colloidal particles in a platy or fibrous structure similar to that of clay. This gel has an enormous surface area and even in its densest possible state has a porosity of about 28%, its very low permeability being attributable to the narrowness of the micropores. These fine capillaries are the remains of the original water-filled spaces in the fresh paste.

8.7   STRENGTH OF CONCRETE

Compressive strength

The compressive strength of hardened concrete is, rightly or wrongly, used as an index of concrete quality and for the estimation of tensile strength, elastic modulus and durability. Compressive strengths of trial mixes are also used as an indication of the concrete-making properties of the ingredients. In designing concrete mixes, the specified strength is not the mean compressive strength, but rather the *characteristic strength* that is exceeded by 95% of test results. Hence where a compressive strength of 20 MPa is specified, it means that only 5% of test cylinder results may fall below this value. The concrete supplier must therefore aim to achieve a mean strength of 27.4 MPa, which is 20 MPa plus 1.65 times an assumed standard deviation of 4.5 MPa.

Tensile strength

Tensile strength is actually more relevant to structural design, though it is much less commonly measured. Instead, tensile strength is assumed – with justification – to vary from about 10 to 20% of compressive strength. This ratio diminishes with increasing strength; in other words tensile strength increases at only about half the rate of compressive strength. Tensile strength can be measured by the flexural (bending) test or by the splitting (Brazilian) test, the latter values being somewhat lower.

Other factors affecting strength

For a particular WCR, the higher the *aggregate/cement ratio*, the greater the strength achieved. In other words, the leaner the mix, the stronger the concrete, provided no voids are present. As the volume occupied by aggregate increases, the space available for cement diminishes; in order to maintain the WCR, so does the water content, and with it the workability of the mixture.

Strength development accelerates at higher *curing temperatures*, but slows down drastically as it falls towards freezing point. Temperature is therefore a factor as important as time in determining the set and strength of concrete.

If concrete is not fully *compacted* a considerable reduction in its strength and in other desirable properties must be expected. The presence of only 5% of air voids in the hardened concrete may decrease compressive strength by a third.

8.8   DEFORMABILITY OF CONCRETE

The deformability of concrete has three aspects: its elastic properties, its volumetric stability (especially the tendency to shrink on drying), and its tendency to creep under sustained loading.

Elastic properties

The elastic properties of concrete comprise its Young's modulus, which is generally in the range 20–50 GPa, and its Poisson's ratio (usually 0.15–0.20). Within the elastic range there does not appear to be much difference between the compressive elastic moduli and those measured in tension. Modulus ratios (E/UCS) vary from 1250 for 20 MPa concrete to 600 at 60 MPa strength. In other words, concrete becomes proportionally less stiff with increasing strength, the opposite trend to that of intact rock.

High-modulus aggregates (50 GPa and more) are used in the production of high-strength (hence high-modulus) concrete, but the tenacity of the cement–aggregate bond appears to be more important than the strength and stiffness of the parent rock. Rough-textured medium-grained acid igneous rocks are reported to achieve much better bond strength than fine-grained basic aggregates of similar UCS but smoother surface texture. Lightweight aggregates and scoria may produce concrete with moduli up to 80% of that containing much harder mineral aggregates, despite having compressive strengths only 1–10% of these rock types.

Shrinkage

Shrinkage affects concrete during both setting (plastic shrinkage) and hardening (drying shrinkage). Because the shrinkage of the cement paste is an order of magnitude greater than that of most dense aggregates, the influence of the latter is only indirect. Concretes with a high aggregate/cement ratio, which usually means coarse and harsh mixes, shrink less than richer and sandier ones. Even poor-quality aggregates like clay-rich volcanic breccias shrink less than cement paste. Typical shrinkage values for six months old concretes vary from about 0.05% with dense, high-strength aggregates to 0.1% with weak and porous stone.

Creep

The tendency of hardened concrete to creep (dilate or extend under constant loads in a structure) is primarily a problem of concrete used in flat slabs, such as floors, or heavily loaded beams. Specific creep (measured in microstrains per megapascal of load) for low-strength concrete – say 25 MPa or less – may be three to four times greater than for high-strength mixes. Marginal-quality aggregates will contribute to this time-dependent deformation, but creep, like shrinkage, is related more to the percentage of cement in the mix than to the physical properties of the aggregate. Nonetheless a high proportion of strong, angular and rough-textured rock chips will exert a degree of frictional resistance and confinement on the cement paste, and thereby limit both the short-term shrinkage and long-term creep.

8.9 DURABILITY OF CONCRETE

Durability of concrete is second only to strength as a design consideration and can be more important in some applications, though high strength, low permeability and good durability generally go together. Durability encompasses the resistance of hardened concrete to external and internal weathering agents throughout the working life of a structure. These cause inferior concrete to crack, then disintegrate, usually over several years.

This deterioration also manifests itself as surface spalling, rusted reinforcement, fretting (crumbling faces) and 'popouts' (cratering around expansive aggregate particles). External agents include chemical penetration by chlorides in salt water, abrasion by traffic, wave impacts, and ice or salt crystallization. Internal degradation is mainly caused by aggregate swelling, sulphate attack on the cement paste, and expansive aggregate–paste reactions.

Porosity, permeability and corrosion

Durability is achieved primarily by minimizing the porosity and permeability of the hardened concrete. This porosity is of three types:

- *Capillary porosity* (or microporosity) due to the hydration reaction, which is minimized by low WCRs and hence by high-strength concrete.
- *Fine cracks* caused by shrinkage and poor aggregate–paste bonding (also minimized in strong mixes).
- *Void space* or visible macropores resulting from bleeding or inadequate compaction, especially of lean or gap-graded mixes; these cavities are by far the most permeable and therefore most likely to result in corrosion.

Corrosion can also be inhibited by ensuring that reinforcement bars are clean when placed and adequately covered by concrete, to a depth of 20–70 mm. The natural pH condition within concrete is slightly alkaline, so, provided this chemical state is maintained, steel will not deteriorate. However, no concrete is totally impermeable and some moisture, as well as oxygen and carbon dioxide, will inevitably diffuse into the mass. The trick is to prevent corrosion by so reducing the concrete's permeability, and by providing sufficient cover, that the wetting front will dry out before it can penetrate to the reinforcing bars.

Salt water and aggressive groundwater

Durability is of particular importance where the concrete is exposed to salt spray in marine applications, to chemical attack in industrial areas, or to mineralized groundwater.

Salt spray causes both physical disintegration by salt wedging, and chemical deterioration by chloride- and sulphate-ion penetration. This type of damage can be resisted by presenting the densest possible surface – hence the highest-strength concrete – to seawater and saline aerosols, and by using chemically resistant cement.

Aggressive groundwater is usually taken to mean sulphate-rich, though bicarbonate and acid waters will also attack concrete. The sulphate–paste reaction results in the formation of calcium sulphoaluminate, with a large increase in volume and consequent cracking. Sulphate-resisting cements are available with low tricalcium aluminate (C_3A) content, and blended cements with a proportion of pozzolan are also effective for this purpose.

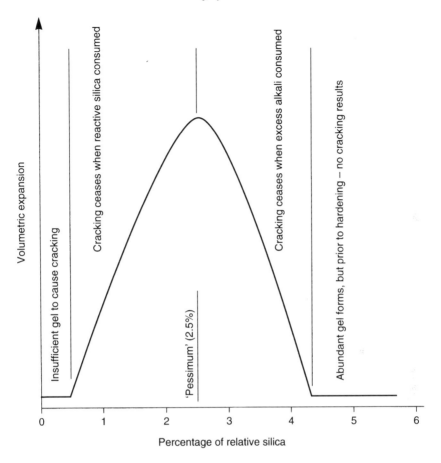

Figure 8.7 Concrete expansion due to alkali–silica reaction and the mechanisms acting. The maximum expansion could be 1–2% in mortar bar tests, less in actual mixes.

Alkali–silica reaction

The alkali–silica reaction (ASR) between cement paste and certain siliceous aggregates needs at least 0.6% by weight of Na_2O equivalent to be present in the cement. This can cause an expansive gel to form, eventually resulting in cracking and disintegration of the concrete. ASR is difficult to predict, since not all of the vulnerable lithologies turn out to be reactive. Reactivity is inversely proportional to grain size; hence silcrete sand is more subject to ASR than silcrete gravel. Unfortunately, the standard mortar bar test is unreliable and slow to produce results, but no widely accepted alternative is presently available.

ASR is counteracted by using relatively coarse-grained and low-alkali cement, preferably blended with 25–35% fly ash pozzolan and – paradoxically – by adding powdered silica to the mix. The reason for this is that the additional reactive silica, ideally microsilica or 'silica fume', mops up the available alkali and forms a non-swelling gel. The influence of reactive silica content on mortar bar test specimens is illustrated in Figure 8.7; note in particular the swelling peak at 2.5% SiO_2 and the diminishing volume instability at higher concentrations. This concentration at which most expansion occurs is called the 'pessimum' (i.e. the opposite of optimum).

REFERENCES AND FURTHER READING

Day, K.W. (1995) *Concrete Mix Design, Quality Control and Specification.* E & FN Spon, London.

Fookes, P.G. (1980) An introduction to the influence of natural aggregates on the performance and durability of concrete. *Quarterly Journal of Engineering Geology,* **13**, 207–29.

Hobbs, D.W. (1988) *Alkali–Silica Reaction in Concrete.* Thomas Telford, London.

Neville, A.M. (1981) *Properties of Concrete,* 3rd edn. Longman, London.

Neville, A.M. and Brooks, J.J. (1987) *Concrete Technology.* Longman Scientific and Technical, Harlow.

Orchard, D.F. (1976) *Concrete Technology,* vol. 3, *Properties and Testing of Aggregates,* 3rd edn. Applied Science, London.

Ryle, R. (1988) Technical aspects of aggregates for concrete. *Quarry Management,* April, pp. 27–31.

Teychenne, D.C. and Blakey, H. (1978) Crushed rock aggregates in concrete. *Quarry Management and Products,* **5** (May), 122–43.

US Bureau of Reclamation (USBR) (1975) *Concrete Manual,* 8th edn. US Bureau of Reclamation, Department of the Interior, Washington, DC.

CHAPTER 9

Asphalt and bituminous surfacing

Bituminous materials used in road-making include a variety of petroleum products, which act as waterproofing seals on road pavements, tack coats for overlays, curing membranes over stabilized bases, protective skins on exposed soil, and binders for asphaltic mixes. However, it is the surfacing and binding functions that are most important, and are the focus of this chapter.

Bituminous road surfaces are of two main types (Figure 9.1): either a thin *sprayed seal* topped with a one-stone-thick mat of coarse aggregate, or an asphalt *wearing course* about 25 mm thick. In sprayed seals the bitumen acts as a flexible membrane and as a glue for the aggregate chips, which in turn protect the bitumen skin. An asphaltic wearing course, on the other hand, is a mixture of coarse and fine aggregate with bitumen binder, which is laid and rolled to form a waterproof veneer over the pavement courses. Sprayed seals tend to be used for remote and rural roads on top of a compacted natural gravel base, while asphalt wearing courses are usually laid onto crushed rock pavements on trunk roads and urban streets.

The functions of the surfacing material are, however, the same in both cases:

- To dustproof the road pavement and prevent it being eroded by wheel abrasion, wind and running water.
- To maintain the shape and smoothness of the road crown for the safety and comfort of road users.
- To prevent infiltration of rainwater into the pavement layers and to keep their moisture content within a narrow range.

This bituminous surfacing membrane or veneer therefore protects the load-bearing pavement courses (base and sub-base), but does not significantly add to their strength. Thicker 'deep-lift' or bound layers and asphaltic overlays (Chapter 11) do, however, increase the bearing capacity or extend the working life of pavements (Figure 9.2).

This chapter is mainly concerned with the materials used in

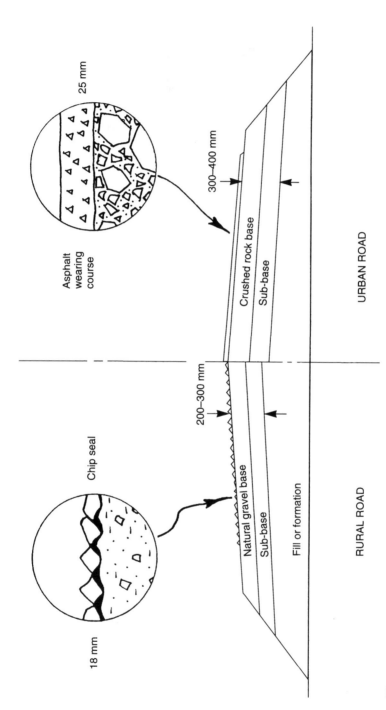

Figure 9.1 Pavement and bituminous surfacing terminology, as applied to light-traffic rural roads and medium-traffic urban roads. Note the differences in materials and thicknesses.

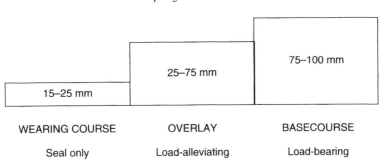

Figure 9.2 Functions and approximate thicknesses of asphalt layers in road pavements.

bituminous road surfacing, particularly mineral aggregates, and only incidentally with the design and construction of these seals. Further information on the procedures used in spray sealing and asphaltic construction are to be found in NAASRA (1980, 1984), Dickinson (1984) and – best of all – in the Shell manual (Whiteoak, 1990).

9.1 SPRAYED SEALS

Sprayed seals, also known as *chip seals*, contain less aggregate and less bitumen than asphalt, and are consequently cheaper and more flexible. Hence they are better suited to marginal-quality natural gravel basecourses, which have relatively low stiffness, and to providing economical all-weather road surfaces in remote areas. In places like inland Australia and southern Africa, this also allows a cost trade-off, whereby low-cost flexible pavements on strong, dry subgrades and topped with a thin coating of bitumen can be used instead of asphalt on crushed rock basecourses. However, sprayed seals demand more skill from the plant operators to ensure good stone coverage and adhesion, and in the choice of bitumen viscosity (too stiff and binding is poor, too thin and penetration is excessive).

Primer

This method of seal construction (Figure 9.3) is usually carried out in two or more stages. The first step is the application of a low-viscosity bituminous primer to the finished pavement surface. 'Finished' in this context means compacted, shaped, swept and moistened just before spraying. The primer coats and binds loose particles, dustproofs and waterproofs the surface, penetrates and fills the pores in the upper 20 mm or so of the basecourse, and provides a good bonding surface to which the subsequent seal coat can adhere. Thinned or 'cut-back' bitumen is the most commonly used primer these days, though coal tar was formerly a cheaper alternative.

The main pavement material properties affecting the success of priming are basecourse compatibility and its surface texture.

- *Basecourse–primer compatibility* depends on the surface chemistry of the gravel and of the bitumen, with the most common causes of seal debonding being excessive clay or salt at the top of the basecourse. The salt may be inherent with calcrete pavement materials (Figure 9.4), or introduced with saline compaction water (bore water), or drawn by capillary rise and evaporation from a shallow saline water table; in any case this is mainly a problem in arid areas.
- *Surface texture* of basecourse is related to its compacted density and hence to its porosity (or voids ratio). Well-graded and densely compacted crushed rock basecourses, which 'ring' when struck, have a tight surface texture and require only light applications of low-viscosity primer. At the opposite extreme are open-textured surfaces, which are highly absorbent. These are said to be 'hungry', needing heavy spraying with a high-viscosity primer to ensure that the upper surface of the basecourse is adequately bound. This condition results from the use of gap-graded or fines-deficient

Figure 9.3   Spreading cover aggregate (chippings) on sprayed hot bitumen, Hamilton, New Zealand. The one-stone-thick layer was subsequently rolled before being re-opened to traffic.

Figure 9.4 Debonded primerseal, near Broken Hill, New South Wales. The cut-back bitumen has penetrated about 10 mm into a calcrete gravel basecourse; seal detachment is probably due to salt in this material. (Photo: I.R. Wilson.)

bases, or natural gravels containing porous aggregates (usually duricrust nodules).

Primerseal

A primerseal is used instead of a primer where traffic is to be allowed onto the road for some time, say 3–12 months, between priming and final sealing. This consists of a medium-viscosity bitumen topped with a blinding layer of sand or crusher grit. The purpose of a primerseal is to allow time for weak areas in the pavement to become apparent and to be patched, or to await hotter or drier weather for final sealing. It may also be delayed until a specialist sealing contractor arrives, since more skill and better equipment are required for this stage of the sealing process.

Surface dressing

The final seal coat or surface dressing consists of a sprayed film of bitumen about 1 mm thick with monosized aggregate rolled into its surface, so that stones protrude above and below the seal (Figure 9.5). The bitumens used for sealing have to meet a higher specification than primers, remaining flexible and waterproof through a working life of 10 years or more. The covering aggregate is sprinkled evenly across the hot bitumen so that the stones are nearly touching each other; in time, loose particles are either broomed away or rearranged by traffic to create a stone mosaic surface. By locking aggregate chips together in this way, both the wear resistance and skid resistance of the seal are maximized. The aggregate mosaic generally covers only about 50% of the seal area when first spread, but this increases to about 70% with rolling and to 80% after a few months of traffic. Usually only one application of bitumen and cover aggregate is required, but 'hungry' seals may need an additional light surface enrichment coat. An extra application of finer aggregate is also sometimes necessary where the stone mosaic surface fails to develop properly due to stripping.

Resealing

With good maintenance and only moderate traffic growth, a natural gravel pavement may last 20–30 years, and be resealed two or three times during this period. Resealing is necessary because bitumen oxidizes and becomes brittle with age, especially in bright sunlight. This allows fine cracks to develop and moisture to enter the basecourse, causing it to deform. One of the chief aims of pavement maintenance is, therefore, to ensure that these cracks are plugged as soon as possible. Thicker bitumen coats oxidize more slowly, so that – other things being equal – a viscous grade of bitumen is likely to prove more durable. The resealing procedure involves gap-filling with successively finer aggregates, such that 10 mm aggregate is used to fill the spaces between the 20 mm stones of the original seal (Figure 9.5).

9.2   ASPHALT

Asphalt, asphaltic concrete (AC), bituminous concrete, hot mix, cold mix and plant mix are names for mixtures of coarse and fine aggregate, inert filler and a bituminous binder. For convenience here, all varieties are collectively referred to as *asphalt*, though terminology varies greatly between British, American and Australian usage. The main functions of asphalt in road-making are summarized in Figure 9.2 and below:

- *Wearing course*   Asphalt is used in Australia primarily as a wearing

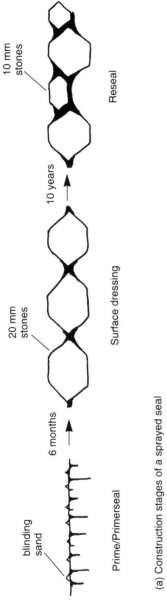

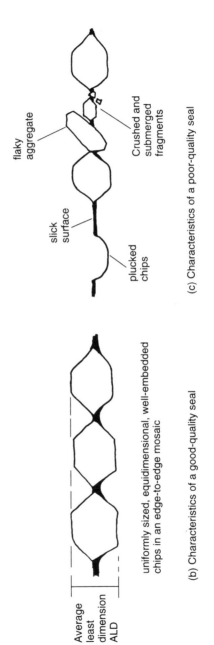

Figure 9.5 Design features and construction stages of a sprayed seal. Note the characteristics of a poor seal and compare with Figure 9.7.

course over unbound fine crushed rock (FCR) pavement layers. These are designed to be about twice the nominal particle diameter in thickness, so that 10 mm stone requires a layer thickness of 20–25 mm.

- *Overlay* The overlay is two to three times thicker and performs a load-sharing or load-alleviating role for unbound pavements approaching the end of their fatigue life. It also corrects surface irregularities created by pavement deformations.
- *Basecourse* Pavement course asphalt (base and sub-base) differs from wearing course in having a higher proportion of coarse aggregate, larger maximum particle sizes, less fines and less bitumen. Asphalt is also widely used as a patching and filling material for road surface maintenance. Cold mix, an open-graded mixture of coarse aggregate bound with slow-setting bitumen emulsion, is commonly used in this application.

Although the names and some of the functions are similar, the differences between asphaltic concrete (AC) and Portland cement concrete (PCC) are worth noting. AC design aims for high shear strength and flexibility, with a proportion of air-filled voids – usually about 4% – left to accommodate thermal deformation; PCC is designed for high compressive strength, rigidity and negligible porosity. Cement provides both the binder and filler in PCC, but in AC they are separate ingredients. The coarse and fine aggregates used are similar in both cases, though AC is more sensitive to their surface properties. The bituminous binders used in AC deteriorate with time, limiting its useful life to about 3–10 years; PCC would normally be expected to perform satisfactorily for at least 10 times as long.

Like PCC, asphalt can be produced in large central batching plants or on construction sites using mobile equipment, with the central batching offering better quality control but greater transport costs. A conspicuous difference between AC and PCC plants is the presence of a rotary drying kiln in the former (Figure 9.6). Aggregate used in AC has to be perfectly dry, dust-free and hot to ensure thorough bitumen coating of all particles, and the kiln performs all three tasks. Some of the dust loosened by drum rotation and drawn off with the hot exhaust air may be later blended back into the asphalt mix as $-75\,\mu m$ filler.

Whereas spray sealing is something of a skilled craft, asphalt-making and laying is more of a manufacturing process (and hence more amenable to quality control). Asphalt is spread and levelled in a similar manner to wet concrete, but is compacted by rollers like roadbase, rather than by vibratory compaction as with PCC. Further compaction is induced by traffic.

Figure 9.6 Aggregate drying kiln in asphalt plant, Kuala Lumpur. Cold stone is fed in from the left-hand end and hot coated stone emerges onto the conveyor (bottom right).

9.3 ASPHALT MIX DESIGN

Asphaltic mixes are designed, like other pavement materials, to ensure adequate stiffness and durability within the constraints of the expected traffic volumes. However, additional requirements may be imposed in urban situations, chief among these being skid resistance, low tyre noise and the capacity to be quickly laid and opened to traffic. A smooth surface that is kind to pedestrians and street-sweeping equipment is also desirable. Typical AC mix designs to meet these requirements are compared in Table 9.1, and their characteristics are summarized below.

Dense-graded mixes

Dense-graded mixes have a continuous particle size distribution from gravel-sized chips (usually 10–14 mm maximum) through all sizes of fine aggregate down to silt-sized filler. When compacted, they present a smooth, tight (close-textured) surface. Complete void filling is nevertheless undesirable, since this makes no allowance for further compaction under traffic or for binder expansion when heated; the lack of this may cause the asphalt to deform. Dense-graded mixes require careful proportioning of materials and thorough blending, with particular attention paid to uniform coating of all particles. A relatively

Table 9.1 Typical asphaltic mix designs

Component	Dense-graded mixes (vol%)	Open-graded mixes (vol%)	Gap-graded mixes (vol%)
Coarse aggregate	44	65	25
Fine aggregate	32	5	46
Filler	4	2	8
Bitumen	12	8	18
Air voids	8	20	3
Total	100	100	100

fluid ('soft') bitumen is used as the binder, and this is stiffened by the filler; together they perform the same load-bearing function as the cement paste in PCC. In another echo of PCC, it is essential that the mix be spread, compacted and shaped to its designed profile while it still retains sufficient heat to be workable.

Dense-graded mixes are the general-purpose asphalts used for most surfacing, pavement strengthening and bound basecourses. Although relatively cheap because of their low bitumen content and its low-viscosity grade, they offer poor skid resistance due to the shallow macrotexture and are very reflective when wet. The lack of binder also reduces their flexibility and makes them more prone to eventual fatigue cracking. Some of these deficiencies are remedied – at a cost – by open-graded and gap-graded AC mixes.

Open-graded mixes

Open-graded mixes are blends of coarse aggregate sizes that are deficient in sand- and silt-sized fines. The result is a kind of bituminous macadam or 'tar macadam' analogous to no-fines PCC. A similar material used in pavement courses, but with coarser aggregate, is known as coated stone. These materials depend for their shear stiffness ('*stability*' in asphalt terminology) on the mechanical interlock of coarse angular stones. The bitumen, which is relatively deficient and a soft grade, essentially acts as lubricant during compaction and as a waterproofing medium in service, rather than as a binder.

The surface texture of open-graded mixes is characterized by visible air-filled voids and resembles flattened popcorn. This makes them free-draining and when combined with non-polishing aggregates imparts good skid resistance. They are also non-reflecting and non-spraying in wet weather, and generate less tyre noise than chip seals. These qualities make such mixes attractive for surfacing urban roads where quietness and skid resistance are at a premium.

There are several disadvantages with open-graded asphalts, the chief one being their short service life. The coarse aggregate, being relatively exposed, wears quickly and is subject to rounding. The binder also oxidizes faster for the same reason, though its content is lower than in other types of asphalt because there is only a small proportion of fines to be coated. Furthermore, the voids eventually close up under traffic, free drainage ceases and the road surface deforms.

Gap-graded mixes

Gap-graded mixes are essentially sand asphalts with a proportion of coarse aggregate added to reduce the cost and provide some surface friction. They require more binder and filler than other mixes, and their stability is much more dependent on the stiffness of the bituminous binder. For this reason 'hard' (low-viscosity) bitumens are used and they are best suited to cool climates, where bitumen stiffness is greatest. The filler – fly ash, cement or stone dust – acts as both a stiffener and an extender for the bitumen. The advantages of gap-graded mixes include being hard-wearing, smooth-surfaced (and therefore quiet) and fatigue-resistant (promising a long working life under heavy traffic). Their skid resistance is poor, though this is enhanced by using angular or 'sharp' coarse sand in the mix and by rolling coarse aggregate chips into the surface (hence 'hot rolled asphalt' or HRA). Gap-graded asphalts are used for low-speed but heavily trafficked suburban streets and parking areas.

9.4   SEALING AGGREGATES

Preferred properties

The mat of stones topping a sprayed seal or an AC wearing course performs the most demanding service of any aggregate. (Deep-lift asphalt can tolerate lesser-quality stone, especially where its only shortcoming is a tendency to polish.) In addition to the dynamic forces generated by the passage of thousands of wheels every day, the stones are subjected to wetting and drying, abrasion by road grit, shearing forces due to braking tyres, freezing and thawing, and the thermal stresses of day/night surface temperature variations of up to 50°C. Although the required quality of aggregate will vary somewhat depending on expected traffic volume, vehicle speed, braking behaviour, climate and seal design life, the preferred properties include the following:

- *Toughness* or resistance to repeated impact loading and crack penetration, as opposed to brittleness. Some siliceous rock types, such as chert, are hard but very brittle and hence unsuitable as sealing aggregate, though they may be satisfactory in roadbase mixes.

- *Hardness* or resistance to surface abrasion and attrition. Some hard rock types are very wear-resistant, but tend to polish (become rounded and smooth-surfaced) under traffic.
- *Durability* or resistance to physicochemical breakdown during the working life of the seal. Non-durable lithologies usually have a proportion of moisture-sensitive secondary minerals, or are relatively porous.
- *Particle shape* that is angular but roughly equidimensional. Flaky or elongate chips are more easily stripped off the seal or fractured by wheel impacts, and require more bituminous adhesive per unit weight of stone than do cubic particles.

Rock strength

Note that rock strength is not a primary requirement for sealing aggregates since the tough, dense and durable lithologies that are most suitable for this purpose are generally of more than adequate strength, with UCS values typically in the range 200–300 MPa, particle SGs above 2.7 and porosities below 1%. In fact, most of the geomechanical properties of good aggregates are closely interrelated. The preferred source rocks are usually fine- to medium-grained, non-glassy and unaltered igneous rocks, or high-grade, non-foliated and non-micaceous metamorphics. Sedimentary rocks, even strongly indurated ones like recrystallized limestone or silicified sandstone, are generally unsuitable as cover aggregates on sprayed seals. Some screened but uncrushed river gravels, though composed of tough and durable clasts, may be rejected because of a lack of skid resistance (because of their roundness), or a tendency to strip (because of their smooth surface texture).

Size

The *nominal size* of cover aggregate used in sprayed seals is the diameter of the coarsest particles, and specifications usually call for a narrow grading range. This preference for single-sized aggregate allows for greater economy with bitumen, more uniform wearing characteristics and better skid resistance. Even more important than the nominal top size is the *average least dimension* (ALD) (see Figure 9.5), since this controls both the seal thickness and the stone spreading rate. In general, coarse sizes (18–20 mm) are specified for new seals because these provide the best armouring, but finer sizes may be used where the coarse fraction is unsound or flaky, where traffic flows are small, or where seal design life is short. Finer aggregates are also preferred on urban streets for pedestrian comfort, minimization of tyre noise and to make future resheeting with asphalt easier. Resealing also makes use of 7–10 mm aggregate to infill between surviving 20 mm particles.

Frictional properties

Finally, and most important in the context of road safety, are the frictional properties of the sealing aggregate. These are required for skid resistance and can be considered at two scales. The *microtexture* (roughness) of single aggregate chip faces is measured in terms of square millimetres, while the visible *macrotexture* refers to the asperity of the seal as a whole. Some of the ramifications of these frictional properties are examined in the next section.

9.5   STRIPPING AND POLISHING

The two main aggregate problems encountered in bituminous surfacing are stripping of cover aggregate from sprayed seals (Figure 9.7) and polishing of stone chips in wearing-course asphalt. Stripping results in a loss of seal macrotexture due to particle detachment from the bitumen membrane, leaving a smooth or slick ('fatty') surface, which is less skid-resistant and may 'bleed' (liquefy) in hot weather. In addition, the bitumen becomes more exposed to wear and to deterioration by oxidation under sunlight. Polishing also reduces skid resistance, but without loss of stone; instead, particles become smooth and rounded due to buffing by tyres impregnated with road grit.

Stripping

Stripping is inhibited by strong stone-to-bitumen adhesion, which dictates that the aggregate must be as clean and dry as possible when pressed into the seal. Some lithologies, such as limestone and certain river gravels, have such an affinity for water that they are said to be 'hydrophilic' and hence adhere poorly to bitumen. Surface charge on aggregate chips (positive in the case of some silica-rich rock types, negative with basic types) may cause similarly charged bitumen to be repelled. Anionic or cationic bitumen emulsions, as well as precoating and adhesion improvement agents such as quick-lime, are used to overcome these problems. Stripping can also be caused by overfluxed (excessively thinned) bitumen, which also leads to seal bleeding, or by rain or a sudden cool change during sealing operations.

Polishing

Polishing is by contrast a microtextural problem, closely related to aggregate petrology and grain size (Figure 9.8). The rock types most prone to polishing are monomineralic, fine-grained and tightly cemented – such as limestone, some fine quartzites, some glassy basalts and

Figure 9.7   Crushed and stripped aggregate in chip seal. Note the disparity between the nominal stone diameter (14 mm) and the fine disintegration products. (Photo: I. Stewart.)

hornfels. Unfortunately, rock types with the opposite qualities – mixtures of weakly bound soft and hard minerals of medium grain size, and those with a rough and finely porous surface – tend to exhibit poor abrasion resistance. The most non-polishing aggregates are synthetic rock chips (see Chapter 17), such as calcined bauxite, set in epoxy-modified bitumens; however, these are also the most expensive. Hence skid-resistant asphaltic wearing courses are used only where required for safety, such as close to traffic lights and pedestrian crossings, and need to be replaced frequently.

9.6   BITUMINOUS MATERIALS

Bitumen

Bitumen is the heavy residue obtained from petroleum refining, though it also exists in natural deposits ('tar sands'). Its mechanical behaviour is both thermoplastic, meaning that it softens on heating, and viscoelastic, meaning that it creeps under prolonged loading and at elevated temperatures, but deforms elastically at normal air temperatures. It can be made fluid for spraying, coating and binding, but behaves as a solid on cooling. Bitumens are classified in terms of their viscosity at 60°C and

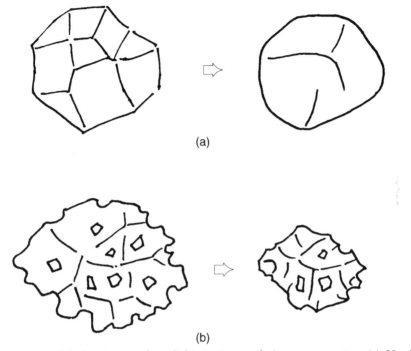

Figure 9.8 Mechanisms of polishing in surfacing aggregate. (a) Hard monomineralic or glassy chips become rounded without much size reduction. (b) Multimineralic or scoriaceous stone becomes smaller but remains rough-textured.

their penetration into a granular base under standard conditions. A 'hard' bitumen has a high viscosity and small depth of penetration, while a soft bitumen is the opposite.

Cut-back bitumen

Residual bitumen is often diluted or 'cut' using kerosene or similar fluxes (thinners) to reduce its viscosity at low temperatures. The cutter oil evaporates with time and the rate at which this occurs depends on the boiling point of the oil used. Cut-back bitumens may contain 3–11% kerosene if used for sealing, and up to 56% if used for priming. They can be produced at the refinery or on site, and tailored to suit different aggregates and air temperatures.

Bitumen emulsion

Bitumen emulsion, the most versatile of the bituminous materials, is a suspension of bitumen globules averaging 1–10 μm diameter in an

Figure 9.9   Spraying with cold bitumen emulsion to extend the life of an old chip seal by filling cracks, Turramurra, New South Wales. In colour, the emulsion changes from light to dark brown then black in about 10 minutes, as it 'breaks'.

aqueous solution. This fine dispersion is made possible by mixing at high temperature with charged emulsifying agents (surfactants) similar to detergents. The bitumen content is typically 60–70% and may be either positively or negatively charged. *Anionic* emulsions are negatively charged, with the globules suspended in an alkaline medium. In *cationic* emulsions the aqueous phase is usually acid, and these are claimed to suit a wider range of aggregate lithologies. Coalescence of the bitumen globules is termed 'setting' or 'breaking' and is brought about by evaporation of the water, or by addition of oppositely charged ionic solutions. This may occur prematurely, so stabilizing agents are added to the emulsion to resist this tendency.

Emulsions are used as tack coats (glues) prior to asphalt-laying, for precoating cover aggregates, for chemically stabilizing granular soils, as curing membranes and for crack-filling of old seals (Figure 9.9). They also act as the binder in 'cold mix' asphalt used for road maintenance patching. Emulsions can be formulated to set rapidly (3 min or less) or slowly (8 min or longer). They can be used hot or cold, but the latter is more common.

REFERENCES AND FURTHER READING

Asphalt Institute (1989) *The Asphalt Handbook.* The Asphalt Institute, Manual Series MS-4.

Broadhead, G.E. and Hills, J.F. (1990) Aggregates for bituminous materials, in *Standards for Aggregates,* ed. D.C. Pike, pp. 142–241. Ellis Horwood, London.

Dickinson, E.J. (1984) *Bituminous Roads in Australia.* Australian Road Research Board, Melbourne.

Hartley, A. (1974) Mechanical properties of road surface aggregates. *Quarterly Journal of Engineering Geology,* **7**, 69–100.

Lay, M.G. (1990) *Handbook of Road Technology,* vol. I. Gordon and Breach, New York.

Lees, G. and Kennedy, C.K. (1975) Quality, shape and degradation of aggregates. *Quarterly Journal of Engineering Geology,* **8**, 193–209.

NAASRA (National Association of Australian State Road Authorities) (1980) *Principles and Practice of Bituminous Surfacing,* vol. I, *Sprayed Work.* NAASRA (now Austroads), Sydney.

NAASRA (National Association of Australian State Road Authorities) (1984) *Principles and Practice of Bituminous Surfacing,* vol. II, *Asphalt Work.* NAASRA (now Austroads), Sydney.

Whiteoak, D. (compiler) (1990) *The Shell Bitumen Handbook.* Shell Bitumen UK, Chertsey, Surrey.

CHAPTER 10

Earthfill and compaction

Earth or soil fill is the cheapest and most widely used of all construction materials, yet so commonplace that it scarcely rates a mention in engineering textbooks. It is used for levelling construction sites; as a foundation for light buildings, roads and railways; and in embankment dams, canals and sea-walls. Its functions may be load-bearing, water-retaining (or water-excluding), providing lateral support, or simply as a sound and sight barrier. Sources of earthfill include the complete range of weak near-surface geological materials, from transported sediments to residual soils and weathered rock, plus waste materials.

The differences between *rockfill* – which is the subject of Chapter 12 – and *earthfill* are summarized in Table 10.1, but in essence they are that earthfill can be easily excavated, can develop porewater pressures (both positive and negative), and is amenable to laboratory testing. It is classified on the basis of its particle size distribution, the plasticity of the fines and its geological origins.

The earth materials discussed in this chapter are those built into engineered fills, which are designed structures, as opposed to more or less random dumped fills. *Engineered fills*, as exemplified by highway embankments, are uniformly compacted in layers of constant thickness at specified moisture contents, usually with some structural function in mind. *Dumped fills* are intended primarily for the disposal of such things as industrial wastes, domestic garbage, building rubble and dredger tailings. Their secondary purpose is land filling and reclamation, but control over layering, compaction and moisture content is minimal.

Dumped fills consolidate over years or even decades by a combination of pore drainage (primary consolidation) and creep due to plastic deformation (secondary consolidation). This may be accelerated by surcharging, artificial drainage, dropped weights, vibro-replacement and so forth. These processes operate through the full depth of fill, usually several metres, rather than in thin layers as in engineered fills. As a result the standard of void reduction achieved is inferior to true compaction, but may be adequate for founding light structures or low embankments.

Another important type of earthfill excluded from this discussion are

Table 10.1 Comparison of earthfill and rockfill characteristics

Earthfill

Disintegrates in water; can be dug by hand or by light excavators without prior blast loosening

Bulks up 10–15% on excavation, but compacted density usually greater than *in situ*

Fine-grained, maximum particle size 200–300 mm, but 80–90% generally finer than 2 mm

Compacted by medium- to heavy-weight rollers at specified optimum moisture content (OMC) to target densities of 1.8–2.2 t/m^3 (to 'end-product' specifications)

Reduced porosity and very low permeability when compacted

Layer thickness 200–300 mm when compacted

Typical design slopes (V:H) 1:2 (clayey sands) to 1:4 (clay-rich soils)

Post-construction settlement usually negligible

Amenable to laboratory testing and field compaction control

Rockfill

Requires blast loosening (in softer rock) and/or blast fragmentation (strong rock), and heavy-duty loaders and trucks

Bulks up 30–80% on excavation (dilation increases with strength); compacted rockfill density only half to three-quarters *in situ* density

Coarse-grained, maximum particle size 300–1000 mm, only 5–10% finer than 2 mm

Compacted by heaviest available rollers at high moisture content, to densities of 1.2–1.8 t/m^3 (in situ rockmass densities are typically 2.3–2.5 t/m^3)

Compaction specified in terms of roller weight and number of passes (to 'procedural' or method specification)

High bulk porosity when loose, and remains free-draining when compacted

Layer thickness 500–2000 mm when compacted; maximum allowable particle dimension about 67% of layer thickness

Typical design slopes 1:1.5 (hard rockfill) to 1:2.5 (soft rockfill)

Settlement 1–3% of fill height, usually takes many years to complete

Density and particle size distribution can only be measured with great difficulty and expense

the materials used in zoned *embankment dams*. These are the largest and most complex of all earth structures, but the variety of soils used – impermeable clay core, permeable filters and random outer shell zones – and the need to design for seepage and control of porewater pressures put them outside the scope of this chapter. Interested readers are referred to Fell *et al.* (1992) for the current state-of-the-art in earth dam design and materials. A useful summary of earthworks design and practice, emphasizing British methods, is given in Horner (1988). A somewhat more dated American equivalent is the *Earth Manual* (US Bureau of Reclamation, 1974). Much of the latest practice in compaction, drainage and earth reinforcement is also described in Hausmann (1990) and Clarke *et al.* (1994).

10.1 DEVELOPMENTS IN EARTHWORKS

Australian road construction practice over the past two decades has generally been to strengthen pavements and subgrades, rather than to improve the standard of fill compaction. Nevertheless, there have been some significant changes in earthworks engineering:

- Cuttings 40–50 m deep are now common on freeway routes, often penetrating into fresh rock below the base of weathering. This generates a proportion of angular boulders, along with the saprolitic earthfill from upper levels in the excavation.
- Deeper cuttings also mean that the water table is now more often intersected, requiring drainage layers beneath the pavement to dissipate uplift pressures.
- Fills are becoming correspondingly higher (10–30 m), to the point where time-dependent settlement (creep) now has to be considered in their design. Previously, fill settlement was assumed to have been completed during the construction period.
- The variety of earth and rock materials obtained from these deep cuttings is such that selective placement and zoning of highway embankments is becoming necessary. These materials include select fill, common earthfill, rockfill, transition layers between rockfill and earthfill, stone facing and free-draining layers.
- Fill lifts have tripled in thickness, to 300–500 mm (and more, for rockfill layers). This has been made possible by larger rollers, though the degree of compaction achieved tends still to be assessed in terms of the standard Proctor test. The 'heavy' or modified AASHO compaction test is mainly used for select fill and pavement courses.
- The use of fill confinement, lime stabilization, subsurface drainage, tensile reinforcement and processing to remove or crush oversize is becoming more common, particularly in the upper layers of embankments.

10.2 EARTH EXCAVATABILITY

As the terms were originally used in eighteenth-century canal earthworks and nineteenth-century railway construction, 'earth' was considered to be material that could be dug by pick and shovel and moved by wheelbarrow, while 'rock' had to be drilled and blasted. Modern earth-moving equipment has blurred this distinction, since weak and weathered rock, which would once have required blasting, can now be ripped by bulldozer tines prior to being excavated, disaggregated and recompacted in fills. This has created a contractual problem in specifying the difference between earthfill, including ripped rock, and harder rock requiring blast fragmentation. Drill, blast and load costs can be three times as great as in rip and push-load excavation, but the difference narrows as the ripping limit is approached.

Some of the criteria that are used to estimate the excavatability of soil, rock and intermediate materials are summarized in Table 10.2. It is worth emphasizing that there is no clear-cut *limit of rippability*, but rather a transition zone depending on the following:

- The weight and power of the *bulldozer*, its mechanical condition, and the skill (and determination) of the operator.
- The strength and toughness of the *intact rock*, and the degree of *rockmass fissuring*, due either to natural jointing or to blast fracturing.
- Other *operational factors*, such as the availability of drill-and-blast equipment, the contractor's experience and prejudices, the maximum block size allowable in the earthworks, and the stringency of blast vibration limits in urban areas (which may dictate that rock be ripped, even though blasting would be cheaper).

Seismic velocity is widely used as a criterion of rockmass rippability, but the effects of geological inhomogeneities – such as zones of secondary cementation, boulders, corestones, duricrust horizons and concretions – must also be taken into account. Closely spaced jointing can make hard igneous rock rippable, while weak but widely jointed massive sandstone and conglomerate may not be economically rippable. Seismic velocity also increases in wet rock, so a boundary between rippable and unrippable layers in apparently uniform rock may only be the water table!

Table 10.2 Excavatability criteria for earthfill and rockfill

---

Earthfill
  Can be free dug by hand, backhoes and small front-end loaders (FELs)
  In situ seismic velocity $V_P < 1500\,m/s$
  Intact UCS < 1–$2\,MPa$, PLSI $= 0$
  Loose bulk density $= 1.5$–$1.8\,t/m^3$, compacted $= 1.6$–$2.2\,t/m^3$

Earthfill/rockfill transition material
  Requires ripping, with or without blast fracturing
  Seismic velocity typically $V_P = 1500$–$2000\,m/s$
  Intact UCS $= 2$–$40\,MPa$, PLSI $< 2\,MPa$
  Particle SG $= 2.1$–2.5, *in situ* bulk density (of rockmass) $= 2.0$–$2.4\,t/m^3$

Rockfill
  Requires blast fragmentation and loosening
  Seismic velocity typically $V_P > 3000\,m/s$
  Particle UCS $> 40\,MPa$, PLSI $> 2\,MPa$
  Particle SG > 2.5, *in situ* bulk density (of rockmass) $> 2.4\,t/m^3$

10.3   EARTHFILL MATERIALS

The simplest classification of earthfill is into cohesive (clay-rich) materials and granular, sandy or non-cohesive ones. The engineering properties of the first group are implied by their *Atterberg consistency limits* (liquid limit, LL; plastic limit, PL; and plasticity index, PI), while the granular soils are classified in terms of their particle size distribution. The most widely used engineering soil classification, originally developed by the US Army Corps of Engineers but now known as the *unified soil classification* (USC), is based on these criteria.

- *Granular soils*   In general, the granular soils provide superior fill because they are easily compacted to high densities at low moisture contents, can sometimes be 'dried back' when too wet for compaction, and become more homogeneous with working due to the kneading action of the gravel clasts.
- *Cohesive soils*   The cohesive soils are difficult to compact uniformly because water cannot penetrate their low-permeability (but high-microporosity) clods. Therefore a given compactive effort can only achieve relatively low densities. The abundance of platy clay minerals causes them to become slippery when wet and to develop shear planes ('laminations') when excessively compacted.

Having been devised in North America, the USC is more applicable to the transported (alluvial, glacial, fluvioglacial and aeolian) surface sediments that are prevalent there, rather than to the *residual soils* and *saprolite* (weathered rock) that dominate the surficial geology of warmer regions. In particular, the USC ignores the strengthening influences of particle aggregation in residual soil fabrics, and of self-cementation in duricrust (lateritic and calcrete) soils.

However, the bulk of earthfill used in Australia and other non-glaciated countries is derived from highly to extremely weathered rock, which has become disaggregated by excavation, spreading and rolling. In a typical cutting the top 1–3 m is residual soil, usually clay-rich, which is unsaturated and often lateritized. This is underlain by 10–20 m of saprolite down to the base of weathering (the 'weathering front' or fresh rockhead). The saprolite becomes less weathered and more rock-like in its properties with depth, and usually becomes unrippable within the 'highly weathered' grade.

As initially broken out, this material is typically a rock–soil mixture with 70–80% of angular or slabby weak rock fragments up to 4 m in maximum dimension. With further cross-ripping and comminution beneath bulldozer tracks, this might break down to a maximum particle dimension of 1 m, and a soil content around 50%. Additional size reduction occurs during scraper loading, spreading and compaction, to the point where the +200 mm percentage becomes negligible. Any oversize blocks can be fractured by grid rollers, or simply pushed to the edge of the fill to act as slope armouring.

Because the engineering requirements for earthfill are so flexible, it is more sensible to think in terms of the few materials that are unsatisfactory rather than the many that are – more or less – suitable for use in embankment construction. These 'problem' soils include the following:

- *Expansive (cracking) clays*, which are subject to large volumetric changes on wetting and drying. These are difficult soils to compact uniformly and may wet up in fills by capillary suction over several years.
- *Dispersive clays*, which erode easily on exposure and which may also undergo internal 'tunnelling' erosion. These soils contain aggregated clays that lose their cohesion (i.e. they *deflocculate* or 'disperse') due to reduction in porewater salinity. This is primarily a problem of small dams and can be controlled by better compaction, or by lime infusion.
- *Silt soils (loesses)*, which are difficult to compact, are highly susceptible to capillary rise (causing waterlogging and frost heave), and may liquefy during prolonged seismic shaking. These soils may be very sticky just below their optimum moisture content (OMC), yet turn into a slurry just above it. They are also prone to piping and surface erosion.
- *Some highly micaceous soils*, which are difficult to compact because of the springiness of the platelets. Where the mica is too fine to be visible, these soils behave like silts, and polished shear planes may develop where the soil is overcompacted.
- *Some andosols* (allophane-rich volcanic ash soils), which have an extremely porous microfabric. These soils may appear dry yet have a very high moisture content, often well above their LL. If overworked during compaction, the cellular fabric collapses, releasing large amounts of water and turning the fill surface into a morass. They may also shrink irreversibly on drying.
- *Halloysitic soils* also develop on volcanic ash, though they may also result from tropical weathering of other parent materials. Like the andosols, they have unusual properties for clay-rich soils, notably a tendency to granulate and become non-plastic on drying. Like the andosols, but unlike swelling clays, this granulation is irreversible.
- *Peat and organic-rich soils*, which are highly compressible, shrink on draining and may even catch fire on desiccation due to spontaneous combustion. Peat itself would never be used as fill, since it contains as little as 5–15% solids, but its presence in estuarine deposits is a major cause of foundation settlement beneath embankments.
- *Some very weak and porous rocks* break down excessively and hence require careful handling when used as fill. They have very low *in situ*

density ($< 2.2 \, \text{t/m}^3$) and low intact strength (UCS $< 10 \, \text{MPa}$), and include chalk, overconslidated clays, diatomite, some pyroclastic rocks and lateritic pallid zone leached clays.

Nevertheless, the most common reason for rejecting soil as fill is simply that it is *too wet*. Acceptance criteria for wet fill include moisture contents less than 1.1–1.3 times PL, or undrained shear strengths above 35–50 kPa for clays. Saturated clean sand and gravel drain quickly, but sands with even a small proportion of silty fines can remain in a semi-liquid state after spreading and are useless as fill. Unsaturated soils can be carefully moistened up, but drying back is not often practicable. Good construction practice with clay fills requires that they be lightly rolled to a smooth finish with a surface crossfall at day's end, to limit possible infiltration by overnight rain.

Moderately wet clay soils can sometimes be satisfactorily compacted by using lighter rollers, since a small compactive effort implies a higher OMC. Conversely, heavy compaction lowers the OMC; the reasons for this will be discussed later. Where wet fill has to be used, lime stabilization of upper layers can provide a working surface (at a cost). Alternatively, settlement of weak embankments may simply have to be accepted, and compensated by stronger pavements and flatter fill slopes.

10.4 EARTHWORKS DESIGN

Design principles

The first principle of earthworks design is simply to balance cut and fill volumes, allowing for discarded topsoil or other reject material and for volumetric swell (bulking) from the *in situ* (bank) state to the final compacted condition. Any fill shortfall has to be made up from borrow pits, while rejected earthfill incurs the triple expense of transport offsite, disposal and replacement. Some typical *bulking factors* (bank to loose) and *compaction factors* (bank to compacted) for earthfill and rockfill are given in Table 10.3. Note that these are larger for rock than for soil, and higher for strong rock than for weak. The nett dilation or shrinkage from *in situ* to recompacted states is small for earthfill, but is +20–30% for rockfill. Typical wastage due to fill rejection, embankment overfilling and haul road construction is 5–15%.

The second principle of earthworks design is to bury the least-satisfactory material as low in the embankment, and as far in from the edges, as possible. Conversely the best soil – usually the most granular – is reserved for a capping layer at the top of the formation. 'Least-satisfactory' in this context generally means the wettest, most clayey, most bouldery and/or most micaceous soil excavated from nearby cuttings.

Table 10.3 Typical earthfill bulking factors

Soil	Bulk density (t/m³)		Bulking	Compaction
	In situ	Loose	factor (%)	factor (%)
Gravel, sandy and clayey (GW)	2.20	1.91	15	95
Sand, well-graded (SW)	2.10	1.83	15	95
Sand, uniformly graded (SP)	1.60	1.45	10	100
Clay, sandy (SC)	1.75	1.46	20	90
Clay, silty (CL)	1.50	1.15	30	–
Clay, heavy (CH)	1.40	1.00	45	120
Fly ash (PFA)	–	–	–	–
Sandstone, highly weathered	2.30	1.80	30	110
Sandstone, unweathered	2.50	1.67	50	130
Shale, unweathered	2.55	1.85	35	120

Design features

Some of the design features of Australian freeway-standard earthfill embankments are illustrated in Figure 10.1. The purpose of the gravelly and/or granular capping layers of *select fill* is to reduce the thickness of high-cost road pavement needed above subgrade level. This is compacted to a higher standard than the rest of the fill and in thinner layers – say 150 mm instead of 300 mm. The uppermost lift may be crushed to improve its grading, or strengthened by stabilization with about 2% cement or 3% lime. Material designated as select fill is expected to have a minimum soaked CBR value (see Chapter 11) of 10–20, with a maximum particle size of 50–100 mm. It is typically a gravelly sand with clay binder, similar to the standard required of a scaled roadbase for a rural main road. Processing of select fill was formerly confined to skimming off oversize, but single-stage crushing is becoming more usual.

The underlying *common fill* can be of lower standard, with a minimum CBR of only 5 after 10 days' soaking. It is typically compacted to only 97–98% of standard Proctor density in thicker (300 mm) lifts than the select fill. The choice of layer thickness is a compromise between the higher cost of compacting thin lifts and the greater fill strength and reduced settlement conferred by these. Heavier rollers make thicker lifts possible, especially with granular soils, although the effect diminishes exponentially with depth, regardless of the equipment used.

Where the fill material is 'harsh' (i.e. stony and bouldery), layers up to 500 mm thick may be allowed, provided the contractor can demonstrate that the compaction procedures are effective to the full depth of the layer. Especially harsh fill material, normally from the bottom of deep rock cuttings, has to be treated as *rockfill*, for which a separate specification applies (see Chapter 12). A wedge of 'inferior fill' such as swelling clay or mining waste is also shown in Figure 10.1, buried deep within the

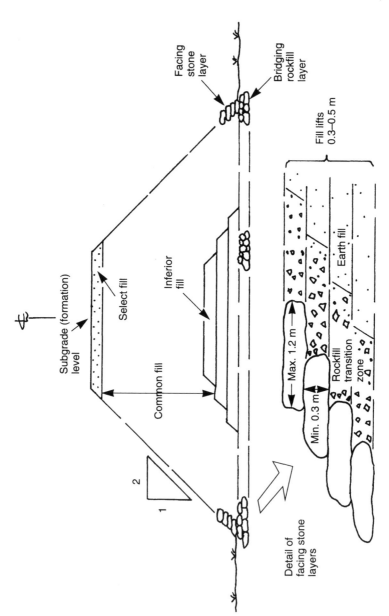

Figure 10.1 Design features of a typical high freeway embankment, not all of which will be present at any particular site. Not to scale; see text for explanation.

embankment. At the base is a *bridging layer*, essentially a construction platform for working on soft ground. This should be free-draining rockfill, placed loose by end-dumping and with minimal compaction.

Figure 10.1 also shows the details of a *rock facing layer*, similar to rip-rap on an embankment dam but with a different purpose. Originally this was a convenient means of disposal for oversize boulders, which made freeway embankments look more natural in a bushland setting. Facing layers are now also expected to provide a degree of confinement to the fill, increasing its shear strength, and to prevent rill erosion. A granular *transition zone* and geotextile sheet act as a graded filter between the clay-rich common fill and the facing layer, to prevent internal erosion.

10.5   COMPACTION THEORY – THE PROCTOR TEST

Earthworks specifications are written so as to make the best use of the materials available, and the simplest way of improving the engineering properties of soil is by compacting it. Essentially air is expelled, voids collapse and particles are forced into closer contact, but the water volume remains the same. Compaction therefore:

- Reduces *compressibility* and hence post-construction earthworks settlement.
- Reduces *permeability*, especially on the wet side of optimum moisture content (and also reduces capillary movement, hence frost susceptibility).
- Increases *shear strength* and stiffness of road subgrades.
- Improves *dimensional stability* (resistance to shrinkage and swell).

Although loosely dumped fills eventually consolidate, the process may take many years and result in settlements of 1–4% of the fill height, causing cracking and loss of shape in the overlying pavement. It is not the movement itself that causes most damage, but rather its inhomogeneity: the final surface is uneven and wavy due to differential settlement. Unsealed roads and low-speed railways are more tolerant of this movement than surfaced roads, since additional gravel or ballast can be added to compensate for settlement.

The 'standard' compaction test was developed in the 1930s by R.R. Proctor, and hence it is commonly known as the Proctor test. In its present form, a soil sample is subjected to a compactive effort of $600 \, \text{kJ/m}^3$ (25 blows of a 2.5 kg rammer on each of three layers in a cylindrical mould), at three to five successively higher moisture contents. Mechanical rammers are now replacing the hand-held sliding hammer for Proctor tests performed in larger laboratories.

At each stage the moisture content is plotted against the oven-dried density of the compacted soil, the result being expressed as a moisture–density curve similar to those in Figure 10.2. The peak of this curve is the *maximum dry density* (MDD) achievable with that particular compactive effort, while the amount of water required to achieve this density is the *optimum moisture content* (OMC). Because the size of rollers and their energy output have increased greatly over the past 60 years, the standard Proctor test has been partly superseded by a 'modified' or heavy compaction test, in which the energy input is more than four times as great (125 blows of a 4.5 kg rammer).

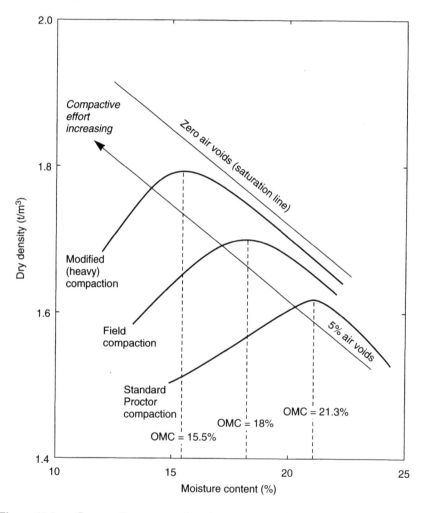

Figure 10.2   Compaction curves showing moisture–density relationships for a clay soil. Laboratory results from standard and modified Proctor tests are compared to the field compaction curve for the same material. Note the increasing density and diminishing OMC with higher compactive efforts.

The moisture–density curve tells us that adding water during compaction increases density (and strength) up to a point, but beyond this 'optimum' both diminish once again. It is believed that initially the added water reduces negative pore pressure (soil suction) and thereby facilitates void collapse. It may have a secondary role as a lubricant, causing fine particles to fit into the gaps between the coarser grains. Eventually the moisture content becomes so great that positive pore pressures develop, although the soil remains just below complete saturation, apparently due to the narrowest voids remaining air-filled. The addition of more water merely results in rearrangement of the mineral grains into a laminated fabric and lower densities, without eliminating these air voids.

The point at which densification ceases and pore pressures become positive, at about 5% air voids (or 95% water saturation), is the best or 'optimum' moisture content for that particular compactive effort. Paradoxically, maximum strength of a clayey soil is developed dry of OMC because of the contribution of soil suction; however, this cannot be relied upon, because soils do not remain permanently dry. Figure 10.2 also demonstrates that even higher densities can be achieved with increasing compactive effort, at decreasing OMCs and void ratios. The 'zero air voids' line corresponds to complete saturation, assuming a particle SG of 2.65. Note also that the on-site compactive effort may be greater than that in the 'standard' laboratory test, though it is usually less than that in the modified test.

Although the same dry density can be obtained from a soil compacted 'dry' or 'wet' of optimum, the other resulting properties may be quite different, as illustrated in Figure 10.3. Soils compacted slightly dry of the OMC are stronger but more brittle in their behaviour, and hence they are more likely to crack in high embankments. They are also more likely to swell or internally erode as water infiltrates along these fissures. Those compacted wet of optimum are less permeable (as much as a hundredfold), but weaker in shear due to the parallelism of their 'laminated' clay platelet fabric. Wet compaction will suppress the tendency of expansive clays to swell, but at a cost of lower bearing capacity and greater slipperiness during construction. For all these reasons, compaction at 1–2% dry of OMC is sometimes favoured for highway fills, while wet of optimum is preferred for embankment dam cores.

10.6   COMPACTION PRACTICE – ROLLERS AND VIBRATORS

Although the Proctor tests provide a necessary benchmark for assessing earthworks compaction, the differences between field and laboratory conditions should be appreciated. In the first place the on-road materials,

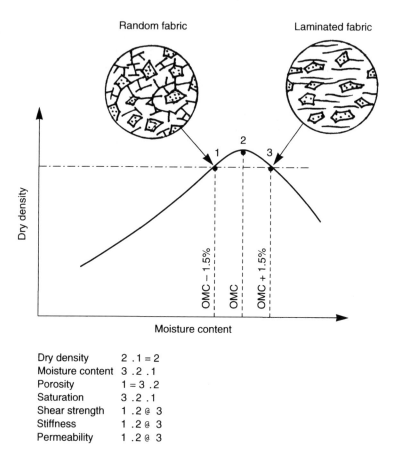

Figure 10.3   Effects of small variations in moisture content close to optimum on the geotechnical properties and microfabric of a clay-rich soil.

even though from the same pit as the laboratory samples, will generally be coarser-grained and less homogeneous. Secondly, the roller compactive effort may be greater or less than that in the laboratory, and hence the achievable MDD and OMC will differ. Finally, the confinement provided by the steel Proctor mould and the low-frequency impulsive loading are quite different to field conditions of negligible confinement and vibratory compaction.

There are three main types of rollers in use for earthworks compaction: steel drum, sheepsfoot and rubber-tyred. All are available in a variety of sizes, may be self-propelled or tractor-drawn, and the first two may also be vibratory or not. None is ideal for all purposes, but some of their advantages in particular materials are pointed out below.

Drum rollers

Drum rollers are up to 15 t in weight and are usually smooth-faced (Figure 10.4), though they may also be cleated or ribbed for breaking down weak rock fragments. Vibration frequencies range from 8 to 85 Hz. Low frequencies (around 12 Hz) and high amplitudes (1.5–2 mm) are most satisfactory for thick fill layers and cohesive soils; while the opposite (60 Hz, 0.4–0.8 mm) is true for thin lifts, granular materials and asphalt courses. Satisfactory compaction can generally be achieved with four to six passes; additional passes achieve little improvement with granular materials, though cohesive soils may continue to become denser. Non-vibratory drum rollers are mainly used for surface finishing.

Sheepsfoot rollers

Sheepsfoot rollers are tamping machines with rod-like feet protruding from a cylindrical drum, which is usually of the vibratory type. These rollers are the most effective in wet clayey soils, and the softer and wetter the soil, the larger the 'footprint' required. The compacting action works from the base of the lift upwards, so sheepsfoot rollers eventually 'walk out' of the fill as maximum density is approached. The surface finish is distinctly pockmarked, but this provides a key for the next layer – hence

Figure 10.4 Typical vibrating smooth drum roller, mainly used for compaction of granular pavement courses.

sheepsfoot rollers are used for common fill and never for select fill or pavement courses. *Padfoot rollers* (Figure 10.5) have shallower and broader tampers, and are replacing sheepsfoot rollers for general-purpose fill compaction. Neither type is suitable for rockfill.

Rubber-tyred (pneumatic) rollers

Rubber-tyred (pneumatic) rollers differ from the previous two types in being non-vibratory and depending on their static weight (up to 100 t) and tyre inflation pressure for effectiveness. They have a kneading action, which works well in finely granular materials; this mixing is accentuated if the wheels are allowed to wobble as the roller moves across the embankment surface.

Other rollers

Three other types of roller have more limited applications. *Grid rollers* are heavy steel mesh drums that are used for breaking up oversize but weak rock fragments, this action being facilitated by high forward speeds. *Vibrating plates* are small, hand-propelled vibratory rammers used for compacting against hard objects (such as bridge abutments) or for densifying backfill. *Impact rollers* are a new development, with a rounded

Figure 10.5 Vibrating padfoot roller. Note imprint on soil – this roller is for earthworks compaction, not for finishing or sealing layers.

square end-section (hence 'square wheel rollers'). The flat faces strike the ground with some violence, and hence these machines are best suited to deep compaction, say in lifts thicker than 0.5 m, though this also makes them unsuitable for use in urban sites.

10.7   SPECIFYING AND CONTROLLING COMPACTION

The degree of compaction required in embankments can be specified in terms of the properties of the end-product, the construction procedures to be used, or the performance of the fill in service.

End-product specifications

End-product specifications are by far the most common in earthfill contracts, generally defining the minimum density ratio (field density/laboratory density) to be achieved. Alternatively, a minimum shear strength may be specified for soft clay fills, or a relative density for clean sands (relative, that is, to the densest state achievable in the laboratory by flooding and vibration).

Method specifications

Method specifications dictate the minimum compactive energy to be applied and the procedures for carrying this out (for example, four passes of a 10 tonne roller vibrating within a nominated frequency range). Such specifications would be employed where *in situ* testing is difficult, say where fill has to be placed very rapidly, or where a high proportion of gravel makes the laboratory procedure unrealistic. In addition, this is the usual practice in rockfill construction (see Chapter 12).

Performance specifications

Performance specifications are rare because fill imperfections – mainly differential settlement – take months or years to develop, so that redress from contractors is hard to obtain. Performance specifications are most common where loose dumped fills (for example, quarry backfill) are being densified prior to site redevelopment. Typically, a minimum settlement under proof loading by surcharge or plate bearing test is specified.

Compaction control testing

Compaction control testing involves measuring the *in situ* compacted density of each lift. The test usually begins with a visual inspection and surface 'drumming' – with a dropped pick handle, for example – of the area to be approved (the test 'lot', say 100 m long and the full width of the fill). The purpose of this is to locate any soft patches within the test area. A small hole, about 150 mm in diameter and the same deep, is then dug in the softest spot, and all the loose soil carefully removed, weighed and sealed in a plastic bag. The hole volume is measured by inserting a water-filled balloon, or by filling it loosely with a standard sand whose poured density is accurately known. Knowing the hole volume and soil weight, the *in situ* density can then be calculated.

The soil recovered may also be subjected to laboratory compaction, but since this takes 24 hours (remember: the contractor is waiting on a result) a rapid test is sometimes used instead. One such is the *nuclear density meter*; in the past this has been criticized because of poor result reproducibility and work safety concerns, but it is now widely used for identifying wet and therefore soft patches at the fill surface.

10.8   COMPACTION BEHAVIOUR

The responses of different soils to standard compaction are shown in idealized form in Figure 10.6 and Table 10.4. In general, granular soils are much easier to compact than cohesive ones; they achieve higher densities and require fewer passes of lighter rollers. In fact, clean sands can be compacted by flooding alone (excess water quickly drains away), and even dirty sands show little improvement with heavy compaction. High-plasticity clays, on the other hand, respond best to heavy rollers and to many passes.

Well-graded sands and gravels with low-plasticity fines – SW and GW soils in the unified soil classification – produce the highest MDDs and lowest OMCs. These are usually the best natural materials available, and should therefore be reserved for pavement courses or select fill. There is a definite tendency for MDDs to diminish, and for OMCs to rise, with increasing clay content (hence increasing PI). Note also that some compaction curves in Figure 10.6 are asymmetrical, with densities and strengths falling off more rapidly on the wet side of optimum; with these materials it is better to be just under, rather than just over, optimum. Some curves, such as the loess sample, are sharply peaked – these have a distinct OMC – while other curves are broad and low, indicating that they are less sensitive to variations in compaction water content.

Some unusual curves are also presented in Figure 10.6: such anomalies are generally due to low or high particle SGs, microporosity, rounded grains, or other peculiarities of grading or mineralogy. The low-density trough to the left of OMC for the uniform dune sand shows the effects of 'bulking', caused by moisture films

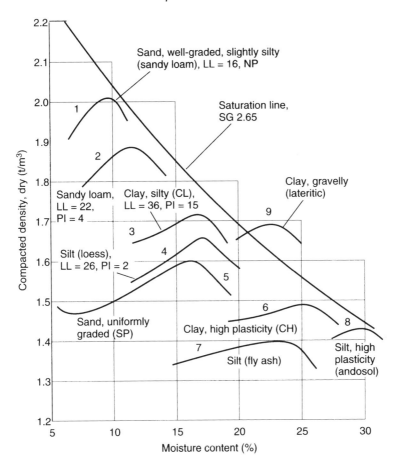

Figure 10.6   Comparison of typical compaction curves (standard Proctor) for a variety of soil materials. See text for further explanation and Table 10.4 for typical properties of these materials.

keeping grains apart in the damp state; this disappears on drying or saturation. The very low MDD for fly ash results from its porous particles and from a very uniform size distribution. The laterite gravel is displaced to the right of the saturation line because this assumes a particle SG of 2.65, while many of the ferricrete nodules are haematitic (SG 5.2). Its high OMC is probably due to absorbent particles, a common feature of duricrusts. Finally, the broad low curve for the andosol (volcanic ash soil) indicates that this sample was compacted at close to field moisture content. Oven-dried samples are much denser and non-plastic, and in fact this type of pretreatment is totally unsuited to these peculiar soils.

Table 10.4 Compaction characteristics of soils

No.[a]	Standard Proctor					Description
	MDD (t/m³)	OMC (%)	LL (%)	PI (%)	USC class	
1	2.01	9	16	NP	SW	Sand, well-graded, silty fines (sandy loam)
2	1.89	12	22	4	SW	Sand, well-graded, silt and clay fines
3	1.72	17	36	15	CL	Clay, silty, low plasticity (lean clay)
4	1.65	17.5	26	2	ML	Silt, low plasticity (loess)
5	1.60	16	–	NP	SP	Sand, uniform medium grading, no fines (dune sand)
6	1.50	26	80	50	CH	Clay, high plasticity, stiff and fissured (swelling black soil)
7	1.39	23	–	–	–	Silt, uniformly graded (pulverized fly ash, PFA)
8	1.45	30	–	–	MH	Silt, high plasticity (andosol, volcanic ash soil)
9	1.69	23	–	–	CL–GM	Clay, silty and gravelly, gap-graded (lateritic gravel)

[a] Number in left hand column refers to compaction curves on Figure 10.6.

10.9 EARTHFILL REINFORCEMENT, DRAINAGE AND CONFINEMENT

Probably the most significant developments in earthworks technology over the past 30 years have been the introduction of strengthening techniques such as reinforced earth and the widespread use of geosynthetics. These have made construction possible in tight locations; have accelerated settlement of embankments, but more especially of soft foundations beneath them; and have speeded up construction generally.

Geosynthetics

Geosynthetics are the most versatile materials used in earth construction. The family includes a variety of woven and non-woven geofabrics (or geotextiles); impermeable geomembranes; sandwich-type geo-composites; and tough meshes and mats (geogrids). Their functions are equally diverse, but the principle ones are as follows:

- *Drainage* at the base of fills on saturated ground (intercepting upward seepage), or below free-draining sub-bases (diverting downward infiltration away from moisture-sensitive subgrades).
- *Filtration*, inhibiting the movement of suspended silt and clay particles in groundwater moving between fine- and coarser-grained soils. The

geofabric filter acts as a sediment trap, allowing clear water to pass through but preventing internal erosion.

- *Separation* of coarse and fine layers, such as preventing railbed ballast or macadam basecourse being pressed into a soft subgrade. This also prevents fines contamination of the ballast, which accelerates its weathering. Usually a cushioning sand layer is laid beneath the stone, but some geosynthetics are robust enough to dispense with this.
- *Reinforcement* of fill layers, but more particularly of pavement courses, by mobilizing horizontal friction. This imparts a degree of tensile strength to the soil and has the same effect as increasing confinement. One interesting application is geofabric-reinforced sprayed seals, which are flexible enough to be placed over very poor-quality bases.

In many situations geosynthetics perform more than one function, and they have numerous secondary applications. These include lining silt traps on construction sites, providing impermeable barriers beneath landfills and tailings dams, and allowing embankments to bridge across small sinkholes in ground subject to mining subsidence.

Reinforced earth

Reinforced earth is a method of construction by which vertical-sided fills are built within segmented concrete walls, allowing the volume of a 10 m high embankment, for example, to be reduced by about a half. The secret lies in the confinement provided by tensile elements embedded in the granular fill – very little is provided by the concrete segments, which function primarily as cladding. These tensile elements are usually galvanized steel strips (Figure 10.7) laid out on the surface of each fill layer, but geosynthetic mats, steel mesh and bars are also used in a variety of commercially available systems.

The backfill for most reinforced earth systems is specified as a free-draining, non-saline and preferably uniformly graded granular material. Dune sand is most widely used for the purpose, but alluvial sands and fine gravels with less than 15% fines are acceptable. The main consideration is developing sufficient frictional resistance between the sand and the tendons to confine the fill. Light compaction in thick lifts is possible because of the non-cohesive nature of the fill material. The main drawback of reinforced earth construction is the risk that in the long term – several decades perhaps – the tendons will rust through and confinement will be lost. The corrosion problem is especially important in marine spray and saline groundwater environments; one countermeasure is to provide extra-thick galvanizing.

Confinement can also be provided by crib walls (box-like earth-filled structures assembled like bookcases from steel and concrete elements),

Figure 10.7   Reinforced earth tendons lying on compacted dune sand fill. Note the ribbing to increase friction between the tendons and the sand, and the cladding units at the rear.

and tied-back gabion walls (stacked cubical ballast-filled baskets). Both are primarily used as gravity retaining walls but have a secondary function confining embankments.

REFERENCES AND FURTHER READING

Bell, F.G. (1993) *Engineering Treatment of Soils*. E & FN Spon/Chapman and Hall, London.

Clarke, B.G., Jones, C. and Moffat, A. (eds) (1994) *Engineered Fills*. Thomas Telford, London.

Fell, R., MacGregor, P. and Stapledon, D. (1992) *Geotechnical Engineering of Embankment Dams*. Balkema, Rotterdam.

Hausmann, M.R. (1990) *Engineering Principles of Ground Modification*, Chs 2–6. McGraw-Hill, New York.

Horner, P.C. (1988) *Earthworks*, 2nd edn. Institution of Civil Engineers, Works Construction Guide Series. Thomas Telford, London.

Institution of Civil Engineers (ICE) (1985) *Failures in Earthworks*. Thomas Telford, London.

Johnson, A.W. and Sallberg, J.R. (1960) *Factors that Influence Field Compaction of Soils*. Highways Research Board Bulletin 272.

US Bureau of Reclamation (USBR) (1974) *Earth Manual*. US Bureau of Reclamation, Department of the Interior, Washington, DC.

CHAPTER 11

Pavement materials and design

The pavement courses of a road, usually the uppermost 200–600 mm, are its main load-bearing component. They dissipate the dynamic forces generated by moving vehicles, so that the stresses imposed on the road foundation – its *subgrade* – are only a small fraction of those at the surface, and thus well within its bearing capacity. Because these transient stresses diminish rapidly with depth in the pavement, its courses (layers) are stiffest and strongest at the top and may be of lesser quality below. The aim of pavement design is to combine adequate strength, hence resistance to deformation, with a satisfactory riding surface and economy in materials.

The true cost of a road pavement comprises the initial capital outlay plus the annual maintenance expenditure; in other words it is a whole-of-life cost. Thus an initially expensive concrete pavement may be justified where, for example, it requires much less patching over its working life than a cheaper unbound gravel topped by an asphalt wearing course. On the other hand, a low-cost pavement can be strengthened by overlaying as traffic gradually builds up over the years.

Two main types of pavement, based on the stiffness of their courses, are recognized.

- *Flexible pavements*   These are by far the more common and are made up of compacted dense-graded crushed rock or natural gravel. These may be either *bound* with cement, lime/fly ash or bitumen, or *unbound*. As indicated later in Table 11.1, unbound flexible pavement courses have moduli in the range 200–500 MPa. Cemented materials are about 10 times stiffer and thus constitute 'semi-rigid' pavements, although this term is not used. Asphalt moduli vary greatly with mix design and surface temperature, from about 12 000 MPa at 10°C to only 600 MPa at 40°C.
- *Rigid pavements*   The other, less common, pavement type is the concrete slab or *rigid pavement*, which may be reinforced or not.

11.1   PAVEMENT DESIGN FACTORS

The variables that are significant in pavement design include the subgrade strength and available roadbase materials, the predicted volume and type of traffic, the climatic (especially soil moisture) environment, and construction and maintenance considerations. Subgrade strength and pavement materials are the main concern of this chapter, but the other design factors will be dealt with briefly now.

Design traffic

The design traffic is the principal consideration influencing the thickness and material quality selected for the pavement. A pavement usually fails by excessive deformation, indicated by trough-like wheel tracks and surface waviness, caused by load repetitions (*fatigue*). A working life of 20 years, with a yearly traffic growth factor of up to 10% cumulative, is generally assumed. The basic design parameter is the permanent strain caused by an *equivalent standard axle* (ESA) carrying a load – in Australia – of 8.2 tonnes (Figure 11.1), in other words by the passage of a single medium-sized truck. New pavements are designed for 10^5 to 10^8 ESAs, but cars and light commercial vehicles are assumed to have negligible effect on the pavement, while a large truck might count as several ESAs. Hence it is important to be able to predict not just the volume of traffic using a particular section of roadway, but also the proportion of trucks and their sizes.

Climatic environment

The climatic environment considerations in design are primarily those of moisture movement into and out of the pavement. Pavement heaving due to freeze and thaw affects only a few roads above 1500 m altitude in Australia, though it is a much more important factor in other countries. Daily surface temperature fluctuations of up to 50°C can, however, greatly affect the deformability of asphaltic pavements on inland trunk roads (hence the choice of concrete for inter-city freeway pavements).

Construction considerations

Construction considerations in pavement design are mainly concerned with practicality and economy, which in turn are conditioned by locally available materials and local engineering practice. For example, some Australian states have favoured crushed rock roadbase, whereas cheaper but less-durable natural gravels are preferred elsewhere. Some road-building authorities favour staged construction, with pavements strengthened by overlays as traffic increases over time.

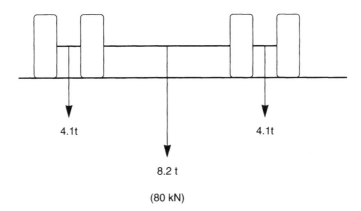

Single axle,
dual tyres at
each end

4.1t   4.1t

8.2 t

(80 kN)

In terms of pavement deformation

1 car   . 0.0001 ESA
1 small truck . 1 ESA
1 semi-trailer . 2 ESA

Figure 11.1 Definition of the equivalent standard axle (ESA), the basic unit in pavement design for assessing fatigue life. Note the approximate vehicle equivalences.

Pavement maintenance

Pavement maintenance is labour-intensive and therefore expensive, so construction using high-grade materials may eventually reduce whole-of-life costs. This is particularly true on heavily trafficked urban roads, where the costs of delays due to lane closures or out-of-hours work for routine maintenance are taken into account. Elsewhere it may be cost-effective to allow a pavement that is to be bypassed, or otherwise replaced, to deteriorate for several years with minimal maintenance.

Much of this chapter is based on Australian practice of flexible pavement design, as summarized in NAASRA (1987), Austroads (1992) and Lay (1990). The wider aspects of pavement design and materials testing, but emphasizing American practice, are covered in Yoder and Witczak (1975). British practice is described in Croney (1977) and in Croney and Croney (1991). A number of cement- and asphalt-promoting organizations have also published design guides for their products, of which the most widely used appears to be Shell's *Pavement Design Manual* (Shell, 1978).

11.2 THE PAVEMENT STRUCTURE

The elements of a flexible pavement structure, which comprises all those parts of the road affected by traffic-imposed loads, are illustrated in Figure 11.2. As well as the pavement proper, this includes the surfacing and the upper portion of the road subgrade or 'formation'. These elements vary greatly in bearing capacity; the stiffness of the basecourse, for example, being typically 10–25 times that of the subgrade and up to twice that of the sub-base. The result is that stress intensity at the top of a weak subgrade may be only 2–5% that at the road surface, though the vertical strains induced in it by each load repetition could be half those at the surface.

The *surface course* is usually either a sprayed bitumen chip seal or an asphaltic wearing course about 25 mm thick. It is assumed to contribute little strength to the pavement structure, but simply to function as a waterproofing and wear-resistant membrane. The *basecourse*, or 'base', is the chief load-bearing element of the pavement structure. The functions of the *sub-base* or sub-basecourse layer are more varied. In addition to a load-spreading role, which is secondary to that of the basecourse, the sub-base may:

- Reduce the overall cost of the pavement by allowing cheaper materials to be used in its less-stressed lower part.

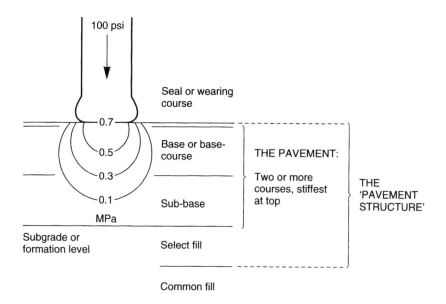

Figure 11.2   Layer definition for a flexible road pavement and for the whole pavement structure. Note how tyre load dissipates with depth in the pavement.

- Provide a barrier against moisture rising from the subgrade (a 'capillary break') and also act as a thermal insulator (to prevent freezing, and therefore heaving, of damp subgrades in cold weather).
- Resist 'pumping' or upward surging of saturated subgrade soil through cracks in concrete pavements.
- Provide a firm working platform on which the basecourse can be laid and properly compacted, free from the risk of contamination by subgrade fines.

The working platform function can also be performed by a select fill or *capping layer* at the top of the formation. Such layers are common on very soft subgrades, both as a construction expedient and as a means of thinning the pavement. The main difference between a capping layer and a sub-base is that, while the latter is usually only slightly out of specification as base, the capping layer is merely the best of the available fill materials.

11.3 SUBGRADE ASSESSMENT

Road foundation strength is assessed for design purposes in terms of the thickness of a standard pavement material, usually *fine crushed rock* (FCR), required to ensure that the subgrade does not become overstressed. Dry granular subgrades, as might have been expected, have smaller pavement cover requirements than soft wet clays. However, subgrades are naturally variable – in the degree of compaction achieved as well as in their particle size distribution – and it is necessary to design for the 'worst case' (Figure 11.3).

California bearing ratio

The most widely used means of evaluating subgrade strength is the *California bearing ratio* (CBR) test. This can be carried out on laboratory specimens compacted to approximate their field density, or – much less commonly – on *in situ* subgrades. The CBR is essentially a penetration test that responds to vertical stiffness in granular soils and shear strength in clays. It measures the force needed to push a 50 mm diameter plunger very slowly 2.5 mm into a compacted cylindrical soil sample. The test is usually carried out on saturated specimens from which +20 mm stones have been removed, as these will impede ram penetration.

The 'ratio' is a percentage of the penetration resistance exhibited by a standard limestone FCR sample, taken as 100% or 100. Well-drained granular soils may have CBRs of 15–20%, while wet and highly plastic clays are below 5%. The main problem in CBR testing is selecting a

Figure 11.3 Pavement failure on a rural road near Wyong, New South Wales. The cause appears to be a combination of insufficient pavement thickness (about 120 mm of crushed sandstone) and a wet shale subgrade.

suitable subgrade moisture content, since the CBR value drops as this increases. Many authorities assume the worst possible subgrade condition – at the end of prolonged flood submergence – and pre-soak the test sample for 2–10 days. Since many subgrades in elevated positions and semi-arid areas are never likely to be flooded, this can be a very conservative assumption.

Empirical methods

A number of empirical methods of pavement thickness estimation have been developed by road-making authorities, but these are falling into disuse except for rural road construction. Some use correlations between measured CBR values and soil classification test results, while others simply presume CBR values from design charts.

11.4   MOISTURE AND DRAINAGE

One problem that is emphasized here and in Chapter 4 is that of water entering and weakening the pavement courses. This concern arises because dense-graded roadbase, unlike free-draining ballast or macadam courses, is often moisture-sensitive. This is especially true of porous and clayey natural gravels, but FCR from some altered basic rocks also contains swelling clays and other secondary minerals. Water entering the pavement after sealing will reduce its effective strength and stiffness due to loss of soil tension (negative pore pressure). In extreme circumstances this can cause complete saturation (development of positive pore pressure) and greatly accelerated failure. The sources of this moisture and its movement in pavement systems are summarized in Figure 11.4 and comprise the following:

- *Direct infiltration* of rainfall through cracks in the seal, and lateral transfer from shoulders, fill batters and table drains.
- *Rises in the water table* and, more importantly, in the capillary fringe above it. Capillary moisture and water vapour can be drawn up to 6 m above the water table in fine-grained soils, and 1 m in sandy soils. Additional damage can be done where this groundwater is saline, since salt can accumulate beneath the bitumen and eventually cause the seal to lift off.
- *Groundwater discharging* at the base of deep rock cuttings excavated below the water table. This can lead to complete saturation of the pavement and to water bubbling up through cracks in the seal.

The most commonly used remedies against pavement infiltration are careful seal maintenance (to plug cracks), sealing shoulders, and increasing crossfall to accelerate runoff. Rising moisture is inhibited by providing capillary breaks or free-draining sub-bases, and by raising formation levels as high as possible above the natural ground surface.

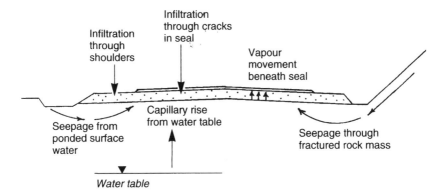

Figure 11.4 Sources of moisture entering a road pavement. See text for further explanation.

11.5 EXPANSIVE SUBGRADES

Problems of soil moisture and movement

These problems are most acute in the case of swelling clay subgrades, which are widely distributed throughout the dry tropics, including much of inland Australia and Africa. Development of these areas requires low-cost, all-weather roads built with local materials, despite their shortcomings (Figure 11.5). The road design is typically a low embankment, 0.3–1 m high, topped with a natural gravel pavement about 0.3 m thick and finished with a sprayed bitumen seal.

The embankment material is the clay soil itself, obtained from roadside borrow pits; there are few cuttings on these steppe-like plains. Because of soil fabric disturbance during excavation and on-road drying, the moisture distribution in embankments built of cracking clays is very inhomogeneous. In particular, much of the fill is likely to be looser and drier than its natural or *equilibrium moisture content*. The result is that expansive clay fills tend to 'wet up' over several years. This process is complicated by annual (wet and dry

Figure 11.5 Experimental geofabric seal placed directly on an expansive clay subgrade (i.e. no pavement courses) in western New South Wales. This is an attempt to provide an all-weather surface for a lightly trafficked road at minimal cost. The aim is to keep the subgrade at a constant moisture content with a good-quality seal; the geofabric both reinforces the seal and resists any heaving or shrinkage that may occur. (Photo: M. Sutherland.)

season) moisture fluctuations, and by longer-term (5–10 year) drought and flood cycles.

Cyclic movement of the subgrade causes the pavement layers to flex and crack, especially close to the seal edge, admitting more water to the subgrade in a self-accelerating process. Natural gravel basecourse offers somewhat greater flexibility in this situation than does more brittle FCR, but this has to be balanced against its own moisture sensitivity.

Control measures

A number of measures have been devised to control pavement distress on expansive soil subgrades. The simplest is, of course, to use granular fill instead, but this is rarely an economic option. A subgrade capping layer of select granular fill about 0.3 m thick may be a sensible compromise. Alternatively, the top 100–200 mm of the subgrade may be lime-stabilized. Good compaction can also reduce the permeability of the fill, though this may simply reduce the rate at which water is taken up. Other suggested palliatives include keeping the embankment as low as possible, to reduce the volume of clay available for swelling, and pre-wetting to approximate equilibrium moisture content.

The purpose with all these measures is to keep the moisture content at the top of the formation as constant as possible, and to move the zone of pavement moisture fluctuation outwards beyond the trafficked part of the road. Specific design features to achieve this include the following:

- *Wide seals*, extending across the road shoulders, with relatively steep crossfalls. The aims here are to widen the zone of more or less constant pavement moisture content from the vicinity of the road centreline to beyond the outer wheel path, and to shed runoff quickly.
- *PVC membranes* on top of the subgrade, which may be pre-wetted or not, to prevent moisture moving downwards. This measure does not appear to have been very successful, probably because of membrane perforation during construction.
- *Geotextile mats* on top of the subgrade have been more successful, since the aim here is to resist subgrade heaving rather than to form an impermeable barrier. Alternatively, placing the geofabric beneath the seal (Figure 11.5) increases its resistance to reflective cracking from below.

Good seal maintenance can also inhibit pavement damage, even on expansive subgrades. Regular respraying or local applications of bitumen emulsion can stop the growth of small cracks and thereby limit infiltration. More severe pavement cracking can be fixed by cold mix patching.

Pavement materials and design

11.6 PAVEMENT MATERIALS

The choice of materials for any particular pavement application is governed by cost, once requirements for strength, stiffness and service life have been decided. There are four main options: unbound granular materials, cemented granular materials, asphaltic concrete (AC) and Portland cement concrete (PCC). Typical properties for these material categories are summarized in Table 11.1.

Unbound granular materials

Unbound granular materials can include natural gravels, coarse crushed macadam courses and fine crushed rock. Only FCR will be considered here, as the other materials are described in Chapters 4 and 12, respectively. FCR is a dense-graded mixture of broken stone, crusher grit and low-plasticity fines, with a maximum particle size of about 40 mm and a compacted CBR of at least 100%. Its vertical resilient modulus E_v is typically 300–500 MPa, decreasing with depth in the pavement courses. Unbound FCR is distinctly anisotropic, with the horizontal modulus E_h being only about half the vertical value.

Both moduli increase significantly with confining pressure but diminish over time with load repetitions, hence the applicability of

Table 11.1 Typical properties of pavement materials

	UCS[a] (MPa)	E_v[b] (MPa)		Anisotropy ratio[c]	Poisson ratio
		Standard compaction	Modified compaction		
Fine crushed rock (unbound FCR) basecourse	2.5	350	500	2	0.35
Well-graded natural gravel–sand–clay (basecourse quality)	2.0	300	400	2	0.35
Natural gravel, sub-base quality	1.5	200	300	2	0.35
Bound FCR (2–3% cement) basecourse	20	–	5500	1	0.20
Bound natural gravel (4–5% cement), basecourse quality	15	–	5000	1	0.20
Bound natural gravel (4–5% cement), sub-base quality	10	–	2000	1	0.20
Asphaltic concrete basecourse	–	–	600–12000	1	0.40
Portland cement concrete (PCC) basecourse	30		20000	1	0.15
Portland cement concrete (PCC) basecourse	45		25000	1	0.15

[a] UCS = Unconfined compressive strength.
[b] E_v = Elastic modulus, vertical loading (E_h = elastic modulus, horizontal loading).
[c] Anisotropy ratio = E_v/E_h.

triaxial testing under cyclic loading conditions. These moduli are also sensitive to the compacted density and *in situ* moisture content of the material, increasing with the first and diminishing with the second (hence the importance of predicting the in-service moisture content of the pavement, and of keeping this as low as possible).

Cemented granular materials

Cemented granular materials are heavily stabilized FCR mixtures that are bound by lime, cement or fly ash blends. They are intermediate between *modified* materials, which have their plasticity reduced by small quantities of these additives but otherwise behave as unbound FCR, and strongly cemented *lean-mix* concrete. 'Bound' granular mixtures are considered to be not only stiffer and more brittle than FCR, but also to behave in a linear elastic and isotropic manner (i.e. vertical and horizontal moduli roughly equal), and to possess measurable tensile strength. Their fatigue behaviour is said to be strain-limited (they fail by cracking), while unbound pavements are stress-limited (they fail by plastic deformation).

Asphaltic pavement materials

Asphaltic pavement materials, as opposed to the surfacing asphalt mixes described in Chapter 9, tend to be dense-graded, coarser (up to 40 mm), lower in bitumen content, and are laid in thicker courses (75 mm and more). Open-graded mixes of 'coated aggregate' or 'tar macadam' are sometimes specified, especially where a free-draining basecourse is required. Polishing resistance is not necessary, so some hard but glassy basalts, dense limestones and quartzites are acceptable.

The main priorities in mix design for asphaltic basecourses, as for all pavement materials, are layer stiffness and fatigue resistance. In addition, an asphaltic pavement has to be sufficiently thinner than the alternative unbound FCR one to be cost-effective. The shear stiffness (*stability*) of asphaltic courses depends to varying degrees on the frictional interlock between the coarse aggregate fragments and the cohesive properties of the bituminous binder.

However, two characteristics unique to bituminous mixtures are that their stiffness is temperature-dependent and that it increases with age, as the binder oxidizes and hardens. At normal temperatures bitumen behaves as a linearly elastic solid, but on very hot days and under heavy truck traffic it may deform plastically. The resilient modulus of asphaltic mixes may reduce twentyfold between 10°C and 40°C, and high temperatures also accelerate the rate of oxidation. The stability of bituminous binders in hot climates is enhanced by using 'hard' (low-viscosity) grades and by blending in a high proportion of fine aggregate and filler.

Portland cement concrete

Portland cement concrete (PCC) used for road pavements varies between low compressive strength (5–7 MPa) for sub-base and medium strength (30–45 MPa) for basecourses. Sub-bases are not expected to contribute to the structural capacity of rigid pavements, but instead to provide a working platform during construction and a stable surface on which the load-bearing basecourse can be laid. The sub-base may even act as a filter, for which a low-fines concrete is specified, to prevent upward movement of subgrade fines by 'pumping'. Other sub-base mixes contain substantial proportions of fly ash for economy, or are simply lean mixes with aggregate/cement ratios up to 20.

Basecourse concrete differs little from conventional mixes, but is usually specified in terms of flexural (bending tensile) strength rather than compressive strength; 3.5–4.5 MPa is the typical range. Strength is also specified at 90 days rather than the conventional 28 day criterion. Skid resistance may be improved by the use of angular ('sharp') fine aggregate, or by grooving the finished surface. The basecourse concrete may be lightly reinforced or plain (unreinforced).

11.7 FLEXIBLE PAVEMENT DESIGN

Structural design of road pavements draws upon elasticity theory and numerical modelling, the details of which are beyond the scope of this book. However, three approaches to pavement design can be distinguished:

- Mechanistic (or 'analytical', or 'rational') methods founded on elasticity theory.
- Semi-empirical methods, based on a combination of experience (in the form of design charts) and testing such as repeated-load triaxial.
- Empirical or prescriptive methods, also experience-based but using only simple index test data or even presumed values.

Mechanistic approaches

Mechanistic design procedures are based on finite-element modelling of a multilayered pavement system subjected to repetitive loading. They are mainly used for heavily trafficked freeways and urban trunk roads, plus airport runways and taxiways. The design models generally assume that pavement layers are homogeneous and behave more or less elastically, although some can simulate anisotropic unbound layers and subgrades. A computer program then calculates the accumulated non-recoverable strains for so many million loading cycles of a standard 8.2 t axle load. The cumulative strain generated by ESA loading cycles is illustrated in Figure 11.6, which also shows the effects of single vehicles

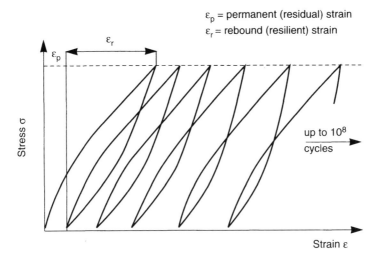

(a) Cumulative pavement strain under traffic

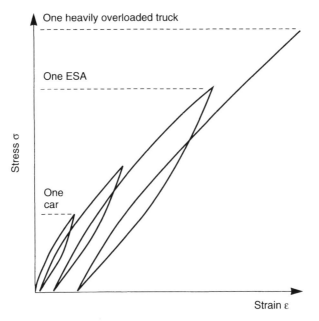

(b) Pavement deflections due to single vehicles

Figure 11.6 Cumulative pavement strains due to the passage of many equally loaded trucks (above) compared with the strains caused by single vehicles of different sizes (below).

of different weights. The critical responses generated by the model are as follows:

- For *bound layers* (asphalt and cemented materials), which are assumed to fail by tensile cracking, the cumulative horizontal strains at the base of the layer.
- For *unbound layers* (chiefly subgrade), which are assumed to fail by excessive deformation, the cumulative vertical compressive strains at the top of the layer.

The input data for the model are usually the resilient moduli (E_h and E_v), Poisson's ratio and thickness for each layer. The model represents a cross-section through the pavement system, in which the subgrade is assumed to be infinitely thick and the pavement infinitely long. Real materials are, of course, inhomogeneous and their elastic properties in bulk are somewhat different – usually less stiff and more anisotropic – than those indicated from tests on small laboratory specimens.

This is especially true of subgrade soils, which are most likely to be non-uniform in degree of compaction and moisture content, and for which the test data available are generally the most scanty and the least reliable. Fortunately, the model sensitivity to material properties diminishes rapidly with distance from the loading point, in this case the road surface. The simulation can therefore tolerate some imprecision in subgrade properties, and in any case test values are scaled downwards to allow for material variations in bulk. Predicted stresses at the base of the pavement are usually so low as to be within the narrow range of subgrade elastic behaviour.

In practice, a lot of the input data are estimated rather than obtained from testing. For example, subgrade moduli are commonly estimated from soaked CBR values, such that:

$$E_{subgrade} = 10\text{–}16 \text{ times CBR} \qquad \text{(range 10–300 MPa)}$$

Moduli for unbound granular basecourses are up to 50 times greater than this, and for cemented materials up to 500 times greater. Nonetheless, measured values are preferred by designers, and the favoured method for determining moduli of unbound pavement materials is by means of a triaxial test under conditions of repetitive loading. For bound materials, flexural tests, also with repetitive loading, are more relevant. In some models the pavement courses have to be divided into sublayers 50–150 mm thick, to allow the moduli – whose values are stress-dependent – to diminish realistically with depth.

Semi-empirical approaches

Semi-empirical design procedures are used for unbound FCR pavements topped by a thin asphaltic wearing course or sprayed seal. Roads of this

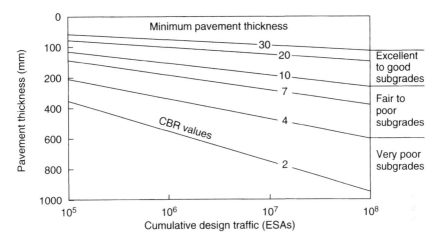

Figure 11.7 Design chart for flexible pavements. Required thickness of specified granular basecourse (minimum 100–200 mm) plotted against whole-of-life design traffic. Note large increase in cover requirement for weak saturated subgrades with CBR < 4.

standard could include single-carriageway rural highways or urban main roads, plus runways at secondary airports. Mechanistic design procedures can be used in this application, but they offer little advantage in return for substantially greater effort.

The essence of the semi-empirical method is the use of design nomograms such as that illustrated in Figure 11.7. The first step is to measure or estimate a subgrade CBR value and a whole-of-life design traffic for the proposed pavement. A subgrade cover requirement, in terms of a standard pavement material (in this case FCR), can then be read off the *y*-axis. Other materials can be substituted for FCR by invoking *layer equivalences*; these mean, for example, that 100 mm of asphaltic concrete is assumed to have the strength of 200 mm of unbound FCR. Note also in Figure 11.7 that there is a minimum recommended thickness of 100–200 mm, even for excellent-quality (CBR > 30) subgrade materials, and that for very weak subgrades (say CBR 2) the required pavement thickness approaches 1 m. This is far too thick for practical purposes, so some sort of stabilization or a granular select fill layer would normally be laid on such a subgrade. The required pavement thickness could then be reduced to about 300 mm.

Empirical approaches

Empirical procedures are the simplest of all, based on a series of design rules, charts and tables. They are most applicable to estimating pavement thicknesses for rural and secondary roads, where the material is either unprocessed or minimally treated natural gravel. The criteria for these design rules are the subgrade and sub-base gradings, fines plasticity and estimated CBR. The on-site materials are simply fitted into defined classes based on these index tests and their cover requirement read off an appropriate table. The basecourse is nominally a 'zero cover' material, though a minimum compacted thickness of 100 mm is usually specified. Some latitude in specification limits is allowed, particularly in PI and LL, for dry climates and well-drained subgrades.

Empirical pavement design procedures formerly had wide currency in Australia, being used up to highway-standard roads. They presuppose the existence of 'standard' materials whose behaviour can be predicted from their long-term performance in existing roads; hence there is little scope for introducing unorthodox new materials. However, their use is now in decline, owing to a move away from prescriptive methods and towards analytical design. This has resulted from a wider understanding of the significance of traffic and repetitive loading on pavement life, and because of a trend towards much greater use of processed and even manufactured pavement materials. Pavement thicknesses calculated empirically are nonetheless conservative in most situations, so failures are more likely to be due to inhomogeneous materials, poor compaction or inadequate drainage than to insufficient subgrade cover.

11.8   RIGID PAVEMENT DESIGN

Rigid concrete (PCC) pavements are undergoing a revival in Australia, after several decades in disfavour because of their high labour component – many of the older concrete pavements were built as 'make work' schemes in the 1930s depression – and their tendency to crack severely with age. This cracking resulted from loss of subgrade fines caused by 'pumping' at expansion joints. Eventually the concrete slab failed by arching ('hogging') and a new focus for pumping was created at the fresh crack. The renewed interest from highway engineers in rigid pavements, for freeways in particular, has arisen out of several developments:

- *Low maintenance*   There is a need for low-maintenance pavements to handle very heavy design traffics (up to 10^8 ESAs) with a high proportion of multi-axle trucks. Flexible pavements meeting these requirements would be unacceptably thick, or unable to match the low maintenance costs of rigid pavements.
- *Climate*   There are climatic limitations on deep-lift asphaltic pavements imposed by large day/night summer temperature variations on inland freeways.
- *Continuous reinforcement*   The availability of continuously reinforced concrete pavement technology has resulted in greatly reduced joint

Figure 11.8 Concrete paving machine, Mittagong Bypass, New South Wales. The machine is advancing from left to right; wet concrete is spread by the front section and compacted and levelled by the rear unit. Note the light continuous reinforcement for the basecourse and the 200 mm thick unreinforced lean-mix sub-base on which the machine is working.

frequency and hence much improved riding quality. This was a major cause of public dissatisfaction with the older unreinforced type, which needed expansion joints at about 6 m intervals.

- *Continuous operations* Developments have occurred in continuous placing, spreading and vibrating techniques (Figure 11.8) for use within slipforms with minimal hand finishing. Paving operations of this type also offer very precise level control using laser guidance.
- *Cost increases* Large increases in the cost of bituminous materials have taken place over the past 20 years, without corresponding rises in cement costs.

In addition to these high-speed and heavily trafficked roads, rigid pavements have two other areas of application. They offer a smooth and hard-wearing surface for urban streets, though access to underground services (gas, water and telephone cables) is somewhat impeded compared to flexible pavements. Concrete pavements are also unrivalled for load spreading on very soft subgrades, such as wet clay fills and swampy ground.

Subgrade strength for rigid pavement design is normally assessed in terms of CBR, as for flexible pavements, but the modulus of subgrade reaction (a type of plate bearing test) can also be used to estimate slab thickness. Lean-mix sub-bases are preferred to cheaper unbound granular materials because they are stiffer and therefore deflect less under load, ensuring a longer service life for the overlying basecourse. Bound sub-bases also resist pumping better.

The design strength and thickness of concrete basecourses are selected by an iterative process. First, a nominated slab thickness, say 200 mm, is related to the permissible axle loading and sub-base CBR using nomograms. This gives a predicted tensile stress at the bottom of the slab due to the passage of one standard axle, which must be less than half the flexural strength of the concrete. Where this is not so, either a stronger mix or – more commonly – a thicker basecourse slab is required. The minimum allowable slab thickness is in any case 150–200 mm. Steel mesh reinforcement, where used, is light by the standards of structural concrete – only about 0.6% by volume. This is not considered sufficient to contribute to the strength of the base, but functions simply to hold the concrete together as shrinkage cracks develop.

11.9   PAVEMENT PERFORMANCE AND STRENGTHENING

Pavement life

It has been emphasized in this chapter that modern road pavements, especially flexible pavements, are designed for a finite service life – typically to carry between one and 20 million heavy vehicles over a period of about 20 years. At the end of this period – or, more strictly, at the end of this design traffic – it is assumed that the pavement will be at or close to fatigue failure. *Failure* in pavement design terms means that it will have deformed so severely that driver comfort will be seriously impaired and vibrations at high speed may damage or endanger vehicles.

Visible indications of impending failure include rutting (trough-like depressions along wheel paths), corrugations (short-wavelength transverse ripples), edge disintegration and a variety of cracks (see Figure 11.3). Note that seal failure is a separate problem, though its cracking will certainly accelerate pavement deterioration. Pavement life can therefore be extended by careful seal maintenance, by the addition of strengthening overlays or by renovation through deep-lift stabilization (see Chapter 18).

The first step is to recognize when the pavement is approaching its fatigue limit. The signs of pavement distress are obvious to an experienced observer, and rating schemes based on extent of patching, degree and type of cracking, loss of seal and so forth have been devised to identify pavement sections at risk. These sections can then be investigated more thoroughly, using deflection testing equipment such as the Benkelman beam, which measures very small road surface movements under standard loading conditions.

Overlays

Overlays provide a means of extending the life of roads whose geometrical design (gradients, curve radii, formation width) is adequate for present and near-future traffic, but whose pavement is failing. They are usually designed on the basis of deflection test results, supplemented by *in situ* testing of the existing pavement courses and subgrades. In some cases they are part of a process of *staged construction* whereby a cheap pavement, usually natural gravel, is laid initially and strengthened – say after 10–20 years – to accommodate the increased traffic.

Asphalt is probably the most common overlay material, because it can provide the same strength as nearly twice its thickness of unstabilized FCR. In addition, it presents a smoother surface and can be effectively compacted in thin layers, since overlays are usually less than 75 mm thick (any thicker and complete pavement reconstruction would be justified). A further advantage with asphalt overlays is that they can be quickly laid, shaped and compacted, and the road re-opened to traffic with minimal delay.

REFERENCES AND FURTHER READING

Austroads (1992) *Pavement Design – A Guide to Structural Design of Road Pavements*, rev. edn. Austroads (formerly NAASRA), Sydney.

Croney, D. (1977) *The Design and Performance of Road Pavements*. TRRL/HMSO, London.

Croney, D. and Croney, P. (1991) *The Design and Performance of Road Pavements*, 2nd edn. McGraw-Hill, London.

Lay, M.G. (1990) *Handbook of Road Technology*, 2nd edn, Chs 8, 9, 11 and 14. Gordon and Breach, New York.

NAASRA (National Association of Australian State Road Authorities) (1987) *Pavement Design – A Guide to Structural Design of Road Pavements*. NAASRA, Sydney.

Shell (1978) *Pavement Design Manual*. Shell Petroleum PLC, London.

Yoder, E.J. and Witczak, M.W. (1975) *Principles of Pavement Design*, 2nd edn. John Wiley, New York.

CHAPTER 12

Rockfill and ballast

These are coarsely fragmented and free-draining rock materials, whose shear strength and stiffness are due to particle interlock rather than dense packing. Both are *open-graded*, though rockfill has a maximum particle size of 0.5–1 m, while for ballast it is only 75 mm. However, they differ in several ways: rockfill is fragmented by blasting and rolling, while ballast is crushed and screened; rockfill is only specified in a rudimentary fashion, while ballast is tightly controlled in grading and durability; and ballast is hard, aggregate-quality stone, while rockfill is much weaker and less durable.

- *Rockfill* is mainly used in embankment dams and as rubble core in breakwaters, or for scour protection on exposed slopes and in channels (pitching and beaching stone). It is also coming into use for highway embankments, because modern deep cuttings are exposing a proportion of fresher rock that does not break down to conventional earthfill.
- *Ballast* (Figure 12.1) is used primarily for railway trackbeds, but also as fill for slope-supporting gabion baskets and mesh mattresses, and as drainage layers in embankment dams and beneath road pavements. Relatively weak ballast is used, to a limited extent, in macadam road pavements.

Comparison grading curves for several coarse rock materials are illustrated in Figure 12.2. Dam rockfill is not included because of its heterogeneity, but it would normally be somewhat coarser than the highway embankment rockfill.

12.1 ROCKFILL CHARACTERISTICS

The principal characteristics of rockfill, as distinct from extremely weathered rock (saprolite) used as earthfill (Chapter 10), are its coarser particle size, higher permeability, steeper angle of repose and larger post-construction settlement. Although there has been some convergence in recent decades, as earthfill has become more stony and acceptable rockfill has become weaker, the fundamental difference remains: rockfill is pervious and earthfill is not.

Figure 12.1 Three coarse rock materials together: railbed ballast (nominal 75 mm size), gabion stone (100 mm) and free-draining rubble (300 mm) behind the gabion retaining wall.

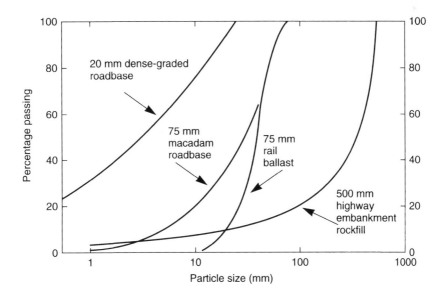

Figure 12.2 Grading curves for rockfill, ballast and macadam compared to conventional roadbase. The rockfill curve represents the upper size limit for this material when used in highway embankments; dam rockfill is coarser still. The other curves are the median lines of their respective grading envelopes.

Being free-draining, positive porewater pressures cannot develop in rockfill, and its slopes can thus be designed steeper than for earth embankments. 'Free-draining' in this context means a bulk permeability greater than about 10^{-5} m/s. These steep fill slopes are also a consequence of the high internal friction generated by coarse, angular, tightly interlocking rock fragments. Consider, for example, two embankment dams each 100 m high – rockfill slopes will stand at about V:H = 1:1.5, while earthfill may require 1:4 or flatter for long-term stability. The earthfill volume will be two to three times that of a rock embankment, and the comparative advantage of rockfill increases with height.

The principal shortcoming of rockfill as an embankment material is its large and protracted settlement. Crest movements of some early hard rockfill dams continued for decades after completion and ultimately exceeded 1.5% of fill height. Modern compacted soft rockfill dams typically settle only about 0.5%, most of which occurs during construction. The settlement consists of two parts: a primary component of about half the ultimate movement, completed shortly after sluicing and vibratory compaction; and secondary consolidation (creep), which occurs over several years. Earthfill usually has only a small creep component, and hence settlement is complete for practical purposes at the end of construction. Furthermore, because of the heterogeneity of rockfill and the difficulty in testing such coarse material, its settlement is much less predictable than that of earthfill.

The void reduction processes include self-weight compression, shearing or crushing of point contacts between rock fragments, and tensile (flexural) fracture of the fragments. These result in smaller particle sizes and reduced fill porosity, especially towards the embankment base, but with sufficient void continuity to ensure that the mass remains pervious.

12.2 ROCKFILL IN EMBANKMENT DAMS

Recent developments

Despite the obvious volumetric economy of rockfill for embankment dam construction, it has only become widely accepted in the past 30 years. This has been mainly due to the following developments:

- *Heavy-duty equipment* The availability of heavy-duty earth-moving equipment capable of loading, transporting, spreading and compacting coarse rock fragments weighing up to a tonne.
- *Vibratory compaction* The implementation of vibratory compaction procedures to ensure rapid and more uniform settlement of the rockfill, while preventing excessive fines generation. About 90% of ultimate settlement is now complete by the end of construction.

- *Faced rockfill*   The introduction of faced rockfill dams able to make use of a variety of low-quality (borderline soil/rock) materials. The face is usually a thin concrete or asphalt slab, made flexible by waterproof joints.

However, settlement can still cause the face membrane to crack; the rockfill behind this must therefore remain pervious to the small leaks that will develop. Furthermore, the embankment must be zoned so as to make best use of the natural materials available, both good and mediocre, and in particular to ensure that the best rockfill is placed where it is most needed.

Zoning in rockfill dams

Examples of two rockfill embankment dams, a concrete-faced and an older central core type, are shown in Figure 12.3. The functions and properties of the rockfill zones are broadly as follows:

- *Zone 1 – select fill*   This is the soundest, least-compressible rockfill, with a maximum particle size of perhaps 150 mm. It must resist deformation to protect the concrete face or clay core, yet be able to drain small leaks. It would usually be produced by scalping oversize from the least-weathered run-of-quarry rock.
- *Zone 2 – drainage layers and filters*   These intercept any seepage from the face, abutments and upstream shell, and divert it away from the core or random fill. Hence these are clean and durable processed granular materials, not rockfill.
- *Zone 3 – transition zones*   This rockfill is intermediate in size and quality between the select and random fill. It prevents intrusion of the coarse outer shell layers into zone 2 (on the upstream side) and piping from the filters into zone 4 (downstream side).
- *Zone 4 – random fill*   This is the lowest quality of the available rockfill (i.e. the coarsest, least-sound, most fines-rich material). The fill layers are thickest here (1–1.5 m) and the largest blocks (0.6–0.9 m) are permitted.
- *Zone 5 – slope protection*   This is partly a means of disposing of oversize and partly an armouring layer. It provides erosion and wave protection on the downstream and upstream faces respectively.

Zones 1–3 normally include sublayers (zones 1A, 1B, etc.), with average particle size decreasing towards the core or concrete face, and outwards from the centre of drainage layers. The complexity of this layering is largely dependent on the availability and properties of the materials. The extensive use of filters and drainage layers has permitted inferior rockfill such as weak sandstone, schist and slate to be used as random zone material (typically making up around 60–70% of the embankment volume). Older homogeneous hard rockfills were sluiced, but compacted

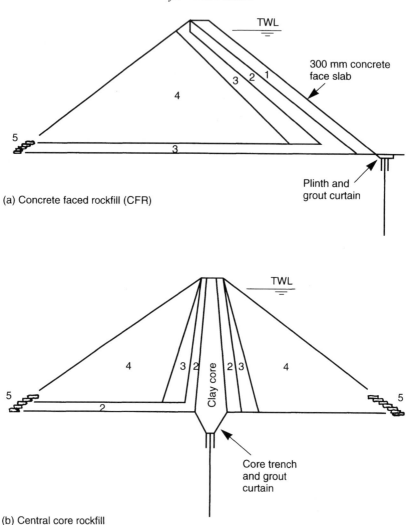

(a) Concrete faced rockfill (CFR)

(b) Central core rockfill

Figure 12.3 Zoning in typical concrete-faced rockfill and central core rockfill dams, somewhat simplified and not to scale. See text for explanation of zone functions and materials.

only under self-weight. These harsh fills had very large initial void ratios and underwent large settlements as a result, though there was never any doubt about their permeability.

Design and construction

Materials selection and embankment dam design are discussed in some detail by Fell *et al.* (1992). Recent British experience with weak rockfill dams is summarized in Reader *et al.* (1993), while aspects of concrete-faced rockfill (CFR) dam design and construction are related in Cooke and Sherard (1985).

It is desirable that oversize be restricted to perhaps 1–3% of the run-of-quarry material. Fragmentation of this order necessitates closely spaced and narrow blastholes, and sometimes deck loading within stronger beds. Rudimentary processing – removal of excess fines or boulders – for select layers may be performed by passing blasted muck over a grizzly with 0.1–0.3 m spaced bars (Figure 12.4), though rockfill is generally placed unscreened as well as uncrushed.

Most size reduction occurs when the rockfill is spread, watered and rolled on the embankment. The roller drum may be cleated or ribbed to improve rock breakage, and strengthened to withstand impact damage. Nevertheless, it is necessary to avoid excessive handling and compaction of weak rockfill, since this will result in fines generation and void clogging. The aim is to achieve a degree of layer compaction, but short of maximum density. This ensures that clasts do not 'float' (become unlocked from each other) in a fines matrix, retains a pervious fabric, yet minimizes long-term settlement. Spray irrigation of quarry stockpiles, in addition to on-site sluicing, allows the fill to be spread in its weakest condition (that is, thoroughly wet and – ideally – saturated).

During vibratory compaction some fines invariably work their way to

Figure 12.4 Rockfill being passed over a bar screen ('grizzly') to skim off oversize. The chains allow the grizzly to be towed a few metres as each pile of undersize is completed.

the upper surface of the fill layer, giving a false idea of its permeability. Though the surface voids are filled in this way, plenty of open space remains in the compacted rockfill beneath. Light tining of this surface between lifts will prevent permeability layering, whereby horizontal permeability is up to 10 times the vertical. The embankment face is protected from erosion by a covering of boulders, mostly winnowed out of the dumped rockfill by bulldozer blading, which also help to confine the fill (Figure 12.5). Better-quality rock is required for the upstream face, which is subject to wave action, than for the downstream face.

12.3 TESTING AND SPECIFICATION OF ROCKFILL

Preferred characteristics

A range of typical rockfill properties is given in Table 12.1. The present-day notion of good rockfill sources is that they should have the following characteristics:

* A rock substance of *medium strength*. Strong rock bulks more on excavation, resists roller compaction, produces higher initial void

Figure 12.5   Slope protection boulders on downstream face of Mangrove Creek Dam, a soft rockfill CFR dam north of Sydney. The maximum size of this rip-rap is about 1.5 m. (Photo: D.F. Branagan.)

Table 12.1 Typical properties of embankment dam rockfill

Compacted rockfill properties

Maximum particle dimension	0.6–1 m (60–100% of layer thickness)
Maximum layer thickness	1–1.5 m
Allowable fines –75 mm	<10%
Bulk density	1.8–2.1 t/m^3
Void ratio	0.2–0.4
Porosity	17–29%
Permeability	10^{-5} m/s (free-draining) to 10^{-4} m/s (semi-pervious)
Fill modulus	<100 MPa
Angle of repose	40–45°
Embankment slope angle (V:H)	1:1.5 (34°) to 1:1.2 (40°)
Interparticle friction angle	>40°
Bulking factor	30–50%
Compaction water added	10–20% by volume of rockfill
Compactive effort specified	4–8 passes of 10 t vibratory drum roller at 16–25 Hz

Source rock intact properties

Uniaxial compressive strength	20–200 MPa (mostly 40–100 MPa)
Wet/dry UCS ratio	0.4–0.6
Tensile strength	3–10 MPa
Point load strength	1–5 MPa
Elastic modulus	2–40 GPa
Specific gravity	2.2–2.6 (mostly 2.3–2.5)
Water absorption	1–6% (mostly 2–4%)

ratios and more settlement. Weak rock simply breaks down to earthfill.

- Chemical stability and sufficient *durability* not to break down excessively, though nowhere near aggregate soundness. Fortunately, the moist, constant-temperature environment of an embankment dam is ideal for resisting weathering.
- A rockmass that is *thickly bedded* (layers up to 0.6 m thick), but not massive. The latter, even when of low intact strength, causes much the same oversize problems as strong rock and generates compaction fines as well.
- *Closely spaced joints*, to reduce blasting costs and oversize.

There is no ideal source lithology, however, since suitability as rockfill is determined more by availability than by intact rock properties. All of the main igneous, metamorphic and harder sedimentary rock types have at some time been used as sources of rockfill. Interbedded sandstone and shale have even been used, where the shale was moderately durable. Slates and schists are sometimes suitable, although there are problems due to platy blocks and unstable sulphide minerals (Figure 12.6).

Rockfill quarry sites are, wherever possible, located upstream of the

Figure 12.6   Corin Dam, near Canberra, Australia, an example of a 1960s central core rockfill dam. The darkest material is the filter zone between the core and the rockfill shell (light tones). Some of this rockfill was sulphide-bearing phyllite, which subsequently gave problems with acid leachate. (Photo: the late E.J. Best.)

proposed dam and below top water level for concealment, and haul distances are generally less than 1–2 km. Spillway excavations are also major contributors of rockfill. Selection of a quarry site on rock quality grounds is less important than making the most of what it contains. In other words the dam is designed around the material available, rather than the material being specified to fit the design.

The first task of the project geologist is therefore to define the classes of material available and their relative proportions, since this will control dam zoning (Figure 12.3). Some processed filter and drainage materials may be imported, but most of the rockfill will have to be found on site or close to it. Different areas or layers within the quarry will be earmarked for the various rockfill zones. Rockfill quarry sites are investigated mainly by drilling, trenching and geotechnical mapping, backed up by a limited amount of core sample testing (UCS, PLSI, water absorption, density and so on). Outcrop mapping and trench logging should emphasize joint/bedding spacing, since this will affect both the blast pattern and the proportion of oversize to be expected.

Testing

The real testing of a rockfill quarry begins with small-scale trial blasts to assess fragmentation and, sometimes, to measure blast vibration. This is followed by the construction of a series of *trial embankments*, each 20–30 m wide, as shown in Figure 12.7. Using these rockfill pads, layer thickness, compactive effort, water content and maximum particle size can be varied to obtain the best result. The testing programme includes measurement of pad settlement by levelling, compacted density and grading (from pits 2–3 m in diameter and about 1 m deep), and falling-head permeability tests. The durability of proposed rockfill can sometimes be assessed by digging up old embankments composed of material from the same formation. Aggregate durability tests such as sulphate soundness and Los Angeles abrasion loss are, however, too severe for rockfill evaluation.

Because particle size distribution and density tests on rockfill are expensive, perhaps a thousand times the cost of comparable soil tests,

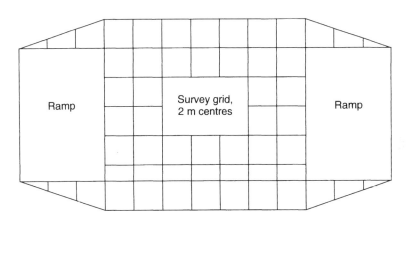

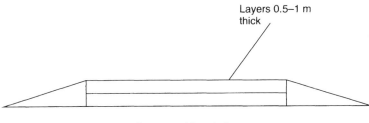

Figure 12.7 Rockfill trial embankment pad, not to scale but typically about 20–30 m wide. A layout would usually comprise about 6–10 pads, testing the settlement and permeability characteristics of different pit materials at varying water contents, layer thicknesses and numbers of roller passes.

few are performed. Usually the results from the trial embankments are combined into a *method specification*, which dictates maximum particle size, layer thickness, size and type of roller, number of passes, volume of sluicing water and sources of rockfill for each zone within the dam. During construction a few density tests, perhaps one per $50\,000\,m^3$ of rockfill, are usually performed, but routine material acceptance is based on simple index tests (PLSI, particle density, moisture content and Schmidt rebound hardness) plus visual impressions.

12.4 ROCKFILL IN ROAD EMBANKMENTS

Rockfill embankments for highways, as distinct from stony earthfill and macadam pavement courses, are a comparatively recent development resulting from the excavation of deeper cuttings. These produce relatively unweathered fill material that is too harsh for compaction to conventional earthworks standards. There were some reservations when rockfill was first proposed for road embankments, since settlements in the order of 0.3 m, which would be quite acceptable in an embankment dam, could cause severe cracking and loss of shape in a road pavement.

A typical highway authority rockfill grading curve is illustrated in Figure 12.2, where it is compared with particle size distributions for rail ballast and coarse macadam. Like the last two materials, highway rockfill is open-graded, consisting mainly of fragments between 100 and 500 mm. However, it is much coarser than macadam, and has a higher percentage of fines than ballast. Unlike dam rockfill, the highway equivalent is specified in terms of grading and intact rock strength (minimum PLSI 1 MPa, roughly corresponding to a UCS of 20 MPa), though compaction requirements are similar. The largest particle permitted is 500 mm, in 750 mm lifts. Rockfill layers must be separated from earthfill by transition zones 0.5 m thick, and no rockfill is allowed within 1 m of the embankment top. The transition zones are finer rockfill (150 mm maximum) in thinner lifts (250 mm).

12.5 RAILWAY BALLAST

The functions of railway trackbed material are: to support the sleepers (ties) and provide a non-rigid base for them; to spread and dissipate the dynamic loads imposed by moving trains; to absorb impact loadings and vibrations by elastic compression and rebound; to resist lateral movement on curves; to maintain track gradient and allow re-levelling; and to act as a drainage layer. To achieve all of this, rail ballast is ideally a mixture of coarse, hard aggregate fragments in the size range

10–75 mm, with no fines at all. However, inferior materials may have to be used for low-speed branch lines.

Several developments in railway engineering over recent decades have affected ballast specifications. First, traffic has now tended towards either high-speed inter-city passenger lines (very large shearing forces on curves) or bulk freight (very heavy loads, slowly applied), rather than the former mixed passenger rail/general freight. Secondly, modern concrete sleepers act as low-frequency vibratory hammers, cyclically loading the ballast layer; whereas old-time timber sleepers had a greater load cushioning effect, though their service life was shorter and their pulldown capacity much less. Thirdly, hand labour for trackbed maintenance is now expensive, so ballast service life (hence durability) becomes more significant. The design life of trackbed material is now about 30 years and it should be suitable for infrequent maintenance using heavy mechanical equipment.

Although many ballast test requirements (such as an even mixture of particle sizes, non-flaky and durable fragments) are similar to those for roadbase, the two materials function in quite different ways. Before describing the specific requirements for ballast, these differences should first be clarified. They are as follows:

- Rail ballast is subject to intermittent and very high loading by 30 tonne axles, four to five times that on highway pavements. It also has to cope with much greater lateral forces, so that interparticle frictional resistance (stability) is at a premium.
- Trackbeds flex significantly with the passage of trains, and hence interparticle rubbing and bending are not negligible, as they are in roadbase.
- Although trackbeds are subject to fatigue due to load repetition, as highway pavements are, this can be fixed by adding fresh stone during routine maintenance. The trackbed as a whole does not require reconstruction, as a failed pavement may.
- The ballast surface is unsealed, so the material is subjected to repeated wetting and drying. This would cause inferior rock to break down (the rail subgrade is also more subject to water softening as a result). A waterproof bituminous seal means that somewhat less-sound aggregate is acceptable for roadbase.

In order to reduce the thickness of high-quality and therefore expensive ballast required for new tracks, it is common practice to top the formation (fill) with a sub-ballast or *capping layer* of compacted sand, inferior crushed rock or select fill. This also minimizes infiltration to the subgrade and hence the tendency of ballast particles to be pressed into it. A geotextile mat is often laid beneath the ballast for the same purpose. The capping layer specification is similar to that for rural highway roadbase, requiring a well-graded mixture between 0.4 mm and 20 mm, a maximum PI of 10 and a compacted density greater than $2\,t/m^3$.

12.6   TESTING AND SPECIFICATION OF BALLAST

Because of these requirements, railway ballast specifications place much emphasis on crushing strength, particle shape, durability and abradability. Crushing strength is assessed by the aggregate impact value (AIV), aggregate crushing value (ACV) and ten per cent fines (TPF) tests, rather than in terms of the intact rock UCS. Wet/dry strength variation in these tests is one measure of durability – a large difference indicating potentially rapid breakdown. Durability is also evaluated in terms of weight loss in the sulphate soundness and Los Angeles abrasion (LAA) tests, though neither is regarded as entirely satisfactory because they fail to duplicate the *in situ* conditions of rock breakdown. Particle shape is important because flaky fragments reduce the compacted density of the trackbed, and hence its internal friction, and are more prone to flexural breakage. Angular, equidimensional and rough-textured particles are preferred because of the greater stability they confer. Where alluvial cobbles are being crushed for ballast, one to three broken faces are required per rock fragment.

Ballast is described as open-graded, meaning that it should have an even distribution of particle sizes at the coarse end of the grading curve, but be deficient in void-filling fines (Figure 12.2). This is especially important on tight curves and steep gradients, where lateral stability is most needed. However, single-sized (uniformly graded) stone is sometimes preferred where *fouling* (void clogging) caused by particle breakdown is a problem, since this maximizes the trackbed permeability. Fouling greatly accelerates the disintegration of the remaining ballast fragments and thereby reduces trackbed life.

Instead of the coarse aggregate particles being cushioned by finer grains, as is the case with dense-graded roadbase, point-to-point contacts predominate in ballast and interparticle stresses are high. Consequently, rock proposed for ballast has to be considerably stronger, tougher (to resist fracture), harder (to resist attritive wear) and denser (to resist lateral movement) than that used for roadbase. In fact, such a rock would produce a 'harsh' roadbase, to which fines (crusher dust or loam) would have to be added to achieve maximum compacted density and minimum void ratio.

The most suitable ballast lithologies tend to be unaltered and unweathered fine- to medium-grained igneous rocks (basalt, acid and intermediate lavas, porphyry, dolerite and diorite). Some coarse-grained granites and gabbros also perform well, but they are often brittle due to microfracturing. Gneiss, granulite and quartzites can also meet specification, but foliated and lower-grade metamorphics are avoided. Carbonate rocks that performed satisfactorily in the past beneath wooden sleepers may not be hard enough to withstand impact and attrition beneath concrete sleepers. Slag has variable properties as ballast – steel furnace slag is denser and more abrasion-resistant than blast furnace slag, and hence is more acceptable.

Coarsely crushed stone is also used for filling *gabion baskets*. For this purpose there need be no particular soundness requirements (though ballast-quality stone is sometimes specified), since the least-durable part of the system is the wire mesh. Coarse fragments, around 75–100 mm, are required on the sides of the gabions, though finer ballast may be used in the centre. For temporary works, crushed concrete and old bricks are also used.

12.7   MACADAM PAVEMENTS

Macadam pavement materials are coarsely crushed rock, or simply broken rubble, with a maximum particle size of about 75 mm (though much coarser fragments are included in the old road pavement in Figure 12.8). They are free-draining, with no optimum moisture contents and low compacted densities. They depend for stability on tight interlock of

Figure 12.8   Unsealed macadam pavement made up of calcrete rubble up to 200 mm in size overlying a clay subgrade. Material as coarse as this would not be acceptable in a modern sealed road pavement.

the angular particles, and in this they resemble ballast rather than conventional dense-graded roadbase. However, macadam differs from ballast in having a small proportion of fines, just sufficient to level the road surface and to provide a tolerable riding surface. Macadam pavements are mainly used for heavy-duty unsealed haul roads, but are also applicable where:

- A source of moderately hard but easily fragmented rock (such as indurated shale) is available and can be worked by ripping, grid rolling or all-in crushing without screening.
- A permeable sub-base is required to provide a capillary break above a shallow water table, or a drainage layer to intercept upward seepage in deep cuttings. Alternatively the sub-base may act as a thermal blanket above a subgrade prone to frost heaving.

Although cheaper to produce than conventional roadbase and available in areas where other materials are scarce, macadam pavements are unpopular because of their poor riding surface. Like ballast, they are unsuitable for swell-prone subgrades without a geofabric separation layer. Another problem lies in assessing particle soundness, since macadams are generally derived from rock that is only moderately strong. Cyclic wetting and drying tests used to predict aggregate durability are too severe for these materials. A more realistic approach is the repeated compaction test, which simulates the extent of particle breakdown during on-road working.

Shale and hard sandstone pavements are variants of the macadam principle, being composed of coarse broken rock that is unscreened except for oversize removal. Note, however, that many poorly cemented sandstones are also used for dense-graded pavements rather than for macadam. Coarse coal washery reject (CWR) (Chapter 17) is another mixture of angular shale and unweathered sandstone fragments in the size range 1–100 mm. Where this material is not slake-prone, it may be used as macadam base for haul roads or, occasionally, as a sub-base for public roads.

REFERENCES AND FURTHER READING

Cooke, B.J. and Sherard, J.L. (1985) *Concrete Face Rockfill Dams*. ASCE, New York.
Fell, R., MacGregor, P. and Stapledon, D.H. (1992) *Geotechnical Engineering of Embankment Dams*. Balkema, Rotterdam.
Goldbeck, A.T. (1948) Mineral aggregates for railroad ballast, in *Symposium on Mineral Aggregates*. ASTM Special Technical Publication No. 83, pp. 197–204.
Hirschfeld, R.C. and Poulos, S.J. (eds) (1973) *Embankment Dam Engineering*. John Wiley, New York.
Kjaernsli, B. (1992) *Rockfill Dams – Design and Construction*. Norwegian Geotechnical Institute, Hydropower Development Series No. 10.

Penman, A.D.M. and Charles, J.A. (1978) The quality and suitability of rockfill used in dam construction, in *Dams and Embankments*. BRE Research Series No. 6, pp. 72–85. Building Research Establishment, UK.

Raymond, G.P. (1985) *Research on Railroad Ballast Specification and Evaluation*. TRB Record, 1006, pp. 1–8. Transportation Research Board, Washington, DC.

Reader, R.A., *et al.* (1993) Recent experience with soft rockfill in dam engineering, in *Engineered Fills*, ed. B.G. Clarke, pp. 379–88. Thomas Telford, London.

Selig, E.T. and Waters, J.M. (1994) *Track Geotechnology and Substructure Management*. Thomas Telford, London.

Stephenson, D. (1979) *Rockfill in Hydraulic Engineering*. Elsevier, Amsterdam.

Transportation Research Board (TRB) (1987) *Performance of Aggregates in Railroads*. TRB Record, 1131. Transportation Research Board, Washington, DC.

Wave protection stone

Coastal wave protection works include marine breakwaters, jetties, sea-walls and groynes. Their purpose is to absorb wave energy, prevent foreshore erosion and enclose reclaimed land. In older breakwater designs this was done by deflecting rather than dissipating wave energy – slabby stones were fitted together to form a relatively tight, smooth surface. Although this technique, known as *pitching*, makes use of relatively small stones, it has been superseded because it requires much hand labour. Breakwaters may also be constructed of concrete blocks, precast armour units, steel sheet piles, caissons, stone-filled gabion baskets, and any combination of these materials. In this chapter, however, we will be chiefly concerned with the most common – and generally cheapest – type of breakwater, the rubble mound embankment. A typical example is illustrated in Figure 13.1.

Rubble breakwaters have to be free-draining, and able to withstand occasional overtopping and the effects of cyclic wave loadings. Furthermore, the rock material has to be durable in a hostile salt spray, wetting and drying environment. Blocks up to 20 tonnes in weight are required to dissipate the impact of the largest storm waves. Rip-rap facing on embankment dams is, by contrast, generally composed of blocks less than 5 tonnes in weight. Modern breakwater design and materials trends are reviewed by CIRIA (1991), McElroy and Lienhart (1992) and in the proceedings of an ICE conference (Institution of Civil Engineers, 1983).

13.1   RUBBLE BREAKWATER DESIGN

Zoning in breakwaters

Rubble mound breakwaters can be homogeneous or zoned, but the latter type is less voluminous and hence more common, since slopes of V:H = 1:2 or even 1:1.5 are feasible instead of 1:3 with random fill. This also makes best use of the available rock, putting the largest and most durable blocks on the outside and the more erodible materials in the centre. A cross-section through a typical zoned breakwater is shown in

Figure 13.1 Typical rubble mound breakwater, near Cairns, Queensland. Note the attempt to lock together the primary armour blocks on the smooth leeward face. The top surface of the central core has been 'blinded' with crusher grit to allow vehicle movement.

Figure 13.2; the three basic materials are the central core, the outer armouring layers, and the transition zones between them.

- *Armouring layers* The *primary armour* layer is constructed from the coarsest blocks available, generally in the range 5–10 t. Because rock of this size is difficult to win, even in massive geological formations, it has to be used economically. Hence it is typically limited to a one-stone-thick strip, 1–1.5 times expected maximum wave height above and below mean sea-level (MSL) on the seaward face, and to a narrower band on the lee side of the mound. The *secondary armour* of 1.5–3.5 t blocks underlies the primary zone and provides the outer layer below the depth of maximum wave energy.
- *Transition zones* Smaller stone is used in filters or transition zones between the relatively fine rockfill core and the armourstone, to prevent the former being sucked out by backwash. These underlayers also have to provide temporary wave protection until the armour is placed, and a sufficiently rough surface for these blocks to key into.
- *Central core* The core is usually made up of fragmented rockfill from which blocks larger than about 1.5 t have been scalped for armourstone. In general, breakwater designs seek to maximize the volume of rockfill core, since this size range is usually abundant in the blasted muck, and thereby to reduce demand for armourstone (which

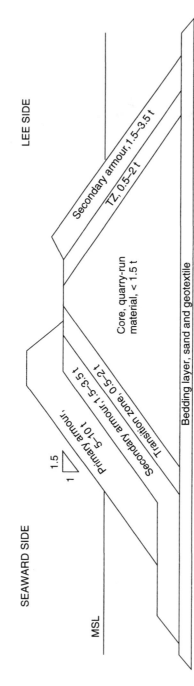

Figure 13.2 Zoned breakwater design, not to scale. The actual dimensions and distribution of the zones would depend on the materials available and the storm waves expected.

may require haulage 100 km or more as individual blocks). The core also provides a foundation for the outer layers and a working platform during construction, which may extend over several years.

Breakwater stone characteristics

The general requirements for breakwater stone quarrying are therefore that:

- Perhaps 5–7% of *large blocks* suitable as armourstone should be present in the blasted muckpile. 'Large' in this context is relative to expected wave height, but usually means 5–10 t (2–4 m^3) blocks and rarely up to 20 t.
- These blocks should be *roughly equidimensional* or at least not too slabby or elongated ('out of aspect'), since this results in bridging and increased flexural breakage. Typically, the ratio of maximum to minimum block dimensions must be less than 2 or 3.
- Blocks should be *sound*, which here means sufficiently free of natural and blast-induced fractures to withstand transportation from the quarry, handling on site and wave impact without excessive breakage. Non-penetrative incipient joints can be especially troublesome, since these may extend as the rubble mound settles under wave action.
- Ideally, armourstone should be *highly durable* aggregate-quality rock, but this requirement often conflicts with that for large blocks. Massive sandstone, for example, is the most widely used armour lithology in New South Wales breakwaters because it can generate 10–15 t blocks when blasted. However, it is relatively porous and usually only moderately durable.
- The parent rock should be of the *highest bulk density* available, preferably greater than 2.60 t/m^3, since dense blocks can be smaller for the same weight (Figure 13.3).

Other desirable characteristics of breakwater stone are high flexural strength, surface hardness, toughness and low water absorption. If these requirements are met, concrete aggregate may be also crushed and screened from the blasted rubble. Core material may be of somewhat lower quality than armourstone, but should nonetheless be an open-graded mixture of durable, tightly interlocking and free-draining rock fragments.

Idealized particle size and weight distribution curves for core material and armourstone are presented in Figure 13.4. In this diagram the blocks are assumed to be cubic, but as-blasted they are usually somewhat elongate. The reality of blasting for coarse sizes is demonstrated in Figure 13.5, where the yield of blocks heavier than 1 t was less than 20%, no 10 t blocks were obtained and only 5.3% were larger than 5 t. The haul of 50 km to the construction site meant that 80% of the blasted rock had

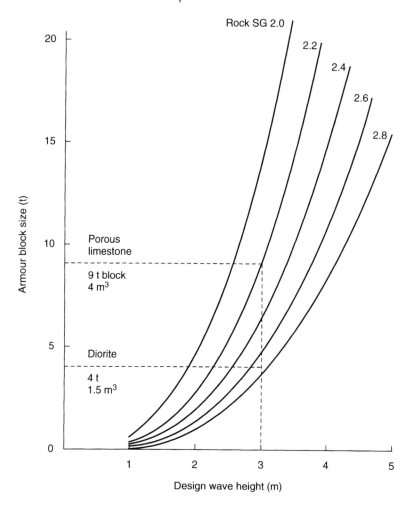

Figure 13.3   Primary armour block size in relation to density and predicted storm wave height. (After Mather, 1985.)

to be wasted, and this result is typical of breakwater quarry sites in Western Australia, where blocks heavier than 1 t are generally only 10–30% of the total muckpile (Mather, 1985).

13.2   DURABILITY OF ARMOURSTONE

This subject has been discussed at length by Fookes and Poole (1981) and Poole *et al.* (1983), with particular reference to limestone armour. They define durability as the ability of the rock material, particularly the most-

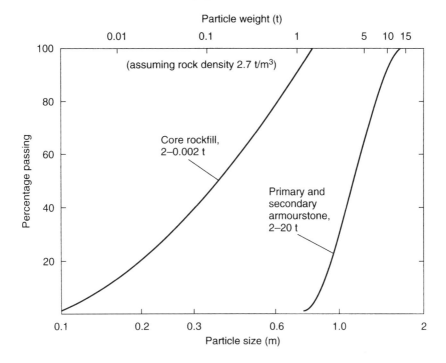

Figure 13.4 Idealized particle size distributions for breakwater armour layers and core rockfill. Note the even grading of core compared with the narrow size range of armour.

exposed primary armourstone, to withstand the processes of weathering in the harshest possible natural environment (Figure 13.6). The mechanisms of rock degradation that are active here include the following:

- *Surface spalling* and crumbling due to salt wedging, thermal stress and freeze–thaw processes.
- *Attrition* due to rocking and jostling of armour blocks by wave action, and *abrasion* by suspended sand in a turbulent medium.
- *Splitting* (tensile failure) of individual stones due to wave-induced fatigue, block-to-block impact or fracture extension.

The consequences of these processes are reductions in block weight, increased rounding and loss of mechanical interlock. Lighter and looser wave-degraded blocks are more free to move around during storms, causing further impact and attrition damage. On the other hand, storm energy may also cause the blocks to settle into a more stable packing, and eliminate the least-sound (which can later be replaced). This is the thinking behind some modern breakwater designs, which are deliberately placed loose to encourage wave compaction and selective breakage.

The durability of proposed armourstone source rocks is assessed in

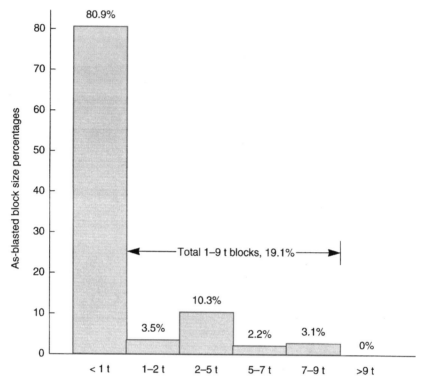

Figure 13.5   Actual armourstone block size distribution, as-blasted, from a granulite quarry. (Data from Mather, 1985.)

the same way as that of aggregate (Chapter 7), except that some lithologies (such as limestone, sandstone and conglomerate), which are not of aggregate quality but are very thickly bedded, may nonetheless have to be accepted. The most important indicators of armourstone durability, apart from performance in service, are the sulphate soundness tests and petrographic examination. Sulphate soundness is considered a severe test for aggregates, but is at least appropriate in situations where salt weathering is the main agency of rock destruction.

Petrographic examination of thin sections should highlight the presence of swell-prone secondary minerals produced by alteration or weathering. Of the two, alteration is the greater problem, since altered igneous rocks can appear quite fresh in hand specimen, and because alteration may be more pervasive and undiminished with depth (while it may be quite feasible to excavate below or around weathered zones in a quarry). Secondary minerals are usually conspicuous under micropetrographic examination as fibrous aggregates, cloudiness in feldspars, dark clays in olivines, and as vugh or vein infillings.

Figure 13.6 Damage to a 30-year-old disused breakwater near Newcastle, New South Wales. The rock used was randomly dumped weathered sandstone overburden from an open-cut mine (maximum block size about 5 tonnes).

13.3   OTHER INTACT ROCK PROPERTIES

Strength

In addition to being durable, it is desirable that breakwater stone be dense, tough and strong. In practice, these properties are interrelated, since dense and strong rocks are usually of very low porosity, resist water and crack penetration, and consequently are both durable and tough. *Compressive strengths* of armourstone are typically between 50 and 150 MPa, about the same as or a little less than those of most aggregate source rocks; *tensile strengths* are usually 5–10% of the UCS value.

Porosity

Porosity increases as the specific gravity of the rock particles falls below about 2.60, since the most common rock-forming minerals (quartz and feldspar) are about 2.65. Porosities of up to (say) 5% may be reasonable in sedimentary armourstone, but above about 1% in igneous lithologies it can indicate weathering or microfracturing. The densest commonly occurring intact rocks are in the SG range 2.65–2.80, and include dolomite, intermediate and basic igneous rocks, and some ultrabasic metamorphic rocks.

Hardness

Surface hardness is most relevant in the case of limestone armour, since calcite (Moh's hardness $M = 3$) is significantly softer than quartz ($M = 7$) and the silicate minerals. It is for this reason that limestone blocks are particularly prone to rounding and loss of weight in breakwater service. Furthermore the indurated and non-porous varieties of carbonate rock are relatively brittle, with modulus ratios (E/UCS) around 700. These limestones, which are otherwise attractive as sources of breakwater stone, may therefore be liable to tensile fracture and a relatively short fatigue life.

Incipient jointing

Incipient jointing – tight, impersistent fractures that can extend during handling and storm buffeting and cause the block to split – is a particular problem with armourstone, since the larger the block, the more likely that it will contain these and other defects. These cracks are difficult to detect by inspection in muckpile blocks, and even where visible may not obviously affect the block's strength or durability. One method of assessing this is a simple drop test in the quarry, though care has to be taken to ensure that the drop height is not unrealistically high or the impact surface too hard!

Promising results have also been obtained from ultrasonic velocity probing of individual boulders, since low V_p can be a very sensitive indicator of fracturing, especially in dry rock. Careful handling and placing – stone by stone, using cranes rather than tipping, and wedging into position with crowbars and smaller stones – is recommended. This can keep incipient joints short of the 'critical length' required for rapid crack propagation according to the Griffith theory of fracture.

13.4   TRANSITION ZONES AND CORE

Transition zone material

In a typical rubble breakwater the core makes up 60–80% of the volume, the armour about 10% and the balance is *transition zone* material between the two. This functions primarily as a filter, preventing movement of the finer core outwards under suction or flushing. It also provides temporary wave protection for the core before the armour is placed, and acts as a foundation (or underlayer) for the armour blocks.

To achieve these ends, the transition zone is intermediate in particle size between the core and the armour, but open-graded like the core rather than single-sized like the armourstone. As an underlayer, it has to provide an even slope of between 1:1.5 and 1:2 for the armour blocks to be laid upon. This surface should be as rough as possible to keep the blocks in place, adding base friction to the lateral confinement provided by tight packing. Because the underlayer is also expected to contribute to wave energy dissipation, its void ratio should approach that of the armourstone layer, about 0.30.

Core material

Core material makes up the bulk of rubble breakwaters and also acts as a working platform during construction, so its top width is dictated by the wheeltrack of dump trucks and cranes. Core is usually either run-of-quarry (random) rubble, the residuum left after primary and secondary armour blocks have been scalped off, or inferior rock from within the quarry. The maximum fragment size is 1–2 t, say 0.4–0.8 m^3, and the smallest particles usually weigh 2–20 kg. This rubble is end-dumped from trucks, or bottom-dumped from barges. Some care is needed to avoid segregation and bridging, which could cause differential settlement. To prevent stone being pressed into soft marine mud, a fine filter layer or permeable mat is often laid on the seabed prior to dumping.

In most breakwaters the core is designed to be pervious and to have a high shearing resistance, contributing in small measure to the attenuation of wave energy. Nonetheless, fine-grained core material, such as carbonate sand, may have to be used instead of rockfill. A similar problem arises where the landward side of the breakwater is to be backfilled with dredger tailings. These situations require carefully designed and placed fine filters, backed by geofabric sheets, in addition to rockfill transition zones. Constructing these can be no easy task, as the filter materials can easily be washed away and geofabric can be ripped or perforated by large sharp stones. As a bonus the geofabric can, by wrapping and enclosing the core, impart significant shear strength to the embankment.

13.5   AGGREGATE FOR MARINE CONCRETE

Where armourstone of adequate size is not available within economic hauling distance – which could be anything up to 100 km or so – precast concrete armour blocks or elaborate shapes (known as tetrapods, tribars, hanbars, dolos and so on) have to be used instead. This is also the case where storm waves higher than 6 m have to be designed for, since rock blocks heavier than 20 tonnes are rarely obtainable. Concrete armour blocks are usually cubic in shape and unreinforced, and are used as replacements for boulders that have degraded. Armour shapes appear to be gaining popularity for new construction where natural rock is uneconomic, since they are much lighter and require less concrete than blocks. They can also economize on core rockfill by standing at slopes of 1:1, are more effective than natural rock for energy dissipation, and pack tighter with time.

Concrete deterioration in the marine environment can have a number of causes, but especially salt penetration along shrinkage cracks, and can be very rapid in warm climates. With reinforced concrete there is the further problem of rust initiating where cracks or honeycomb cavities expose steel rods. Fibreglass reinforcement is one means of increasing tensile strength of precast armour blocks without rust problems.

Aggregates for armour blocks should have high resistance to chemical attack and attrition, hence low LAA and sulphate test losses. Very coarse gradings can be used in the precast cubes because of their size – a 20 t cube is roughly 2 m on each side – and lack of reinforcing cages. This permits a low proportion of cement to be used for a specified strength, though there is a risk of void formation with such harsh mixes. Quartz sand is far superior to carbonate sand as the fine aggregate, in terms of both abrasion resistance and chemical stability, even though the latter may be more abundant at coastal sites.

Mix designs for marine concrete aim for surface toughness and very low permeability, since scour and spalling – even without the effects of attrition and chipping – can remove up to 15 mm of surface per year. To achieve this, compressive strengths in excess of 40–45 MPa and low water/cement ratios (below 0.45) are desirable. Where armour units are reinforced, a minimum depth of concrete cover of 90 mm over the steelwork is recommended.

13.6   BREAKWATER QUARRY INVESTIGATIONS

Quarry sites

Breakwater stone quarries differ somewhat from aggregate quarries in being relatively small (a large job might require $300\,000\,m^3$ of loose rock, equivalent to $200\,000\,m^3$ or 0.5 Mt *in situ*) and short-lived (though they may be re-opened from time to time for maintenance stone). Older quarries were often located almost on the construction site, since bold coastal headlands tend to provide both a degree of storm shelter, which the breakwater could augment, and a source of relatively unweathered rock. In many cases the worked-out quarry has been recycled for shops, ship support facilities, and even for housing. The down side is that these headlands, or other prominent hills in coastal zones, are nowadays often protected against development by planning constraints.

Site investigation procedures for breakwater stone quarries are similar to those for other hard rock sources, except that a proportion of large blocks is required. This immediately limits the number of lithologies worth prospecting to the following: coarse-grained plutonic rocks such as granite, gabbro and diorite; high-grade metamorphics such as gneiss or granulite; intrusives such as dolerite plugs or porphyry stocks; thick lavas and major sills; and some massive and well-cemented sedimentary rocks such as quartz sandstone, conglomerate and limestone. Nevertheless, even these rocks are often unsatisfactory because of weathering and close jointing.

Geological investigation

The geological investigation procedures follow the lines outlined in Chapter 2, but with a few refinements. *Joint mapping* (attitude, spacing and persistence) assumes a greater importance, and rock faces may have to be exposed for mapping, sampling and trial blasts by bulldozer excavation and sluicing. Shoreline cliffs and road cuttings, even at some distance from the proposed quarry, may give an indication of what can be expected in terms of rockmass discontinuity spacing. *Angled drillholes* can be very cost-effective in this situation, particularly where the alternative is extensive bulldozer costeaning. Careful logging of oriented drillcore can give a good idea of joint distribution in three dimensions. Numerical modelling of joint block size distributions, and hence prediction of possible size percentages of armour blocks in the blast muckpiles, is also possible using map and drillhole data.

Trial blasting

Trial blasts give a rough idea of expected particle sizes, but care is needed in setting up the blast and interpreting the results. The face being shot should be truly representative of the whole rockmass, although the tendency is to choose the most-prominent outcrops; these are, of course, its least-weathered parts. On the other hand, trial blast faces are usually only 6–10 m high, and deeper rock will almost certainly be more widely jointed and less weathered. Furthermore, production blasts will eventually be modified to maximize block size, but this is not possible in one or two tests. Another benefit from trial blasts is an indication of the ground vibrations to be expected. This is important because choke blasting methods used in armourstone production tend to generate higher amplitudes than conventional fragmentation blasting, and hence more complaints from nearby residents.

13.7 BLASTING AND SIZING ARMOURSTONE

The principal aim in breakwater quarry blasting is to produce a proportion of oversize, on the assumption that sufficient 'fines' (core material) will be generated in any case. In fact, some aggregate quarries set aside their oversize to serve this market, rather than carry out secondary blasting. Deliberate production of oversize presents both geological and blasting problems.

The geological problem

The geological problem, simply stated, is that no fragments larger than the maximum dictated by the *in situ* joint and bedding spacing can be produced, and even these blocks may be damaged by inappropriate blasting techniques. Muckpile blocks heavier than about 2 t (0.8 m^3) are almost invariably bounded by at least one joint, and two or three joint faces are commonly present. Large fragments in layered rockmasses are likewise nearly always platy, since bedding is more persistent than jointing. A further complication is that the larger the joint block, the more likely it is to contain a number of impersistent joints or fractures, and hence the more likely it is to break up during handling. In these situations, concrete blocks may be required to supplement the largest blocks (Figure 13.7).

The blasting problem

The blasting problem is the exact opposite of that encountered in normal quarrying – here the aim is to generate as much oversize as

Figure 13.7   River-mouth breakwater, Yamba, New South Wales. The primary armour here is sandstone, supplemented by 10 tonne concrete blocks to the extent of about 5%.

possible. To achieve this it is necessary to aim for rockmass fracturing rather than fragmentation, usually by *'choked' blasting* (or muffled blasting) of single-row, widely spaced blastholes fired without delays. The burden is increased by trial and error until the blast fractures just reach the bench face. In this situation there is hardly any move-out (or dilation) of the rockmass – hence the confinement or 'choking' effect on blast energy. Blasthole spacing here is approximately equal to burden distance, compared to about 1.3 times burden in a conventional fragmentation blast. The burden depth is critical – once it becomes too large, the only result of the blast will be annular crushing around each blasthole and intensified ground vibration due to overconfinement of blast energy.

Another approach is to use *low-density charges* – such as loose ANFO or some ANFO/emulsion mixes – to maximize the proportion of heave energy (gas pressure) to strain energy ('brisance') generated. This tends to cause some rockmass dilation, but most of the fracturing is likely to occur by extension along pre-existing joints and bedding. In addition, the degree of internal microfracturing generated within blocks is less than occurs using explosives with higher velocity of detonation (VOD).

A third method is *presplit blasting*, using closely spaced, small-diameter holes loaded with decoupled charges. The aim here is to create tensile fracture planes between blastholes, and in effect to cut out the large blocks bounded by these fractures and natural joints. This is most applicable to very thickly bedded but weak rocks such as sandstone or conglomerate.

Two other blasting methods have been used over the years in breakwater stone quarrying, although they are not common these days. *Tunnel blasting* involves a single very large charge fired in an underground cavern to produce a huge muckpile, from which suitable armourstone blocks can be selected. Because of the high proportion of waste generated, this method is suitable only for remote sites. *Snakehole blasting* makes use of subhorizontal blastholes drilled at the base of high and near-vertical faces, causing rockfalls rather than conventional fragmentation. Again, this is a method to which environmental objections are very likely, since it results in high, bare and only marginally stable faces, and much undersize debris is likely to be left behind.

13.8 QUALITY CONTROL

Testing and specifying breakwater stone can be a difficult balance between what is desirable from a design viewpoint and what is possible from the local geology, with plenty of scope for disputes between the contractor and the client's engineer. In the past the tendency was for the client to draw up a brief specification, mainly concerned with acceptable sizes of primary armour, and let the contract tenderer choose both the quarry – either new or existing – and the method of operation. The present trend is for the client to select and prove up a quarry site and require the contractor to use this, or propose a cheaper alternative meeting the same requirements. Disputes in this case centre more on the method of operation, especially blasting practices, rather than on the rock itself.

Size and shape

Size and shape of armour blocks are probably the most common causes of discord, and can be checked by routine measurement and weighing. With experience in a quarry, nomograms can be constructed so that weight can be estimated from three orthogonal measurements, with sufficient accuracy for construction control purposes. 'Reference stones' of known weight and acceptable shape, often whitewashed to make them more conspicuous, are sometimes set aside for operator training, since most blocks will have to be visually accepted.

Table 13.1 Acceptance criteria for armourstone[a]

Test or measurement	Armour and underlayers	Rock-fill core	Comments
Max/min dimension (aspect ratio)	<2.0	<2.5	May be relaxed to 3.0
Bulk density	$>2.6\,t/m^3$	$>2.5\,t/m^3$	Some limestones acceptable down to $2.4\,t/m^3$
Water absorption	<2.5% by weight	<3%	Total porosity can be higher, up to 5%
Uniaxial compressive strength (UCS)	>85 MPa	>50 MPa	
Aggregate crushing value (ACV) loss	<20% by weight	<25%	Ten per cent fines (TPF) test similar
Wet/dry UCS strength ratio	>75%	–	
Point load strength index	>3.5 MPa	>2 MPa	5–10% of UCS
Methylene blue absorption (MBA)	<0.007	<0.01	By weight; up to 0.015% acceptable
Sulphate soundness test loss	<8% by weight	<16%	Magnesium sulphate test preferred, 5–15 cycles

[a] Data from Fookes and Poole (1981).

Soundness

Soundness generally has to be visually assessed also, although ultrasonic velocity testing offers promise as a means of quantifying this. In this procedure, a number of velocity measurements are made across a block, with the highest velocity taken to indicate intact rock. Soundness is then assessed in terms of the velocity ratio (V_{block}/V_{intact}), low ratios indicating fissured rock. Usually different areas within the quarry, based on prior testing, are designated for particular zones in the breakwater.

Index tests

Index tests that might be used for routine quality control include rock SG, water absorption, Schmidt hammer rebound number, and point load strength index (PLSI) on lump samples. Some suggestions for acceptance test limits for armourstone and core material are listed in Table 13.1. Two recent developments in breakwater stone assessment techniques are a self-abrasion (milling) test to predict rounding and weight loss rates in limestone armour, and the use of fracture toughness tests on notched drillcore. Fracture toughness is not yet a routine procedure, but useful correlations between this property and both aggregate impact value (AIV) and sulphate soundness have been reported by Poole *et al.* (1983).

REFERENCES AND FURTHER READING

CIRIA (1991) *Manual on the Use of Rock in Coastal and Shoreline Engineering*. CIRIA Special Publication 83. Construction Industry Research and Information Association, London.

Dibb, T.E., Hughes, D.W. and Poole, A.B. (1983) Controls on the size and shape of natural armourstone. *Quarterly Journal of Engineering Geology*, **16** (1), 31–42.

Fookes, P.G. and Poole, A.B. (1981) Some preliminary considerations on the selection and durability of rock and concrete materials for breakwaters. *Quarterly Journal of Engineering Geology*, **14**, 97–128.

Institution of Civil Engineers (ICE) (1983) *Breakwaters – Design and Construction*. Thomas Telford, London.

McElroy, C.H. and Lienhart, D.A. (eds) (1992) *Rock for Erosion Control*. ASTM Special Technical Publication 1177.

Mather, R.P. (1985) Rock for breakwater construction in Western Australia – availability and influence on design. *Engineering Geology*, **22**, 35–44.

Niese, M.S.J., Van Eijk, F.C. and Verhoef, P.N. (1990) Quality assessment of large armourstone using an acoustic velocity analysis method. *IAEG Bulletin*, No. 42.

Poole, A.B. (1991) Rock quality in coastal engineering. *Quarterly Journal of Engineering Geology*, **24**, 85–90.

Poole, A.B., Fookes, P.G., Dibb, T.E. and Hughes, D.W. (1983) Durability of rock in breakwaters, in *Breakwaters – Design and Construction*, Institution of Civil Engineers, pp. 31–42. Thomas Telford, London.

US Army Corps of Engineers (USACE) (1995) *Construction with Large Stone*. USACE and ASCE Design Guide No. 13.

Wang, H., Latham, J.P. and Poole, A. (1991) Blast design for armour stone production, Parts I and II. *Quarry Management*, July, pp. 17–21, and August, pp. 19–22.

CHAPTER 14

Dimension stone

Dimension stone is rock that is used in cut or split form, rather than broken as rockfill or crushed as aggregate. It can be subdivided into *structural stone* (solid masonry), where rock blocks or ashlars are stacked to form load-bearing walls, and *facing stone* (cladding), where rock slabs are assembled into curtain walls. Up until the early part of the twentieth century, structural stonework dominated, but with the development of steel-framed buildings, facing stone became much more important.

Dimension stone is also used extensively for paving and tiling, interior finishes, landscaping, monuments and statuary, and building restoration. Different lithologies and rock qualities are used in all these applications. For example, more wear-resistant slate is needed for paving than for landscaping; polished marble is popular for internal facing but may stain, warp or crack if used externally; denser sandstone is required for exposed sills than for plain walling. The stone is supplied as roughly cut blocks; as dressed stone (recut to specified sizes); or in finished form (surface-ground, polished, hammered, etc.).

Facing stone is experiencing a worldwide revival in demand, mainly due to the development of thin *veneer panels* supported by steel or concrete frames (Figure 14.1) in place of the glass curtain walls of the 1960s. There has also been large growth in the market for interior stone finishes as wall facings, floor tiles and bench tops. Computer-controlled diamond saws have allowed thinner stone veneers, sometimes only 10–30 mm thick, to be cut without vibration-induced microfracturing. This has not only economized on the use of high-grade stone, but has halved the weight of stone facades. Furthermore, these panels can be assembled, on or off site, by less-skilled labour than traditional masons.

This chapter is an introduction to stone technology, emphasizing current growth areas in thin slab facing and restoration of historic buildings. A much fuller treatment, emphasizing the mineralogy and petrology of dimension stone, is given in Winkler (1975). Shadmon (1996) gives a broader view of stone production and processing, with particular relevance to labour-intensive methods in underdeveloped countries. The current state of the art in design and construction of stone veneer cladding is reviewed in Donaldson (1988).

Figure 14.1   Sandstone veneer panel attached to a reinforced concrete frame. This is a mock-up assembled in a Gosford, New South Wales, quarry. Note the relatively thick (75 mm) slab thickness required with sandstone.

14.1   DIMENSION STONE PETROLOGY

In stone technology 'granite' includes all coarse-grained igneous rocks, plus gneiss and coarser schists; 'sandstone' also covers quartzite and some granular limestones; and 'marble' comprises all non-porous carbonates capable of taking a polish. In addition, a number of other tradenames are used: travertine (for any porous limestone), ashstone (tuff), and bluestone (usually basalt, but the term can include some slates).

Requirements of dimension stone

Customer requirements of building stone are dictated more by aesthetic considerations than by physical properties. Nonetheless, most of the lithologies used in stone veneer construction are dense, strong rocks such as granite, porphyry and marble. Weaker and more porous rocks are acceptable for solid masonry and landscaping, but must be used – if at all – in thick slabs (around 75 mm) for cladding. Aesthetic factors include fashionable colours, a coarse grain size that shows off mineral variations, an interesting fabric and an attractive finish. These fabric elements could include: bedding, fossils or breccia bands in sandstone or limestone; layering, foliation and porphyroblasts in metamorphics; and vesicles, flow banding and phenocrysts in igneous rocks.

Colour-fastness, dimensional stability, batch uniformity and sufficient reserves in the quarry are also important factors on large building projects. Customers prefer stone with a record of good service, so traditional sources such as Carrara marble remain popular centuries after they were first quarried. The main commercial considerations are prices competitive with – but not necessarily cheaper than – synthetic and composite facings, consistent quality of finish, and assured delivery dates. A cost premium will be paid for certain rare and attractive veneers, since stonework is expected to lend prestige to a building.

Suitability of dimension stone

It is difficult to generalize about the suitability of particular rock types for use as dimension stone, since a single petrological name will include both acceptable and inferior varieties. However, the following comments can be made:

- *Granite* is the preferred rock type for external cladding, since it can be cut thinly yet is nearly impervious to weathering. It is also widely used for interior facing and paving, though it is difficult to split evenly and can be slippery when wet if polished. Some varieties lighten or darken on exposure to sunlight, and a redder tone can be achieved by flame treatment. Kaolinization may increase water absorption and cause swelling. Sulphide traces and biotite flakes may weather and leave rusty stains.
- *Marble* is generally restricted to indoor applications because some varieties are liable to become discoloured, crack or distort when exposed to sun and rain. Atmospheric pollutants can also cause black and white encrustations and solution grooves to form. A wide range of colours, from brilliant white to dark red and green, and varied fabrics are characteristic of marbles.
- *Limestone* has similar applications and problems to marble, coupled with greater water absorption and hence a tendency to stain readily.

On the other hand, some limestones weather to pleasing 'warm' brown tones. Carbonate rocks are the most commonly used dimension stone in the world, because they are widely distributed and easily worked, yet adequately strong.

- *Sandstone* is as widely used for building stone in Australia as limestone is overseas, and for much the same reasons: it is abundant, relatively easy to cut and shape, and performs satisfactorily in many applications. Like limestone, it tends to be porous and absorbent, and hence only moderately durable. Most sandstones wear rapidly under traffic – but some are relatively hard and conveniently thin-bedded ('flaggy') for this purpose. They do not take a polish and are unsuitable for thin veneers, though they may be used for external cladding of lower storeys. Sandstones with a trace of sideritic cement can weather from grey to a warm pink colour, while others develop an attractive buff tone ('yellow block') due to limonite staining.

14.2   QUARRYING DIMENSION STONE

The quarries

Although many of the rock qualities that make for good dimension stone can also produce good aggregates, the two styles of quarrying are incompatible, since blast damage reduces both the size of blocks obtained and their durability. A striking feature of Australian dimension stone quarries is that they are generally small and appear to be abandoned at most times (Figure 14.2). Most operate intermittently, in response to single large orders or to replenish reserve stocks at distant stone dressing factories.

The 30 granite quarries in New South Wales, for example, average only about 700 tonnes of cut blocks per year per quarry. The largest dimension stone quarry has an annual output of only 7000 tonnes of sandstone (compared with 1.5 Mtpa for the state's largest aggregate/roadbase quarry, and 3 Mtpa for the largest gravel and sand pit). Some stone quarries are only single faces in a small outcrop, or even a single huge boulder, say 50 m square.

Large dimension stone quarries become so because of their long working life – decades in Australia, centuries in Europe – rather than because of high annual output. It should be noted, however, that the weight of cut blocks is not a true measure of a dimension stone quarry's production, because of the inherent wastage caused by trimming to obtain regular-shaped blocks and the need to avoid rockmass defects. Wastage may be up to 40% in sandstone and marble workings, but is higher for granite and up to 90% in slate quarries.

Worldwide, it is common for quarrying and processing to be carried

Figure 14.2 Wondabyne sandstone quarry, north of Sydney, New South Wales. Dimension stone quarries have operated intermittently here for over a century, but the worked area covers only 1 ha and is 40 m deep. Note the smooth wire-cut faces and, beneath the crane, presplit blastholes.

out in different countries; this is the one truly international extractive industry. Italy is the world's leading processor, drawing on suppliers as far apart as Scandinavia and South Africa. India is another major producer of cut stone, since its quarrying and processing are still well suited to labour-intensive methods.

Quarrying objectives

The planning objectives in dimension stone quarrying are therefore to ensure that:

- Sufficient proven and accessible reserves are available to fill large orders at short notice. As far as possible, however, stockpiles are held 'in the ground' to save costs. Output can be increased rapidly by working a second shift in large quarries, or by re-opening small quarries.
- Areas of uniform stone (in appearance and properties) have been outlined by exploratory drilling and geological mapping ahead of demand, and quality control procedures are in place. Conservation of especially rare and valuable stone should be included in the extraction plan.
- A variety of colours and rock qualities can be offered, either through selective working of a large quarry, or from small quarries that may be hundreds of kilometres apart (and equally distant from the dressing factory). Transport costs are relatively unimportant with these rock products.

Quarrying operations

Blocks are cut from the rockmass by a combination of line drilling, wedging, splitting, sawing and – occasionally – gentle blasting (Figure 14.3). Typical block dimensions are 1.5 m cubes (9 t) or 1.5 m × 1.5 m × 3 m prisms (18 t). The most common method of outlining blocks is by *line (stitch) drilling*, which requires closely spaced holes, at 100–300 mm centres, with as little as 25 mm of rock bridge left between drillholes. Holes are drilled vertically and horizontally to envelop the block, then wedges are inserted to release it. *Wire saws* use a system of thin steel cables, pulleys and wet sand abrasive to cut sandstone and marble, but diamond-impregnated wire can even cut granite. *Channel cutting* by means of a 2 m diameter pick-studded wheel is also effective in sandstone and carbonates. Mobile *circular saws* and large tungsten carbide *chain saws* are used for cutting soft and non-abrasive limestone.

Hard and abrasive rock such as granite is sometimes *presplit* using low explosives in narrow, closely spaced drillholes. These are decoupled and unstemmed black powder charges, fired instantaneously to maximize tensile strain energy generation while eliminating shatter (brisance). Alternatively, blocks may be broken out using non-explosive bursting agents (synthetic swelling compounds), hydraulic chisels and jacks. Another method used in granite is *flame cutting*, a form of channel cutting using a thermal lance.

To facilitate splitting, the quarry master will take advantage of any natural planes of weakness (*rift*) present within the rockmass (Figure 14.4). These include bedding, foliation, flow banding, cleavage or mineral lineation (*grain*), and are often not visible to the untrained eye in apparently massive rock. Rock without any preferred direction of weakness is sometimes called *freestone*. One type of rift that is not always welcome is stress-relief fracturing, since it is non-systematic and cuts

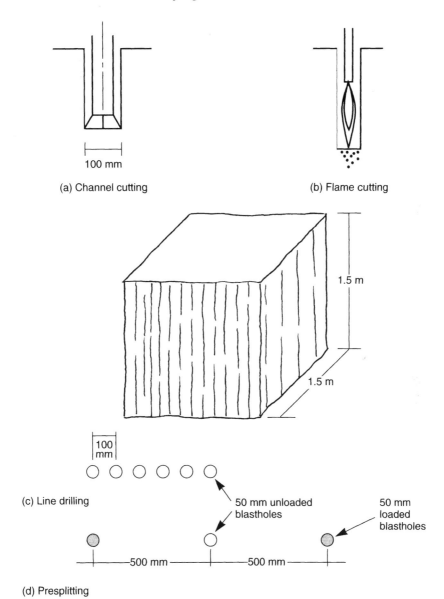

Figure 14.3 Methods of primary block cutting and a typical detached block prior to further processing.

across other discontinuities, causing undersized blocks. However, it can be exploited in granite quarries as a means of horizontal block release. It is most spectacularly displayed in granites, but can occur in any unweathered and massive rock type, even sandstone, with widely spaced joints and locked-in tectonic stresses.

The final stage in quarrying is seasoning of the blocks by exposure to the atmosphere for some weeks or months. This causes pore moisture ('rock sap') to be drawn to the surface and to evaporate, hardening the rock and slightly darkening its colour as iron oxide traces precipitate. Slight shrinkage, generally less than 0.05%, accompanies the drying.

14.3 STONE PROCESSING

The factory processing of roughcast quarried blocks can conveniently be divided into secondary cutting, surface finishing and stone dressing.

Secondary cutting

Secondary cutting is the trimming, sawing and splitting of the quarry blocks into ashlars or veneer slabs of specified dimensions. The blocks

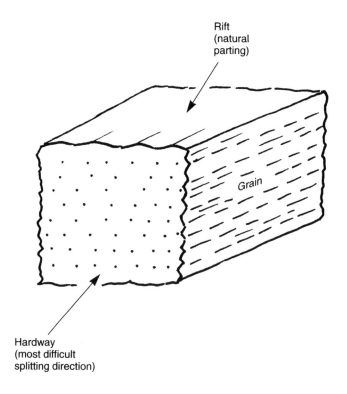

Rift
(natural
parting)

Grain

Hardway
(most difficult
splitting direction)

Figure 14.4   Some terms used in quarrying to indicate the relative difficulty of splitting.

are cut to close tolerances by circular saws edged with fine diamonds or tungsten carbide inserts. *Gang saws* are reciprocating frame saws with a row of thin blades arranged like a bread slicer to convert quarried blocks into slabs (Figure 14.5). These sawn surfaces are adequately planar on soft rocks such as sandstone, but on harder rocks like granite they are often marred by curved striations caused by blade vibration and require additional finishing.

Surface finishing

Surface finishing techniques include honing, grit blasting and polishing. *Honing* is simply grinding to produce a smooth but dull surface in granite, marble and hard sandstone. Though less eye-catching than a polished face, honed surfaces are non-slippery and do not show nicks and scuff marks so easily – hence they are more appropriate for paving stones than for interior panelling. *Grit blasting* can be used with masking templates to create surface patterns of dull and polished granite side by side – to emphasize lettering, for example.

Polishing imparts a glossy, ultra-smooth finish to granite, marble and many close-grained igneous rocks, highlighting the fabric, mineral texture and colour variety within the stone. It is achieved by repetitive grinding and rubbing with successively finer grits, using an automated radial arm polisher. The polishing media include tin oxide, oxalic acid

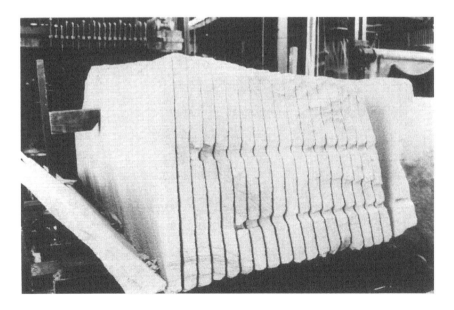

Figure 14.5   Secondary cutting of sandstone block into 75 mm slabs by a gang saw (at rear), Gosford, New South Wales.

and aluminium oxide, which are often softer than the stone and do not only act as abrasives. One theory suggests that surface fusion occurs and a thin film of glass (the Beilby layer) fills in minor irregularities.

Stone dressing

Dressed stone can refer to any type of finishing, but it is commonly restricted to those which require a considerable degree of hand working. This can include *rock-faced* blocks of sandstone and basalt with square edges and very roughly spalled sides. Another coarse-textured finish is achieved by *thermal exfoliation*, especially of granite, which heightens the feldspar colour and increases surface sparkle by 'popping' quartz grains. However, it may also reduce tensile strength and durability in veneers. A number of traditional hand dressing techniques with names like bush-hammering, scabbling, droaving and pointing remain in use, particularly for restoration stonework.

14.4 PHYSICAL PROPERTIES AND TESTING

Even relatively weak sandstone and limestone have adequate compressive strength for most structural stone applications. On the other hand, thin facing slabs are subjected to wind loading, thermal stresses and seismic vibrations, and hence require considerable flexural (bending tensile) strength. The other major problem with the physical properties of stone is its resistance to in-service weathering (durability), or rather the lack of it in some cases.

Bulk density

The bulk density of building stone ranges from about 1.8 to $3.0 \, t/m^3$. Below this, most rocks are insufficiently durable for masonry, though some porous limestones with bulk densities as low as $1.1 \, t/m^3$ are exceptions. Stone in the range $1.8–2.2 \, t/m^3$ is most suitable for hand working or carving. Hard, unweathered, non-porous igneous and metamorphic rocks have densities of $2.6–3.0 \, t/m^3$. In these rocks, variations in bulk density are influenced more by mineral specific gravity (quartz, feldspar 2.6–2.7; mafic silicates 3.2–3.6) than by porosity. The bulk density should be stated as oven-dry, air-dry or saturated, since the wet density can be substantially higher with softer rocks.

Porosity

Porosity is inversely related to bulk density, with total porosity (measured by the vacuum infusion method) values for unweathered igneous and metamorphic rock being generally less than 1–2%, sandstone in the range 5–15%, and limestone up to 60%. However, effective porosity, after soaking for 24–48 hours, is much less (ranging from about 0.1% for tight granites to around 8% for better-quality sandstones and 20% for porous limestones). The ratio of effective porosity to total porosity is known as the *saturation coefficient* and ranges from 0.05 to 0.95. Because water absorption is easier to measure than true porosity, it is more frequently measured, and often misrepresented as porosity.

Permeability

Permeability of saturated rock is related to effective porosity, since both are influenced by average pore diameter and continuity, but also to hydraulic gradient and fluid viscosity. The ability of a rock to draw moisture inwards or upwards by capillary suction is nevertheless more important in dimension stone. This can range between 4 m and 10 m in sandstone, and be much greater in a horizontal than a vertical direction. Capillary moisture movement may be the most important factor in rock decay, particularly when this water is saline, and it is also essential to frost crystal growth.

Strength

Strength of dimension stone is traditionally measured and quoted in terms of *uniaxial (unconfined) compressive strength* (UCS), although failure by bending is much more likely in the case of veneer slabs. Air-dry UCS values are roughly 10–80 MPa for limestones, 30–100 MPa for sandstones, 60–200 MPa for granites and marbles, and around 400 MPa for exceptionally strong quartzites. Soaked values are typically 50–90% of dry strength, and the disparity is greatest in weak and porous rock types.

Tensile strength

There is no standard tensile strength test comparable with UCS, but the two most common methods are illustrated in Figure 14.6. The *indirect tensile strength* (ITS) measured by the Brazilian method averages about 7% of compressive strength (UCS), but ranges from as low as 2% in exceptionally brittle rock up to 20% in plastic shales. For dimension stone the ratio should be between 5% and 10%. However, the *flexural strength* (four-point loading) test is the preferred method of evaluating rock for use in thin slabs. It is higher than ITS, typically between 10% and 30% of UCS. Specimens loaded parallel to the grain of the rock are distinctly weaker than those loaded perpendicular to it, due to natural anisotropy.

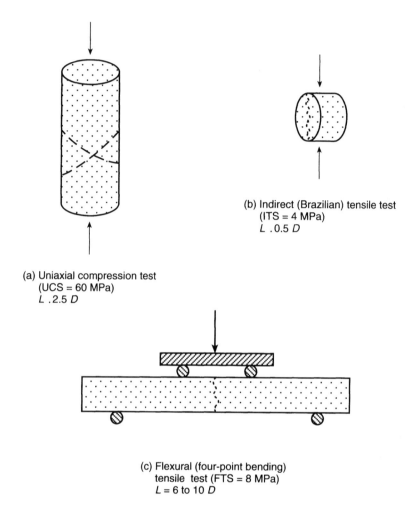

(a) Uniaxial compression test
(UCS = 60 MPa)
L .2.5 D

(b) Indirect (Brazilian) tensile test
(ITS = 4 MPa)
L .0.5 D

(c) Flexural (four-point bending)
tensile test (FTS = 8 MPa)
L = 6 to 10 D

Figure 14.6 Comparison of uniaxial compressive strength, indirect tensile strength and flexural strength test methods and results for a typical dense limestone. Note different specimen length-to-diameter or length-to-depth (*L/D*) ratios.

Elastic modulus

The elastic modulus (E) of stone is used to estimate its relative toughness or brittleness. As a group, rocks are considered to be brittle materials – meaning that cracks propagate rapidly within them, they fail at low strains (generally less than 1%), and they undergo negligible yielding prior to failure. This behaviour is best investigated by *fracture toughness* testing, which measures the energy required for crack propagation, but the procedure is difficult and there is no standard test. Hence several indicators, such as toughness index (= $UCS^2/2E$), are used to identify the less brittle lithologies likely to perform best in veneer panels. 'Less brittle' in this context means those with modulus ratios (E/UCS) of 200–500 and ITS/UCS ratios of 7% or more. Modulus ratios higher than 500 indicate very brittle stone that fractures easily, while weak rocks like sandstone are usually in the range 100–200.

Thermal properties

Thermal properties of rock include its coefficients of expansion and its thermal conductivity. The coefficient of linear expansion does not vary greatly with lithology; it is usually in the range $(5–12) \times 10^{-6}$ per degree Celsius, about the same as concrete. The thermal conductivity of porous rock, however, is 30–50% that of dense rock and only 2% that of metals, and so porous rock is a good insulator.

Abrasivity and abradability

Quartz has the highest *abrasivity* of the common rock-forming minerals, about three times that of feldspar and 30 times that of calcite. As a consequence, quartz sandstone and granite cause much more wear during cutting than limestone, marble or gabbro of the same strength. This should not be confused with *abradability*, which is the tendency of the rock itself to become worn, usually as paving stones under pedestrian traffic. This is measured under standard conditions as specimen weight loss beneath the abrading wheel of a Taber tester.

Typical properties of Australian stone

A range of typical physical and geomechanical properties for some of the better known Australian building stones are listed in Tables 14.1 and 14.2. The *sandstones* range from soft Helidon sandstone from Toowoomba, Queensland, to a relatively hard and dense variety from Carey Gully (Adelaide Hills), South Australia; Sydney (Hawkesbury) sandstone from Bondi, New South Wales, is intermediate in properties. Note the close relationship between strength, sonic velocity and bulk density that is characteristic of sandstones. Sulphate soundness test losses correlate poorly with other properties, but are greatest with the more porous samples.

In contrast, the *granites* have negligible porosity, are almost unaffected by salt crystallization, and have very high abrasion resistance (though the Kingston granite is the weakest and also the most abradable). The relationship between strength, sonic velocity and density is not consistent in this group, possibly due to variations in

Table 14.1 Typical physical properties of Australian stone[a]

	V_p[b] (m/s)	Bulk density (t/m³)	n_e[c]	n_t[d]	SC[e]	Sulphate loss (%)	Taber hardness
Sandstones							
Helidon, Qld	2220	2.18	12.2	18.6	0.66	92	–
Bondi, NSW	–	2.36	–	9.6	–	6	–
Wondabyne, NSW	2710	2.26	8.7	17.9	0.47	5	–
Somersby, NSW	2740	2.23	10.1	19.3	0.52	88	–
Carey Gully, SA	3410	2.49	4.1	7.3	0.56	0.9	6
Granites							
Kingston, SA	5492	2.71	0.1	0.5	0.21	0.02	96
Sedan, SA	5592	2.63	0.5	0.7	0.68	0.10	131
Calca, SA	6213	2.61	0.4	0.5	0.95	0.06	133
Black Hill, SA	7128	2.97	0.2	0.2	0.99	0.03	125
Marbles and limestone							
Chillagoe Marble, Qld	5140	2.70	0.3	–	–	0.01	12
Angaston Marble, SA	5520	2.72	0.2	0.3	0.67	0.04	17
Wombeyan Marble, NSW	5200	2.70	0.4	–	–	–	13
Gambier Limestone, SA	2036	1.21	39	56	0.70	10.6	0.3
Slate							
Mintaro, SA	4995	2.76	0.8	0.9	0.90	0.07	12
Auburn, SA	2990	2.11	11.5	23	0.50	6.6	3

[a] All velocities and densities dry.
[b] V_p = (Ultra)sonic wave velocity.
[c] n_e = Effective porosity (water absorption in 24 h).
[d] n_t = Total porosity (water absorption under vacuum).
[e] SC = Saturation coefficient = n_e/n_t.

mineralogy and microfabric. The Black Hill 'granite' is in fact a gabbro, its high proportion of mafic minerals reflected in a density of 2.97 t/m³ (and the lower Taber hardness than granite, due to the absence of quartz).

The *marbles*, despite being from three states, are remarkably alike – of medium strength, high stiffness and very high modulus ratios (hence very brittle crushing behaviour), similar to high-strength concrete. The Gambier Limestone, on the other hand, is an exceptionally porous, lightweight rock with excellent insulating capacity and a strength that, despite being very low (4 MPa), is nonetheless adequate for load-bearing house walls. Its wear resistance is negligible, as could be expected, but even the dense marbles have only moderate Taber hardness because of their soft calcite content.

The hard and soft extremes in Australian *slates* are represented by the Mintaro slate and the Auburn bluestone, respectively. The first is a true slate, with unusually high flexural strength developed because of the alignment of the metamorphic minerals. The Auburn bluestone is a

Table 14.2 Typical geomechanical properties of Australian stone[a]

	Compressive strength, UCS[b]		Tensile strength		T/C ratio[d]	Elastic modulus	E/UCS ratio[e]
	(MPa)	W/D[c]	(MPa)	W/D[c]		(GPa)	
Sandstones							
Helidon, Qld	20	0.30	–	–	–	–	
Bondi, NSW	42	0.43	11.4M[f]	0.21	0.27	–	
Wondabyne, NSW	46	0.70	13.0M	0.26	0.24	–	
Somersby, NSW	27	0.63	4.6M	0.26	0.17	–	
Carey Gully, SA	91	0.68	15.1M	0.64	0.17	14.7	161
Granites and gabbro							
Kingston, SA	98	0.93	9.3F[g]	0.93	0.10	44.6	450
Sedan, SA	197	0.92	16.1F	1.01	0.08	58.4	296
Calca, SA	150	0.99	10.9F	1.05	0.07	61.8	412
Black Hill, SA	219	0.88	15.5F	1.19	0.07	92.6	425
Marbles and limestone							
Chillagoe Marble, Qld	69	1.09	5.6M	1.55	0.08	55.0	733
Angaston Marble, SA	75	0.99	11.3F	1.00	0.15	53.2	707
Wombeyan Marble, NSW	87	1.02	7.2M	1.78	0.08	64.5	741
Gambier Limestone, SA	4	0.87	1.9M	0.79	0.76	2.5	621
Slate							
Mintaro, SA	183	0.75	37F	0.69	0.20	28.8	157
Auburn, SA	27	0.64	11.4M	0.43	0.42	4.2	156

[a] All strength values are in air-dry condition.
[b] UCS = Uniaxial (unconfined) compressive strength.
[c] W/D = Wet/dry strength ratio.
[d] T/C = Tensile/compressive strength ratio = brittleness ratio.
[e] E/UCS = Modulus ratio.
[f] M = Tensile strength by modulus of rupture (three-point loading) test.
[g] F = Tensile strength by flexural (four-point loading) test.

lightly metamorphosed siltstone, which splits along bedding (it is a 'flagstone') but lacks true slaty cleavage. Rock such as this is suitable for walling and landscaping, but is too soft for paving and cannot be split thinly enough for roofing.

14.5   WEATHERING AND DURABILITY TESTS

By engineering standards stone is a durable material, since it can often perform satisfactorily for centuries. Nonetheless, the accelerated deterioration of urban facades and monuments in industrialized countries over the past few decades has been thoroughly documented. The more common manifestations and causes of dimension stone deterioration are listed in Table 14.3. These decay processes are complex

Table 14.3 Deterioration modes in dimension stone

Discoloration
  Bleaching by sunlight
  Darkening by sunlight or iron oxide precipitation
  Uniform soiling by windborne dust and soot
  Patchy staining by rainwater, rust, chemicals, oil, etc.
  Graffiti or paint penetration, peeling renders and waterproofing agents
  Efflorescences of white salt or mortar exudates (mostly cement–sulphate
    reaction products)

Surface deterioration
  Loss of polish by solution or rain scouring, especially in carbonates
  Etching, pitting and grooving ('fluting') by more prolonged or severe solution
  Loss of surface by abrasive wear, chipping, grooving or mineral plucking
  Spalling, blistering and crazing of surface crusts by ice crystallization or
    swelling minerals, mainly clays and gypsum, on a small scale
  Surface crumbling and powdering due to salt crystallization, followed by
    wind and water erosion
  Honeycomb weathering (= cellular or tafoni weathering), also by salt
    crystallization and erosion, but on a more extensive scale
  Fire damage, mainly extensive cracking and crazing, and discoloration

Internal disruption
  Splitting caused by extension of inherent joints and other planar rock defects,
    root penetration, swelling of rusty bolts
  Tensile failure of blocks due to creep at overhangs
  Shear failure due to compression at arch ends
  Warping due to thermal or moisture stress
  Delamination along bedding, especially of blocks laid 'end on', i.e. courses
  perpendicular to bedding
  Exfoliation of thin slabs of case-hardened sandstone or gypsum-impregnated
    limestone surface

---

and often interdependent, and the agencies are both external (atmospheric) and internal relative to the stone surface.

Atmospheric processes

The atmospheric agents of decay include polluting gases, especially SO_2, SO_3, H_2S and CO_2, which generate 'acid rain', and aerosols such as marine salt and diesel smoke. Figure 14.7 illustrates different rates of stone deterioration in the heart of an industrial city. Salt weathering is probably the dominant mechanism here, but vehicle fumes and steelworks emissions may have contributed. In solution or suspension, these corrode the stone face and are drawn into the pores of the rock by capillary suction during drying. The consequences can be severe for porous marble, limestone and sandstones, especially where wind or water can remove the loosened grains. Ice crystallization and wetting–drying cycles can also cause surface disintegration.

Figure 14.7 Varying degrees of sandstone decay close to the sea shore, Newcastle, New South Wales. This wall has been subject to salt crystallization and wind erosion. The upper courses are of typical Coal Measures lithic sandstone, while the lower courses are of stronger and much more resistant quartz sandstone.

Internal processes

The internal weathering processes of building stone are dominated by the capillary movement of moisture inwards during wet periods, outwards on drying, and upwards below the damp course. In addition to causing feldspars, zeolites, clay minerals and other secondary minerals to shrink and swell, capillary moisture transports salts to the stone face. Salt crystallization loosens surface grains, causes steel pins to rust, and the sulphates may react with mortar. The sources of salt may be shallow saline groundwater, particularly in hot arid areas, windblown sea salt, or connate salt within the rock pores. The main sources of moisture and mechanisms of salt transport are summarized in Figure 14.8.

Disseminated sulphides within the stone may oxidize, causing iron staining along fissures and at surface discharge points. Biotite may 'rust' in a similar fashion and feldspars become cloudy. Other agents of stone decay include biotic weathering by bacteria, algae, lichen and mosses; wetting from interior condensation, pipe and roof leaks; acid runoff from pigeon guano; and harsh cleaning methods.

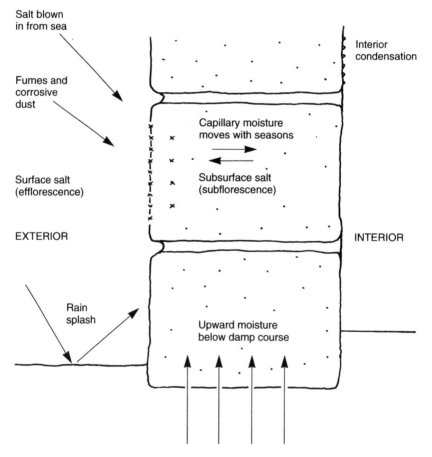

Figure 14.8 Sources of moisture, salt and pollutants, and capillary movement in stone masonry.

'Weatherability' tests

The tendency of stone to break down under the effects of weathering processes can be inferred from the results of routine physical and geomechanical testing. High water absorption, low saturation coefficient, low bulk density, high abradability test losses and the presence of water-sensitive minerals are all indications of low durability. In addition, several 'weatherability' tests have been devised to quantify the rate of stone breakdown under standard conditions, or at least to highlight rock materials that are relatively non-durable. These include the following:

- *BRE crystallization test*   In this, 50 mm stone cubes are put through up to 15 cycles of soaking and oven-drying in a sulphate solution. Non-durable samples lose weight by surface crumbling and may even completely disintegrate. This is a severe procedure, like the sulphate soundness aggregate test from which it was derived. It is most applicable to sandstones, but does not cause significant breakdown in low-porosity marbles, granites and slates.
- *BRE saturation coefficient test*   This measures the absorption of cold water over 24 hours as a proportion of total pore space (that is, the ratio of effective porosity to total porosity). Where this is less than about 80%, the stone is considered frost-resistant, since some air voids remain after water absorption to allow for ice expansion. Again, dense and strong stones with very low total and effective porosities score well in this test.
- *Winkler wet/dry strength ratio test*   This is based on soaked and oven-dry bending tensile strengths. Where wet strength is less than 60% of the dry value, the stone is considered likely to break down. The wet/dry strength ratio of UCS test results is also considered to be an indicator of durability.

14.6   STONE CONSERVATION

Conservation of historic stone buildings and monuments has four main aspects: surface cleaning, waterproofing, preservation of crumbling faces, and replacement of damaged stonework. Because of the architectural significance of these structures, and the variability of natural rock, restoration has to be undertaken with great care, with cost as a secondary consideration. In the past much damage was done by inappropriate techniques, and the ever-present risk is that a new process or chemical treatment may yield a good finish initially, but disfigure the monument after some years of further weathering.

Surface cleaning

The simplest and most generally applicable form of stonework cleaning, though slow, is simply washing and brushing off softened dirt. A good result requires that the surface be gradually 'wetted up' beforehand by means of a fine spray or fogging device. This ensures that the dirt crust is thoroughly softened and therefore can be easily detached from the stone. Reducing the quantity of water applied also eases the problem of dirty water disposal. Steam cleaning is little used today, except for removing deep-seated dirt. Sand or grit blasting is fast but severe, and has also fallen from favour because it can gouge out soft bands and give a patchy result. Where it is permitted, wet rather than dry blasting, low nozzle pressures and non-siliceous abrasives may be stipulated.

Chemical cleansers are usually acid or alkaline mixtures with added surfactants; the most effective ones are based on sodium hydroxide or ammonia. Dilute hydrofluoric acid and ammonium bifluoride are fast and effective cleaners of very dirty stone, but demand great care in use. Poultices of clay with special chemicals are used for severe local encrustations and deep stains, and toxic washes are used against lichens and similar organic growths.

With the exception of water washing and scrubbing, all of these methods in unskilled hands can cause damage to stonework, and pose both a safety risk to operators and a nuisance to the public. Furthermore, the treatment has to be adapted to the problem; the most severe methods should only be used where gentler techniques have been tried and failed. It is good practice to try out any new techniques on small and inconspicuous areas of stonework first, or on slabs replaced during restoration.

Waterproofing

Cleaned stone was traditionally waterproofed by painting, coating with linseed oil, limewashing and waxing. Modern water repellents include various silicone compounds and resins that provide a semi-permeable membrane, allowing the stone to 'breathe'. There is a risk of spalling if moisture can be drawn from within the building, or upwards from foundations, and builds up pressure behind impermeable coatings. Water repellents ideally shed most rain, but are pervious to water vapour moving into or out of the stone. Waterproofing is most needed in lower masonry courses, where rain splash is effective, and also close to the sea, to prevent salt uptake.

Preservation

Fretting of exposed stonework, such as sandstone, can at least be temporarily arrested by the application of consolidants. These are organic liquids that coat and bind loose grains and restore integrity to the surface. Although there are chemical similarities, consolidants differ from water repellents in penetrating to greater depth in the stone, so they can better resist exfoliation. Ideally they are chemically inert, transparent, low-viscosity liquids with high setting strength and excellent durability. Wax solutions have been in use for over a century, but the most commonly used consolidants today are epoxies and acrylics. As with other surface treatments applied to dimension stone, unexpected and unwelcome reactions may occur between the chemical and the stone, so careful laboratory and field trials are recommended before application.

Replacement

Replacement of stone blocks may become necessary where carved work has become blurred or where sections of masonry are structurally unsound. In Australia the main problem is with sandstone masonry, but in Europe limestone is most subject to such decay. Matching the new stone with the old is no easy task, since the original source will usually be unknown or long abandoned. A new quarry working the same formation is one possibility, but whether this can fully reproduce the appearance, carving characteristics and weathering behaviour of the older stone may be questionable. Duplicating the old finishing techniques seems to be less difficult, but attempts to create an 'antique' finish may or may not be successful!

A more common approach is to recycle demolition stone. Even quite small buildings, say a house or a shed, may provide sufficient stone for a large repair job. In Sydney, for example, there are at least two stockpiles of sandstone blocks from demolished buildings and from defunct sandstone quarries available for future restoration work. In large cities an inventory of reserved stone may be needed, though there is usually considerable awareness of this material among trade specialists. In addition, the high price of newly cut stone ensures that most is recycled from demolition sites.

As a last resort, decayed blocks can be replaced with synthetic stone, a mixture of sand filler, colorants and binder. The binder used to be cement, but polymers and resins are now used. The result may be of adequate strength and durability, but match the original appearance poorly. Normally synthetic stone would be restricted to locations that are not easily seen, or for duplicating intricate carvings. Composite stone blocks can also be fashioned by cutting back decayed ashlars to expose the unweathered interior, or simply reversing the exposed and interior faces, and making up the lost stone with concrete or epoxy.

14.7 CASE STUDY: THE AUSTRALIAN STONE INDUSTRY

The Australian dimension stone industry grew strongly during the 1980s, after decades of decline (Robinson, 1992), in response both to the building boom of those years and to renewed architectural interest – locally and internationally – in stone as a building material. Sandstone represents nearly half the local output and granite one-quarter, with the remainder mainly marble. Production probably peaked at about 150 000 tonnes per annum in the late 1980s and has since declined with the building slump. World production at this time was about 23 Mtpa, so Australian output represented less than 1% of the total market.

The local industry is greatly hampered by fluctuating demand, which for specific products can rise or fall by 50% or more in successive years. This is made worse by the industry's dependence on the top end of the market: high-rise buildings, prestige projects and expensive houses. The industry also sees itself threatened by imports and a bias towards overseas stone from local architects. Exports amount to about half the imported tonnage, but also fluctuate greatly.

More than 80% of Australia's dimension stone production comes from New South Wales, Victoria and South Australia. The industry quarries a considerable variety of granites and sandstones and a limited range of marbles, but gneiss, limestone and slate are not produced in significant quantities. Some of this shortfall is made up by importing block stone and cutting and finishing it locally.

REFERENCES AND FURTHER READING

Ashurst, J. and Dimes, F.G. (eds) (1991) *Conservation of Building and Decorative Stone*, vol. 1. Butterworth-Heinemann, London.

Barton, W.R. (1968) *Dimension Stone*. USBM Information Circular 8391.

Bell, F.G. (1992) The durability of sandstone as a building stone, especially in an urban environment. *Bulletin, Association of Engineering Geologists*, **29** (3), 49–60.

Committee on Conservation of Historic Stone Buildings and Monuments (1982) *Restoration of Historic Stone Buildings and Monuments*. National Academy Press, Washington, DC.

Donaldson, B. (ed.) (1988) *New Stone Technology, Design and Construction for External Wall Systems*. ASTM Special Publication 996.

Robinson, M.J. (1992) Forming a stone industry down under. *Stone World*, September, pp. 73–86.

Schaffer, R.J. (1932) *The Weathering of Natural Building Stones*. DSIR/BRE Special Report 18. DSIR (UK) and Building Research Establishment. Reprinted 1985.

Shadmon, A. (1996) *Stone – An Introduction*, 2nd edn. Intermediate Technology Publications, London.

Sims, I. (1991) Quality and durability of stone for construction. *Quarterly Journal of Engineering Geology*, **24**, 67–73.

Wallace, I. (1969) *Building Stones of New South Wales*. New South Wales University, Department of Industrial Arts Monograph 1 (2).

Winkler, E.M. (1975) *Stone: Properties, Durability in Man's Environment*. Springer, New York.

Winkler, E.M. (ed.) *Decay and Preservation of Stone*. Engineering Geology Case Histories 11, Geological Society of America.

CHAPTER 15

Limestone and cementitious materials

The term 'limestone' covers a variety of geological materials that vary greatly in their origins and properties, but which can be most simply defined as sedimentary rocks made up of 50% or more of calcite and dolomite, in which calcite is the more abundant. The proportion of these combined carbonates in commercially mined *high-calcium limestone* is usually 90% or greater. The main criteria used in carbonate rock classifications are:

- *Mineral composition*, meaning the percentages and types of the carbonate minerals present (calcite, aragonite, dolomite, siderite, magnesite), and the extent to which they are mixed with non-carbonates (mainly quartz and clay minerals).
- *Geological origin* of the sediment, whether detrital (clastic), organic (mainly reef material), chemically precipitated, or a combination of these.
- *Grain size and texture*, in terms of porosity, percentage and type of matrix and cement.
- *Geomechanical properties*, mainly strength, permeability, hardness and deformability – all of which are closely related to bulk density and porosity.

Industrial end-uses for carbonate rocks have given rise to terms such as cementstone (a type of argillaceous limestone with the correct proportions of $CaCO_3$, Al_2O_3, SiO_2 and Fe_2O_3 for cement manufacture without additives) (Figure 15.1), kilnstone, fluxstone and glass stone. Some of the specified requirements for these uses are listed in Table 15.1.

Carbonate rocks make up about 15% of the volume of all sedimentary rocks, but the proportion in a single region may range from about 80% to almost nil. This naturally affects the value attached to individual deposits and the extent to which they are used as low-value construction materials, rather than as higher-price raw materials. Construction uses of the carbonate rocks – and in particular limestone – include the following: as aggregate, railway ballast and roadbase; as

Figure 15.1 Interbedded limestone and shale, naturally proportioned for cement raw material ('cementstone'), Hendry Quarry, North Wales. (Photo: J.H. Whitehead.)

breakwater stone and rip-rap; in cement and lime manufacture; and as facing or structural stonework.

The chemistry, geology and industrial uses of limestone are discussed at length in Boynton (1980). Its engineering properties are summarized by Bell (1981) and Dearman (1981). Cement-making materials and processes are dealt with by Neville (1981).

15.1 GEOLOGICAL EXPLORATION

Geological exploration of extractive-industry sites usually comprises two phases – location of the deposit, followed by assessment of its reserves and quality. The main carbonate deposits in settled regions are usually well known, owing to their distinctive landforms and the relics and records of previous small-scale mining.

Table 15.1 Requirements for limestone as a raw material

Cement-making ('kilnstone')
  Preferably 'high-calcium' recrystallized limestone (>95% CaO)
  Low magnesia (<3% MgO)
  Low to nil phosphates, sulphides
  An ideal, though uncommon, raw material is argillaceous limestone with
    uniform $CaO/SiO_2/Al_2O_3/Fe_2O_3$ ratio of about 2.7/0.8/0.2/0.1
    ('cementstone')
  Actual cement kiln feeds are usually proportioned mixes of high-calcium
    limestone, shale and high-ash coal

Steel-making ('fluxstone')
  High-calcium limestone with minimal acid oxides required
  Silica content typically <8%, preferably <2%
  Phosphates, alumina, sulphur also low
  Crushed product should be uniformly graded –30 mm gravel with
    minimal fines (hence a hard parent rock)

Quick-lime
  High-calcium limestone is preferred
  A wide range of carbonate compositions is acceptable

Glass-making ('glass stone')
  Main requirement is for a very low Fe_2O_3 content, <0.1% ideally
  Uniformity in composition, particularly silica content (a requirement common
    to all chemical end-uses)

Locating deposits

Depositional environment is important because it determines the size, shape and purity of limestone deposits. Shallow-water high-energy deposits are likely to contain less non-carbonate material than low-energy micritic limestone environments, where substantial dilution by clay- and silt-sized particles can be expected. Carbonate sediments, being relatively soluble, are also highly susceptible to post-depositional alteration and modification. This is beneficial where recrystallization results in a higher calcite content, better cementation and higher strength. However, it may also cause dolomitization and render the deposit unsuitable as a cement raw material.

Assessment of reserves

Once the quarry site has been selected and its limits defined by geological mapping and preliminary drilling, the assessment of its reserves and their quality variations can begin. The bedded nature of most limestone deposits means that considerable vertical changes in grain size, chemical composition and rock texture can be expected. Lateral facies changes may also be present, but these will have less influence on product quality since they can be isolated into discrete blocks, to be allowed for in product blending.

Special care should be taken when mapping and sampling surface outcrops of soluble rocks, as these may be unrepresentative of the deposit as a whole. Shale or argillaceous limestone beds may be obscured by soil, while limestone blocks may be scattered downslope by mass movement; both processes tend to exaggerate the proportion of carbonate present within the sequence. The surface layers may also be enriched in secondary calcite, dolomite and gypsum; extensive and massive-looking outcrops may turn out to be mere crusts or cappings!

Core drilling on a regular grid will provide the basis for statistically valid sampling of the whole deposit, and also highlight lateral variations. The pattern size naturally varies with the complexity of the deposit; usually coring at 100 m centres is supplemented by non-core (downhole hammer) drilling at closer spacings. Rotary percussion air-flush drilling such as this is the cheapest and quickest method of sampling available, and contamination due to caving may be reduced by casing off above the bit. Air-flush core drilling may also be necessary because of water losses in cavernous limestone.

Downhole geophysics is especially applicable in carbonate rocks. Density and gamma logging of uncored holes and correlation with nearby fully cored drillholes will greatly improve the quality of data obtained. Below the water table, neutron logs are sensitive both to lithological changes (limestone/dolomite/shale ratio) and to porosity variations.

Surface geophysical techniques such as gravity, resistivity and transient electromagnetics can also make use of the distinctive physical properties of carbonate rocks. Their main applications appear to lie in defining narrow and steeply dipping deposits, and in detecting caverns beneath quarry floors.

Sample evaluation

Samples for *chemical analysis* may be only 100 g, though much larger quantities are required for aggregate testing. Core samples are split, crushed, mixed and quartered several times to obtain representative analytical specimens of 1–2 m vertical intervals. Even thinner layers may have to be sampled to determine if they are to be discarded or blended into the run-of-mine product during quarrying. The analytical test results are reported in terms of weight percentages of oxides present, with the best-quality high-calcium limestone having CaO contents above 95% and SiO_2, MgO, Fe_2O_3 and Al_2O_3 each less than 1%.

Aggregate evaluation of limestone deposits usually concentrates on index tests (uniaxial compressive strength, UCS; porosity or water absorption; bulk density; elastic modulus), and on durability (Los Angeles abrasion test loss, sulphate soundness, petrographic examination). The durability tests require large samples, typically

10–20 kg, and this in turn puts a premium on large-diameter coring. The largest core diameter widely obtained in Australia is HQ (61 mm), which is roughly equivalent to 8.2 kg per linear metre of dense limestone core. Where large volumes of core are required for testing, paired HQ drillholes are cheaper than large-diameter single holes.

15.2   GEOTECHNICAL INVESTIGATIONS

Geotechnical investigations for quarry design should be carried out in parallel with environmental impact studies and reserve estimation, since much of the surface mapping and drilling results can be used for all three aspects of the site exploration. The geotechnical factors most relevant to limestone quarry investigations include slope stability, hydrogeology, hazard detection and excavatability.

Slope stability

Evaluation of face and bench slope stability is particularly important in narrow and steeply dipping limestones. Because of the long working life of these deposits – a century is not uncommon – and the need to mine deep to make the best use of the reserves, the steepest possible pit slopes consistent with safety and efficient operation are needed. Information on bedding and discontinuity spacing, joint shearing resistance, orientation and persistence can be collected during outcrop mapping and coring, and supplemented by bulldozer and excavator trenching. One redeeming feature of jointed limestone faces is that they are usually free-draining, thereby removing one major destabilizing factor.

Hydrogeology

Where the quarry floor is located below the water table, *groundwater inflows* incur considerable pumping costs, and can have regional consequences (dewatering of distant wells, groundwater pollution and sinkhole collapse). Even in quarries above the water table, inflows from open joints following heavy rain can be a nuisance. Groundwater investigations at the exploration stage are largely confined to recording water levels in boreholes, but occasionally large-diameter bulk sampling drillholes (150–200 mm) may be re-used for pumping tests.

Hazard detection

There is a risk to workers and equipment from floor collapse into *solution caverns* during drilling, blasting or working. The principal detection technique is probing by rotary-percussion rigs, but gravity and electromagnetic (EM) surveying are sometimes used. Ground-probing radar also shows promise in this application, though all geophysical methods suffer from interpretation ambiguities. Is the anomaly a small void at shallow depth, or a large but deep one? How does the response vary if it is filled (or partly filled) with air, water or clay?

Excavatability

The excavatability of a carbonate deposit comprises its drilling, blasting and crushing characteristics. The principal considerations here are the presence of hard or abrasive layers (chert nodules, silicified bands), depth to the water table (which determines whether or not ANFO can be used for blasting), and the prevalence of cavities and open joints (which cause blasthole gases to depressurize). Tabular or slabby blocks are also common in bedded deposits like limestone, and these can cause jams at conveyor transfer points and in crushers. They may also mean that rolls crushers are more suitable than jaws for primary comminution.

15.3 GEOMECHANICAL PROPERTIES OF LIMESTONE

Typical geomechanical properties of some important British and American limestones are summarized in Table 15.2, emphasizing that they are more variable than any other rock type, from very strong and dense marble to weak and porous oolite. It should be noted that some chalks and aeolianites are weaker still, while marble may have compressive strengths up to 450 MPa. The brittleness of strong carbonate rocks is indicated by their high compressive/tensile strength ratios and high modulus ratios (E/UCS). Limestone decreases in strength by 20–40% with saturation, another consequence of its porous nature (the strongest and densest samples show the least weakening). Deformation behaviour is also strongly affected by porosity, with the Carboniferous Limestone failing suddenly at relatively small strains (0.15%), while weaker limestones tend to yield and behave plastically prior to ultimate failure at strains of about 0.5%.

In practical terms, these results confirm that only the strongest and least-porous limestones are suitable as aggregate, but that these are nonetheless relatively easy to crush because of their brittleness. For use as armourstone, carbonate rocks are inferior to siliceous rocks of the same strength because of their softer surfaces (Moh's hardness of calcite is 3, compared with 7 for quartz). Limestone armour blocks are consequently much more susceptible to rounding and weight loss in a high-energy wave environment. Nevertheless, weak and visibly porous limestones have adequate strength for use as building stone (but not as veneers), though their durability may be suspect.

Table 15.2 Geomechanical properties of typical carbonate rocks[a]

Formation	Carbon-iferous Limestone (UK)	Magnesian Limestone (UK)	Great Oolite (UK)	Niobara Chalk (US)	Salem Limestone (US)	Knox Marble (US)
Geological age	Carbon-iferous	Permian	Jurassic	Creta-ceous	Carbon-iferous	Ordo-vician
Compressive strength (MPa)						
Dry	106	55	16	26	75	322
Saturated	84	37	9	–	–	–
Tensile strength (MPa)						
Brazilian	6.5	5.8	1.9	–	–	–
Flexural	–	–	–	4.1	11.0	26.2
Point load	2.8	2.7	0.9	–	–	–
Elastic modulus (GPa)	69	41	16	–	–	–
Modulus ratio (E/UCS)	657	791	1023	–	–	–
Bulk density (t/m^3)						
Dry	2.58	2.51	1.98	1.81	2.37	2.84
Saturated	2.61	2.60	2.23	–	–	–
Particle SG	2.70	2.83	2.70	–	–	–
Porosity (%)	4.4	11.4	22.0	8.3	11.0	0.7

[a] Data from Carr and Rooney (1975) for US samples and from Bell (1981) for UK samples.

15.4   LIMESTONE AGGREGATES

On a worldwide scale, limestone is perhaps the most important single source rock for concrete aggregate, railway ballast, crushed roadbase and breakwater stone. In the USA it comprises about 75% of all crushed stone aggregate, and in total production it is second only to alluvial sand and gravel. Limestone suitable for use as aggregate has to be dense (around 2.7 t/m^3), and consequently strong and of low porosity (less than 1%, preferably less than 0.5%). Its attractiveness for this purpose arises partly from the fact that the capital cost of equipment in limestone quarries can be offset against other products (lime for cement- and steel-making fluxes, for example), and partly from its inherent virtues as an aggregate.

These include compressive strengths typically in the range 50–150 MPa – that is, just stronger than most high-strength concrete but weaker than most igneous rocks. The moderate strength of carbonate rocks also results in a well-graded roadbase product, which is less 'harsh' (contains more fines) than that obtained from harder igneous rocks. Such a roadbase is consequently more workable – it can easily be spread, compacted and shaped on the road surface. Some weak limestones, such as calcrete and aeolianite, can be broken down by ripping and rolling alone to provide cheap pavement materials for lightly trafficked rural roads. Better-quality rock is crushed for highway roadbase (Figure 15.2), although it may still be inferior to harder igneous rock under really heavy traffic. Limestone is, however, rarely acceptable for surfacing aggregate because of its susceptibility to breakdown under repeated loading and – more importantly – because of its tendency to 'polish' under traffic.

In general, limestone is an excellent concrete aggregate, though some dolomitic and argillaceous varieties may react with the cement paste (alkali–carbonate reaction) to produce an expansive gel. Opaline silica nodules and gypsum veins within limestone may also result in deleterious reactions. Aggregate–paste reactions can nevertheless be beneficial: some limestones bond strongly with the cement, producing concrete of high tensile strength from aggregates of only moderate UCS.

Two geomechanical properties of high-strength carbonate rocks are especially important in crushing:

- They are much less abrasive than quartzose and silicate rocks. In terms of metal wear, calcite has only 3% of the abrasivity of quartz. This translates into extended life for crusher plates, smaller and lighter crushing plant for the same output, less pipe wear where concrete made with limestone aggregate has to be pumped, and so on. It also means that limestone blocks can be easily sawn for dimension stone.

Figure 15.2 Mobile crushing and screening plant in operation north of Adelaide, South Australia. The complex arrangement of processing stages was necessary to obtain highway basecourse from sheet calcrete ('paddock limestone') of marginal quality, and incurred a high proportion of fines wastage. (Photo: G. Harvey.)

- Crystalline limestone is relatively brittle, even by rock standards, and hence does not store 'strain energy'. This means that it can be crushed with low expenditure of energy and that the crusher can have a higher capacity (output in tonnes per hour) than it could achieve with a tougher lithology, such as dolerite or granite.

15.5   MINING AND BENEFICIATION

Limestone quarrying does not require specialized equipment. The main considerations are minimizing the proportion of overburden to product, segregating the thicker waste layers during mining, and carefully blending the rest to obtain a uniform product. Where limestone mining does differ from other types of hard rock quarrying is that it may be economic to upgrade or 'beneficiate' some deposits to obtain a higher calcite content or a more uniform product. These techniques make use of density differences between waste and product, with the product being the denser. Separation is effected by turbulent movement of the finely crushed run-of-pit material in water, with or without a heavy-medium suspension (such as magnetite or ferrosilicon). The reject is floated off, while good stone sinks. In the case of bedded limestone, the main waste materials are shale partings and cavity infills.

Underground mining

Underground mining of limestone accounts for only about 4% of production in the USA, and none at all in Australia and most other countries. Nevertheless, interest in this method is reviving, despite high production costs, mainly due to environmental restrictions on surface mining. The applications and advantages of underground mining include:

- Where there are *uneconomic overburden ratios* in proposed workings due to deeply buried limestone, coupled with a lack of surface deposits within acceptable hauling distances.
- Where there are *slope stability problems* in existing deep quarries, or where the overburden ratio is becoming intolerable due to lateral expansion of the pit walls into waste rock (in this case the underground mine is simply an extension of the surface workings).
- Where there is the possibility of *selectively mining* high-quality seams that command premium prices (and, conversely, to avoid low-grade argillaceous beds).
- Where there are *environmental benefits*, such as concealed workings, and less noise and dust; in this way extraction can be carried out close to urban areas, or in locations where surface mining would be rejected.

Underground workings are excavated by board-and-pillar techniques leaving about 30% (by area) of limestone in the ground as roof support. Because limestone of sufficient quality to justify underground mining is much stronger than coal, tunnels can be driven higher and wider, sufficient for access by large trucks. The underground space created can also be turned into an asset – in the form of waste disposal sites, storage caverns for gas and liquids, and space for other industrial uses.

15.6 ENVIRONMENTAL PROBLEMS

Limestone mining presents a number of environmental problems that are either unique, or more severe than those created by other types of quarrying. Carbonate deposits commonly occur in areas of great scenic beauty, as conspicuous escarpments, gorges and cliff-lines. They frequently enclose spectacular cave systems, which may contain important archaeological and zoological remains (which, nonetheless, would have remained undiscovered without mining!) The terra rossa soils developed on limestone beds may also support unique floral and faunal colonies. Consequently, environmental objections are likely to be even more strenuous than to other forms of quarrying, especially since limestone mining sites tend to be the largest and most long-lived of extractive operations. However, one site where limestone quarrying has had a beneficial effect is illustrated in Figure 15.3.

The conflict between limestone quarrying and the environment is illustrated by the situation in New South Wales. Carbonate rocks are rare here – they cover less than 1% of the state's area – and are mostly located in rugged and picturesque areas. The largest limestone quarry in Australia is located near Marulan, New South Wales, adjacent to a national park, a spectacular gorge and a state recreation area. The quarry has been operating for 70 years and at present produces 2.4 Mtpa, nearly half the state's consumption. This is almost all in the 'essential raw material' category – for cement- and steel-making – and extraction is scheduled to continue well into the twenty-first century.

A quarry rehabilitation plan expected to last 40 years is now in operation, and has as its chief aim the screening of the quarry from a nearby tourist area and lookout. The key elements of this development plan involve, sequentially: dividing the quarry in two, with an earth barrier between the halves; working out the nearer (southern) pit first, so far as possible below eye level; later filling this with overburden from the more distant pit; and revegetating this overburden, while working the now-concealed northern quarry.

In other countries, where limestone is more widespread, the main environmental problems concern groundwater. Dewatering in cavernous limestone deposits is not only expensive for the quarry operator, it

Figure 15.3 Buried karst exposed in a limestone quarry, Kuala Lumpur. This site was formerly a derelict alluvial tin mine, and is being further redeveloped into a housing estate as limestone extraction is completed.

creates problems of disposal and may eventually result in drawdown over a wide area. This, in turn, can cause the collapse of literally scores of sinkholes per square kilometre and dry up local wells. Where the quarry is located above the water table, pollutants – muddy runoff water and petroleum wastes – may be flushed downwards during heavy rains. Once in the saturated zone, these contaminants can move rapidly towards discharge areas (wells and springs) owing to the exceptional permeability of karst terrains.

A point that deserves emphasis is that, while limestone is a fairly common rock type, large deposits of the highest purity (95% or more CaO) are relatively rare and should be preserved for essential chemical uses. Cement can be manufactured from lower-grade calcareous materials – in fact, clay up to a certain proportion is advantageous – and aggregate can be obtained from a variety of other sources (many superior to limestone). Carbonate production in the USA in 1979 totalled 812 million tonnes, of which 65% was used as aggregate and 18% for cement manufacture; the remaining 17% included all other essential chemical uses and some for which substitutes could be found. There is thus considerable scope for economy in limestone utilization.

A number of other environmental problems and constraints, such as noise, dust and blast vibrations, are common to all forms of quarrying. Nevertheless, the dust nuisance from limestone mining is more severe, owing to the high proportion of fines liberated during crushing and the fine grinding processes involved with cement manufacture. These problems and their solutions are discussed in Chapters 19 and 20.

15.7   CEMENT MANUFACTURE

Portland cement is powdered 'clinker', a synthetic calc-silicate rock obtained by sintering finely crushed mixtures of limestone, shale and minor constituents such as iron oxides and gypsum. The calcined product consists of 98% or more CaO, SiO_2, Fe_2O_3 and Al_2O_3 combined into two main mineral constituents, tricalcium silicate (known as C_3S) and calcium orthosilicate (C_2S). Although limestone of purity greater than 90% $CaCO_3$ is preferred as the source of CaO, it has been obtained from chalk, marl, argillaceous limestone, coral sand, shell deposits, carbonatites, lime-rich slag and even calcareous feldspar (anorthosite).

Approximately 1.8 tonnes of raw materials, including coal fuel, are required to produce 1 t of cement clinker. Finely ground limestone (70–75% by weight) is mixed with 6–9% shale or clay, 10–17% pulverized coal, and small amounts of silica sand and iron oxide. The coal used is generally high-ash coal (25–35%), since it is a source both of fuel and of clay minerals; as a bonus, this is also the cheapest grade of coal. For gas-fired kilns, shale is increased to 10–30% of the charge.

Among the minor constituents of clinker are some whose concentrations must be severely limited. Foremost among these is magnesia (MgO), for which a maximum concentration of 3–4% is usually specified; this rules out dolomite as a source of CaO. Other deleterious materials include excessive SO_3, Na_2O, K_2O and P_2O_5. The effects of these oxides include premature setting of the cement paste, expansion and cracking of the finished concrete, and reaction with some siliceous aggregates to form a swelling gel.

The processes of milling, blending, firing (or 'calcining') and grinding the clinker are described as 'wet', 'dry' or 'semi-dry', depending on whether the raw materials are mixed with water during milling or not. The firing temperature in cement kilns is in the range 1400–1600°C, well below the melting point of most of the raw materials and made possible by the action of fluxes in the kiln feed.

Clinker emerges from the kiln after several hours as granulated spherical pebbles, whose diameters range from about 30 mm down to 1 mm (Figure 15.4). These are reduced by ball milling down to a powder of 90 µm (0.09 mm) maximum size before bagging. Fine grinding is essential in both the raw materials (to ensure a homogeneous clinker composition) and in the cement product. The latter ensures rapid and complete hydration, and predictable strength gain, setting time and long-term chemical stability of the resulting concrete.

Figure 15.4   Cement clinker, calcined from marble, Angaston, South Australia.

15.8   CEMENT TYPES

Portland cement is sold in a number of variants, in addition to the 80–90% market share of the 'standard' product (which itself varies somewhat with time, raw material sources and local demands). These involve changes in the raw materials or finer grinding, and generally carry some penalty in terms of strength gain or hardening rate.

Blended cements

Blended cements are mixtures of general-purpose Portland cement with ground granulated blast furnace slag and/or fly ash. They are mainly used for economy, since both admixtures are much cheaper than clinker, but other favourable properties may result – such as reduced bleeding or segregation, improved pumpability and reduced permeability. Ultimate strengths may also be higher than equivalent Portland cement concrete mixes, though hardening is slowed. The admixture may be used as a cement *extender*, which is ground with the clinker, or as a *raw material* prior to calcining.

Rapid-hardening cement

Rapid-hardening (high early-strength) cement reaches about 75% of its 28-day strength in three days rather than seven. Some types achieve this in hours, but at a cost of much heat of hydration being generated. These cements are very fine-grained and are mainly used where concrete formwork has to be made available quickly for re-use. They are also suitable for cold weather concreting, but not for large pours, in which excessive heat generation may cause cracking.

Low-heat cement

Low-heat cement is the opposite to the previous one; cracking is diminished but at a cost of low early strength. This type is used in very large concrete structures, such as dams. Strength at 28 days is only about 60% that of rapid-hardening cement, but equal or greater at the end of five years. The slow hydration is made possible by small changes in the oxide balance relative to normal Portland cement.

Sulphate-resistant cements

Sulphate-resistant cements inhibit post-hardening chemical attack from sulphate ions in groundwater and sea spray. To achieve this, a reduced proportion of tricalcium aluminate (C_3A) is needed. This type of cement develops strength slowly, though finer grinding may compensate by speeding up hydration.

Low-alkali cements

Low-alkali cements have Na_2O equivalents ($Na_2O + 0.66K_2O$) less than 0.6%, making them more resistant to alkali–silica reaction. The drawback in this case is again low early strength, which may have to be counteracted by a higher cement content than would otherwise be necessary.

15.9   CASE HISTORY: THE VICTORIAN LIMESTONE INDUSTRY

By Australian standards, Victoria is well endowed with limestone deposits, ranging from hard recrystallized Palaeozoic carbonates to Pleistocene lime sands (Inan *et al.*, 1992). About 70 deposits with actual or potential commercial value have been identified and investigated, but non-roadbase production is concentrated in just five quarries. Roughly 85% of the 2.7 Mt of limestone extracted annually in Victoria is used for cement-making, 9% is crushed for aggregate and roadbase, with the remainder used for agriculture and as an industrial raw material. Although crushed and broken limestone used for road-making is second only to cement-making in terms of tonnes produced, this represents only 1.1% of the state's consumption of crushed rock. There is at present no significant output of carbonate dimension stone or flux stone.

Victoria has three operating cement works, of which the two near Geelong contribute 93% of production. These mine local Tertiary limestones and marls of variable hardness and $CaCO_3$ content (41–96%). The two other important commercial deposits – both Palaeozoic – are at Lilydale, near Melbourne, and near Buchan in the east of the state. The Buchan limestone is high-grade and present in large quantities, but is distant from urban markets, while the Lilydale quarry is better placed but highly magnesian in places.

The future prospects of the Victorian limestone industry are closely tied to the fortunes of the cement producers. At present these have about 40 years' worth of proven reserves and little likelihood of substantially increasing their cement output. The most promising areas for future exploration are the Buchan area (high-calcium limestone) and the Portland–Warrnambool region in the west of the state (low-grade but easily mined Tertiary coquinite and Pleistocene aeolianite).

REFERENCES AND FURTHER READING

Bell, F.G. (1981) A survey of the physical properties of some carbonate rocks. *IAEG Bulletin*, No. 24, pp. 105–10.

Boynton, R.S. (1980) *Chemistry and Technology of Lime and Limestone*, 2nd edn. John Wiley, New York.

Carr, D.D. and Rooney, L.F. (1975) Limestone and dolomite, in *Industrial Minerals and Rocks*, ed. S.J. Lefond, pp. 833–60. AIME, New York.

Dearman, W.R. (1981) Engineering properties of carbonate rocks. *IAEG Bulletin*, No. 24, pp. 3–17.

Gillieson, D. and Smith, D.I. (eds) (1989) *Resource Management in Limestone Landscapes*. Special Publication No. 2, Department of Geography, University College, Canberra.

Gunn, J. and Bailey, D. (1993) Limestone quarrying and quarry reclamation in Britain. *Environmental Geology*, **21**, 167–72.

Inan, K., Summons, T.G. and King, R.L. (1992) *Limestone Resources in Victoria*. Geological Survey of Victoria, Report 97.

Jefferson, D.P. (1978) Geology and the cement industry, in *Industrial Geology*, ed. J.L. Knill, pp. 196–223. Oxford University Press, Oxford.

Neville, A.M. (1981) *Properties of Concrete*, 3rd edn, Chs 1 and 2. Longman, London.

CHAPTER 16

Brick clays

'Heavy' or *structural clay* products, which include bricks, earthenware pipes and roofing tiles, consume 80–90% of all clay mined. For the sake of convenience, all such clay and crushed shale materials are referred to here as brick clays unless otherwise noted. They are distinguished from higher-value *ceramic clays*, used for pottery and refractories, whose properties are more closely specified and which are subjected to more physicochemical modification. Hence the average ex-pit price for Australian brick clays is only one-quarter that for kaolin (china clay). Their production fluctuates between 5 and 7 Mtpa – about a third of a tonne per person – depending on the fortunes of the house-building industry.

It should be remembered that bricks, tiles and pipes are manufactured objects, and that brick clay is a raw material in an industrial process. As such, its blended properties are altered to suit the process and product, unlike other quarried materials, which are only crushed and screened. Consequently, there is no such thing as a brick clay specification, and brick-making has as much in common with baking as with quarrying. A wide variety of fine-grained mineral compositions, ranging from slurries to lithified mudrocks, can be used for the purpose. The general requirements for brick-making materials, however, are as follows:

- When moistened, the clay will behave plastically – meaning that it can be extruded, shaped and pressed into a mould.
- This formed clay 'body' will not shrink or crack excessively on drying, or during firing.
- The clay body will have sufficient 'green strength' to be handled after drying, and adequate fired strength.

In practice, these requirements are fairly easily met, either by mixing and modifying natural clays, or by manipulating kiln conditions. The three main products are: *facing bricks*, for which appearance is the most important characteristic; *commons*, which are cheaper and are used for interior walls or where looks are unimportant; and load-bearing or *engineering bricks*. Of the three categories, facing bricks are nowadays the most important in terms of numbers produced; commons have largely disappeared, along with double-brick house walls, replaced by brick veneer construction and concrete blocks. Reinforced concrete has long replaced engineering bricks in most load-bearing applications. In this regard, developments in brick technology have paralleled those in dimension stone, changing it from a structural material to a decorative cladding.

This chapter highlights the influence of the raw materials on the brick-making processes, and vice versa. Readers seeking more detail should consult standard reference books such as Clews (1969), Brownell (1976) or – most comprehensive of all – Bender and Handle (1982).

16.1   PROPERTIES OF FIRED BRICKS

The range of brick compressive strengths is wide, 10–130 MPa, though most are between 20 and 60 MPa. Engineering bricks and pavers are at the stronger end, while extruded bricks are stronger and denser than pressed ones. Brick density and strength are closely related, reflecting the greater degree of glass formation obtained with 'hard firing' at elevated temperatures.

One convenient indicator of brick strength and soundness is the water absorption test; values for Australian bricks are generally in the range 6–12% after 24 hours soaking, with the figures for pressed bricks again being about twice those for extruded bricks. Water absorption test results have a number of implications: high values indicate porous and therefore weak batches, which may have to be rejected. Porous bricks, like porous concretes, are vulnerable to surface disintegration by salt attack. Moderately absorbent ('high-suction') bricks require wetting before laying, to reduce their uptake of moisture from the mortar. On the other hand, the capacity of bricks to absorb rainwater and subsequently to give it up to evaporation prevents moisture movement into the building interior.

All bricks expand slightly after removal from the kiln, due to hydration of the minerals formed during the firing process. This irreversible *brick growth* is typically 1–2 mm per metre and occurs over 10 years or more, though the rate diminishes with time. Allowance is made for this by providing expansion joints within courses. Much smaller, but reversible, strains occur due to thermal fluctuations and the effects of wetting and drying cycles.

16.2   PROPERTIES OF BRICK AND TILE CLAYS

Brick-making clays

The heterogeneity of brick clay physical properties is illustrated in Table 16.1, which lists typical parameters for a number of Australian and

Table 16.1   Physical properties of some brick clays[a]

	Dry density (t/m3)	In situ moisture (%)	Plasticity index, PI	Quartz content (%)	Clay mineral (%)
Australian clays					
Bringelly Shale	2.6	1–2	3–17	25–55	45–65
Ashfield Shale	2.5	1–2	1–7	20–45	40–60
Hindmarsh Clay	1.9	30	55–80	30	65
Clay tailings	1.1	60	–	30	70
Lateritic clay	1.5	5	10	35	65
British clays					
Oxford Clay	1.8–2.2	15–25	28–50	–	30–70
London Clay	1.7–2.1	19–28	40–65	45	40–72
Etruria Marl	2.1–2.3	9–22	8–32	–	12–25
Weald Clay	1.6–2.4	5	28–32	–	20–74
Coal Measures Shale	2.2–2.6	8	9–19	–	14–43

[a]Data from various sources.

Figure 16.1   Suggested particle size ranges for brick and tile clays. Note that very clay-rich raw material is usually too shrink-prone to perform satisfactorily, especially where the clay size fraction is mostly clay minerals. (After Bender and Handle, 1982.)

British clays. In consistency these range from soft washery tailings, through overconsolidated silty and sandy clays (London and Hindmarsh Clays), to clay–shales (the Oxford and Weald Clays), and lithified Bringelly Shale and Coal Measures mudrocks. Deeply weathered metasediments are represented by the lateritic clay. Although there is no ideal brick clay, the best materials are likely to have the characteristics summarized below.

- *Particle sizes*   The particle size range should include an even mixture of clay, silt and fine to medium sand. A small percentage of coarser *texture fraction*, up to 10 mm particle size, is included in most clay blends to improve mixing and for aesthetic reasons. This is composed of shale chips and inert 'grog' (ground pre-fired clay). Particles in the size range 50 μm to 1.2 mm constitute the *filler fraction*, whose function is to control shrinkage and lamination, and to provide gas drainage paths during firing. The filler grains are mostly sand-sized quartz and comprise 20–65% of the mix. A high sand content may be used for an attractive texture (hence 'sandstock'), or to balance a very plastic clay, but too much can weaken the brick and abrade moulds. The *plastic fraction*, 35–50% by weight, is made up of clay- and silt-sized particles, not all of which are clay mineral or even sheet silicates. Suggested

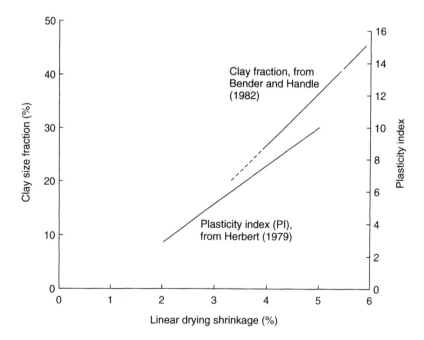

Figure 16.2   Pre-firing drying shrinkage of clay bodies in relation to clay size content and plasticity index of the raw material.

grain size distribution limits for brick and tile materials, based on German experience, are illustrated by the triangular diagram in Figure 16.1.

- *Plasticity* Low to moderate plasticity is needed for moulding and shaping. The *Atterberg plasticity index* (PI) is a rough guide to both clay content and clay mineralogy; however, a low PI may be due either to a non-plastic clay mineral such as kaolinite, or to an excess of fine sand in the raw material. Generally the more plastic a clay is, the more it shrinks when dried or fired, as demonstrated in Figure 16.2. Hence a compromise between adequate plasticity and low shrinkage is needed; for practical purposes a PI in the range 10–20 appears to be most satisfactory.

- *Moisture content* A moderate working moisture content is necessary to facilitate extrusion and pressing. This should be somewhat above

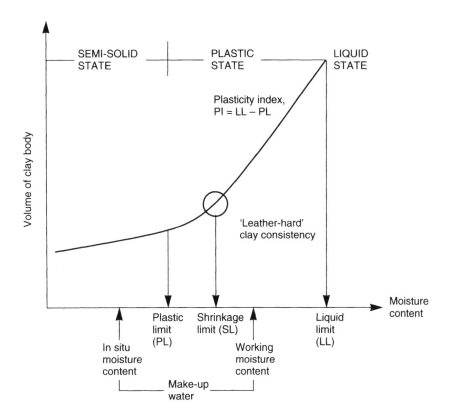

Figure 16.3 Consistency relationships in brick clay raw materials. The ideal firing moisture content is close to the shrinkage limit, at a consistency sometimes referred to as leather hard. Make-up water is that needed to bring the stockpiled clay up to its working (pre-drying) moisture content.

the plastic limit, but well below the liquid limit and close to the shrinkage limit (Figure 16.3), and is typically in the range 15–25%. It is nearly always preferable to wet up a relatively dry shale/clay mixture than to dry back a clay slurry, both to save fuel and to minimize shrinkage.

Nonetheless, the principal market requirements for facing bricks are aesthetic rather than physical. Bricks of adequate strength and durability are fairly easily produced, but public tastes in colouring and surface texture are much more demanding, and vary with both time and place. In Australia a number of 'boutique brickworks' have emerged to cater for the top end of the market, offering decorative products and bricks matching old styles for restoration or renovation.

Tile-making clays

The properties of tile-making clays are similar to those for brick clays, except for the following:

- Tile clays have to be more plastic, but at the same time non-shrinking, to permit thin and complex shapes to be pressed. Hence a higher proportion of clay size (at least 40%) and clay mineral are required. Finer grinding by ball milling and more thorough mixing are also desirable.
- The product must have greater flexural tensile strength. This demands hard firing and more careful selection of ingredients – fibres such as wood pulp can be added to increase tensile strength.
- Low water absorption is essential, and hence a hard vitrified surface is required. Firing temperatures are therefore somewhat higher (1200°C) than for brick-making; this also increases strength.

The fired colour of tiles is less critical than for bricks, since natural terracotta red–browns are still acceptable to customers. Ceramic glazes, which also act as waterproofing agents, are available in a number of colours.

16.3 MINERALOGY

Clay mix and firing

The preferred clay mineral assemblages in brick-making raw materials are those in which disordered (poorly crystallized) kaolinite, illite and mixed-layer illite–smectite predominate. Well-ordered kaolinites tend to be relatively coarse-grained, non-plastic and to have a high melting point; this makes them useful as refractory raw materials, but less suitable for brick-making. Smectites (montmorillonites) shrink excessively and are difficult to use in blends. Illites impart good plasticity but tend to vitrify at low temperatures and, worse, to do so over a narrow temperature range. A good brick-making material might therefore comprise illite/kaolinite/quartz in proportions around 20/25/55. In such a mixture the illite provides plasticity, the kaolinite contributes fired hardness, and the sand-sized quartz grains act as a stabilizer and filler, to resist shrinkage and impart gas permeability.

Other fine-grained sheet silicates, such as the chlorites and micas, are non-plastic but contribute *fluxes* (oxides of Ca, Mg, K, Na and Fe) that lower the fusion temperature. Some fluxes may also extend the *vitrification range*, the temperature interval between the first appearance of glass (at around 1000°C) and the melting point. A narrow vitrification range means that kiln temperature has to be very closely controlled, and this presents practical problems. Too low a temperature, and the product will be weak due to insufficient sintering ('underfiring'); too high, and it will deform as fusion accelerates ('overfiring'). A vitrification range of at least 100°C is desirable, and for most brick clays it is much greater. The *firing range*, on the other hand, is the much narrower kiln operating temperature, which typically fluctuates through only 50°C.

Detrimental minerals

Detrimental minerals can include carbonates, sulphates, sulphides, chlorides and micas in large proportions, plus solid inclusions. Carbonates in moderate proportions, up to about 15% by weight, are beneficial to brick-making; they act as fluxes and lighten fired colours. Larger percentages can be tolerated, provided the grains are evenly disseminated. Eventually, however, a calcareous shale (marl) 'bloats' by giving off copious CO_2 gas. Furthermore, where the carbonate particles are relatively coarse, say 1 mm or more, they may leave CaO residues, which hydrate and swell.

Sulphides oxidize to sulphates in the kiln, and both sulphides and sulphates may generate noxious SO_2 and SO_3 in flue gases. Chlorides can cause scaling on kiln burners, and these and any other soluble salts can be drawn to the finished brick surface as white *efflorescence*. Salts baked into the brick surface are referred to as *kiln scum*; like efflorescence, this has no effect on the physical properties of the brick, but it cannot be washed off. Coarse mica and chlorite flakes may cause *laminations* (weak shear planes) to develop in extruded clay, and hence reduce its green strength.

Fragmented and gritty *inclusions* cause damage to the grinding machinery due to their hardness. They include vein quartz and silcrete, ferricrete and phosphatic nodules. Clays containing a high percentage of quartz sand, particularly coarse grains, act as an abrasive paste and cause rapid wear of extrusion and forming machinery.

Useful ingredients and fired colours

A small proportion of natural *fuel* – coaly, carbonaceous or bituminous matter – is useful within the clay. This will reduce the amount of oil, pulverized coal or gas required to fire the kiln, and may also produce a desirable cream colour. Excess carbonaceous matter can, however, cause 'black cores' in the finished bricks, particularly if the firing has been too rapid.

As mentioned above, moderate amounts of carbonates can be beneficial, as they act as fluxes and lighten fired colours. The fired colours of bricks are due chiefly to oxides of iron, modified by other chemical constituents present, such as CaO and MgO. Ferric oxide (haematite) imparts shades of red and yellow, while Fe_3O_4 (magnetite) yields blue or black tones, and manganese oxides produce purples. Various ratios of Al_2O_3, CaO and Fe_2O_3 give rise to shades of buff, white and pink. Colour is also affected by the kiln atmosphere, since cutting back on the circulating oxygen creates reducing conditions and dark tones (Figure 16.4). Naturally red-burning clays are more abundant in Australia than lighter-firing materials, but market demand has favoured the latter since the 1960s. Brick-makers have met this change in customer taste by blending, adding powdered limestone (stone dust) and altering kiln settings.

Figure 16.4 Light and dark tones obtained from the same raw materials by changes in kiln ventilation. The actual colours are buff to light brown (left, oxidizing atmosphere) and purple-brown (right, reducing atmosphere).

16.4 CLAY PREPARATION, FORMING AND DRYING

The principal manufacturing steps in brick-making comprise: clay mining (or 'winning'); preparatory processing, blending and moistening; forming the clay 'body' (brick, tile or pipe); drying; and firing. These processes vary slightly with the raw materials and the age of the plant, but are summarized in Figure 16.5.

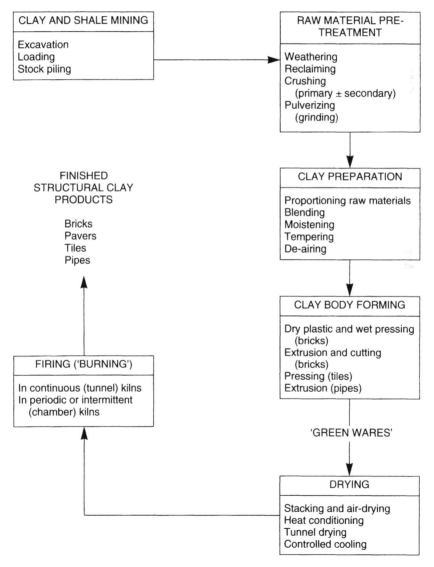

Figure 16.5 Processes in the manufacture of heavy clay products. See text for further explanation.

Mining

Clay and shale mining is a straightforward earth-moving operation. The stronger shales may require blast loosening (not fragmentation), but most can be dug by a front-end loader or a face shovel. More rarely, in very large pits, a bucketwheel or ladder excavator (shale planer) may be used, with the aim of extracting clay evenly from all layers in the sequence. Alternatively, horizontal layers can be peeled off by bulldozer ripping and push-loading scrapers. This allows different seams within the shale formation to be preferentially blended later, or discarded.

At some pits digging is performed only in the drier months, with material stockpiled – sometimes covered against rain saturation – between seasons. This promotes shale breakdown by slaking over several months, plus better mixing of material from different parts of the pit. It also allows moisture to infuse more evenly through the shale and provides time for quality control testing of the raw material, well before it is used.

Processing

Comminution (breakage) of harder shale lumps is carried out using small jaw or rolls crushers, which reduce maximum particle dimensions from about 300 mm to 50 mm. Secondary crushing is by grinders, ball mills or hammer mills, reducing to a maximum size of 10 mm. Oversize is scalped off by screening. Excessively damp material at this stage can jam crusher outlets or block off screen apertures. Ground-up material is stored in surge bins, then later discharged in weighed proportions onto a conveyor belt. At this point the clay mixture is *blended* by the rotating paddles of a pugmill, and chemical additives (colorants, plasticizers and binders) are mixed in, along with make-up water as required.

Forming

Bricks are formed or shaped by one of three processes:

- *Extrusion and wire-cutting*   A square-ended column of wet clay is squeezed through a die and then wire-cut. Usually these bricks are perforated to accelerate drying, firing and cooling time, and to minimize thermal gradients within the body. This also makes them lighter and conserves clay. The consistency of the mix can be damp to wet (18–25% moisture content is typical), but it is usually extruded as dry as possible so that the green product can be handled without delay and to save fuel.
- *Stiff-plastic pressing*   Bricks are formed at somewhat lower moisture contents, typically 14–17%, in box-like moulds. Each clay body is

pressed twice, initially to give it the rough brick shape and subsequently to impart sharp corners and a central depression or 'frog'.

- *Dry pressing*   In this process, bricks are actually damp pressed, at moisture contents around 10%. This process has been largely superseded by extrusion, but it does eliminate the drying stage and associated shrinkage.

In terms of bricks produced, extrusion is the dominant process since it is best suited to firing in modern tunnel kilns.

Drying

The drying process makes use of waste heat from the kiln, usually at temperatures around 110°C. Product wastage due to distortion, warping and cracking can be severe at this stage if drying is carried out too rapidly, or with too high a temperature gradient between the centre and surface of the clay body. Typical drying periods are between 18 and 30 hours.

16.5   CLAY FIRING

The firing process

When a clay body is fired or 'burned' it successively loses porewater, chemically bound (lattice) water, organic matter, sulphur oxides and carbon dioxide. It undergoes partial fusion, becomes harder and less permeable, changes colour and develops some new minerals. The idealized firing behaviour of pure well-crystallized kaolinite and illite under differential thermal analysis (DTA) is illustrated in Figure 16.6. Real heavy clay blends contain a proportion of fluxing oxides, whose effect is to displace exothermic peaks to the left and otherwise to complicate the DTA signature. The main features to be noted in Figure 16.6 are the following:

- An endothermic peak around 150°C in both clays, due to the vaporization of *adsorbed* surface water from pores. This is larger in the case of illite because it is the more plastic, with a higher working moisture content.
- A second peak between 550 and 650°C, coinciding with the maximum loss of *bound* (lattice) water. This is followed by the formation of amorphous metakaolin from kaolinite, but only lattice distortion without mineralogical change in the case of illite.
- An exothermic peak about 950°C in which mullite ($Al_6Si_3O_{15}$) begins to crystallize and the earliest glass phase forms. This is the start of the

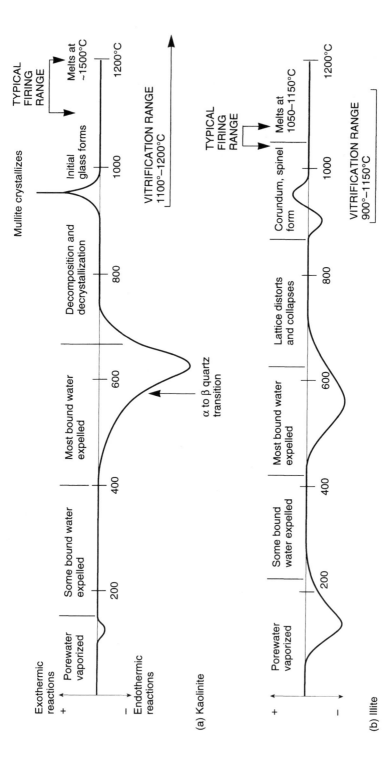

Figure 16.6 Idealized firing behaviour of pure clay minerals; the actual shale and clay blends used in brick manufacture depart significantly from these models.

vitrification range, which continues up to the melting point. The proportion of crystalline mullite is believed to be an important factor controlling finished product strength.

Vitrification has much in common with thermal metamorphism, and brick mineralogy includes species such as wollastonite, spinel and gehlenite, in addition to mullite. It is more prolonged in the case of kaolinite, which finally melts at around 1500°C. Illite melts between about 1050 and 1150°C, lower still if overfluxed. Hence kiln temperatures are commonly kept in the range 1040–1080°C with illitic clays, to avoid overfiring and distortion of the product.

A typical kiln heating, firing and cooling sequence for a brick clay mix is illustrated in Figure 16.7. The total firing duration may vary from 30 to 150 hours, but about five days (120 hours) appears to be most common. The kiln temperature is initially raised at about 100°C per hour, held at approximately two-thirds of peak temperature for 6–24 hours (the *oxidation hold* or 'soaking period'), then slowly raised to a maximum, commonly in the range 1050–1150°C. These temperatures ensure that enough mullite and glass phase is generated to impart strength to the brick. Tiles and paving bricks, on the other hand, require 'hard firing' at temperatures in excess of 1200°C to meet their higher required strengths.

The two most common types of brick ovens used in Australia are the annular or Hoffman kiln and the tunnel kiln. In the older *Hoffman kiln*

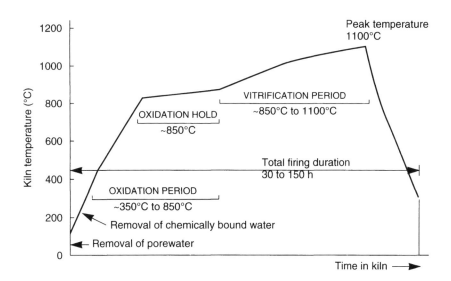

Figure 16.7   A typical firing schedule for a red-burning kaolinitic clay in a brick kiln. Note the gradual and stepped increases in temperature, followed by rapid cooling. See text for an explanation of the terms used. (After Brownell, 1976.)

batches of formed bricks are sealed up in chambers around the periphery of a ring-shaped building. The seat of the fire is progressively moved around beneath these batches, such that they are first dried by exhaust gases and then gradually fired, in a cycle extending over several days. In a modern *tunnel kiln* it is the pallet-loaded bricks rather than the fire that moves, but the principle (i.e. progressively raising the temperature and then holding it constant for an extended period, before gradually lowering it again) is similar.

Product flaws

Product flaws that can develop during manufacture of structural clay wares include the following:

- *Underfiring* occurs particularly where the brick-maker is trying to avoid deforming a narrow-vitrification-range material. The result is a weak product, which may nonetheless be acceptable in non-loadbearing situations. A similar result occurs if the sand content of the mix is too high, but here the aim may be to achieve a pleasing surface texture.
- *Overfiring* is indicated by product distortion and increased water absorption on cooling. Extensive development of gas bubbles is another manifestation of overfiring and is known as *bloating*. This can be turned to advantage in the manufacture of lightweight expanded aggregate. The source of the bubbles is CO_2 given off by carbonates, a high percentage of which is therefore desirable in this application.
- *Limebursting* or lime popping shows up as pits or small craters in the brick surface and is caused by hydration of quick-lime residues from coarse carbonate particles. This raw material problem can be treated in a number of ways, including finer grinding, addition of salt and extended firing.
- *Black cores* (or reduction cores) are due to incomplete burning of carbonaceous matter or prolonged reducing conditions in the centre of the brick. This is often accompanied by bloating and can cause the product to be understrength. The remedy is a longer oxidation hold or a more permeable mix, which allows gas to dissipate faster.
- *Efflorescence* is the result of soluble salts (mainly sulphates) leaching from the interior of the body. The origin of these salts may be gypsum present in the raw material, or reaction products from the fuel or clay. Sulphate efflorescence can be controlled using barium carbonate as a fixer.

16.6 GEOLOGY OF AUSTRALIAN BRICK CLAYS

Because brick clay is a low-value resource with no fixed physical properties, a wide variety of clay-rich (but not necessarily clay-dominant) rocks and soils have been successfully used in its manufacture. In Australia these have included Cainozoic stiff clays and clay–shales, lithified Triassic and Permian shales, deeply weathered metapelites and industrial wastes. Normally a blend of two or three materials is used, including a ground shale plus a plastic clay. Older brickworks may draw most of their materials from distant pits, while modern plants seek to draw the components from different seams in large adjacent excavations.

Alluvial and lacustrine clays

The alluvial and lacustrine (fluviolacustrine) clays are exemplified by the Hindmarsh Clay of Adelaide and clays from other small Tertiary and Pleistocene basins in the vicinity of Sydney, Adelaide, Brisbane and Melbourne. These are stiff to very stiff (but uncemented) fissured clays or 'clay–shales', with *in situ* moisture contents in the range 20–30%. They are mainly reworked lateritic clays from ancient weathering profiles, but often contain a substantial proportion of sand. They may also include lignitic clay seams, though these do not appear to contribute much fuel value. The better varieties are suitable as refractory clays, or for blending with inferior shales and metapelites to raise their plasticity or to improve their cream-burning potential. The main problems with these young clays are their sometimes high proportion of montmorillonite, hence high plasticity. They also tend to be inhomogeneous, particularly in sand content, and may contain unwelcome inclusions such as gravel and ironstone nodules.

Lithified shales

Lithified shales are those with the geomechanical properties of weak rock rather than of dense soils, and represent a further step of induration beyond the clay–shales. The processes of lithification compress and cement clay platelets into compact aggregates, which behave like silt particles rather than clay minerals even when the rock is finely ground; consequently this material has a low plasticity and low shrinkage. Unweathered shales of this type have very low *in situ* moisture content (1–3%), high bulk density (2.5–2.6 t/m^3) and rock-like compressive strengths (10–60 MPa).

They are typified by the Triassic-age Ashfield and Bringelly Shales of the Sydney Basin, whose ceramic potential has been comprehensively described by Herbert (1979). These two formations comprise the most important brick clay resource in Australia, generating one-third of the national output. About 2 Mt of clay is produced annually from them, with the Bringelly Shale proportion about 70% and rising. Both are dark grey, brown-weathering mudrock formations composed of siltstone, claystone and laminite seams. Lithic sandstone occurs in partings, bands and in a few thicker beds.

The 60 m thick Ashfield Shale is the lower of the two and was the main source of structural clay in the Sydney region up to the 1960s, but production from these pits is now in decline due to urban encroachment (Figure 16.8) and obsolete equipment. The overlying Bringelly Shale is thicker, about 250 m, and outcrops in the central western portion of the Sydney Basin, where house-building and hence demand for bricks is greatest. The brickworks here are larger and more modern, built around tunnel kilns.

The two formations are mineralogically similar, with 45–60% clay minerals, 25–45% quartz and the remainder (up to 16%) mostly siderite. The siderite occurs as nodules up to 100 mm in the Bringelly Shale, but

Figure 16.8   Brickworks (lower right) and Ashfield Shale pit, western Sydney. The pit is about 300 m square and has since been backfilled with garbage and recycled as a public park. Note the suburban encroachment on two sides, though the brickworks managed to operate for several decades after these houses were built. (Photo: P. Dupen.)

is more abundant, though disseminated, in the Ashfield Shale. Phosphatic nodules also occur, though apparently not in sufficient quantity to affect brick-making. In both formations kaolinite is the dominant clay mineral, averaging 60% in the Ashfield Shale and 55% in the Bringelly Shale, which is somewhat the more plastic as a result. The remainder of the clay content is a mixture of fine illite and interstratified illite–montmorillonite.

In the Bringelly Shale brick-pits the main concerns are making best use of the more plastic and pale-burning claystones, particularly the slightly carbonaceous varieties, and avoiding sandstone bands, quartz-rich laminites and concentrations of siderite. The Ashfield Shale presents fewer opportunities for in-pit blending, so plastic clays must be imported. This is in any case inevitable, since the few remaining Ashfield Shale pits have little remaining accessible reserves. The higher plasticity of the Bringelly Shale is better suited to the continuous extrusion processes now in use than the older dry press method formerly dominant in Ashfield Shale brickworks.

Deeply weathered shales

Brick clay deposits located within deeply weathered shales and low-grade metamorphic rocks are widespread in Australia, as a consequence of several cycles of long-duration weathering extending well back into the Tertiary and even the Mesozoic. Intensely weathered shales, metasiltstones, phyllites, argillites and schists of Proterozoic to Triassic age provide the bulk of the brick-making materials in the Brisbane, Melbourne and Adelaide metropolitan areas. Kaolinite is usually the dominant clay mineral, but hydrous micas and muscovite are also abundant. Plasticity is improved by blending with Tertiary fluviolacustrine or residual clays.

Industrial wastes

The use of industrial wastes, mainly clay tailings from sand washing plants, as brick-making materials offers possibilities for consuming what are at present environmental nuisances, while at the same time relieving pressure on natural resources and landscapes. Other materials in this category include pulverized fly ash (PFA) from power stations, and coarse and fine coal washery rejects (as raw material and fuel respectively). Small amounts of these materials are used at present, but their wider application is limited by distance to brick plants, problems of dewatering tailings with around 60% moisture content, and some undesirable constituents (such as siderite nodules and disseminated pyrite in colliery wastes). The main drawback from a commercial point of view is, however, their heterogeneity. Where naturally occurring homogeneous, low-moisture clays are available and cheap, these will be preferred by the brick-makers.

16.7 INVESTIGATION AND SAMPLING

The main purpose of geological investigation and sampling in brick-making materials is to ensure that seams with superior fired colours or plasticity are identified and can later be selectively mined. Conversely, seams with high carbonate or sand content, for example, may have to be carefully blended or discarded. In proving-up reserves and subsequent pit layout design, the chief requirement is that a consistent product can be achieved without too much selective mining. The present tendency is to develop large brick-making facilities as close to markets as environmental pressures will allow, and to ensure their long-term viability by proving up 20 years or more of reserves prior to construction.

The principal sampling technique in shale is rotary diamond-bit coring, using triple tube barrels of the largest available inside diameter. Weaker clay–shales can be cored using tungsten carbide (TC) or sawtooth bits up to 100 mm in diameter. The sample quality achieved is somewhat inferior to diamond drilling, but the size is almost three times greater – 20 kg per metre compared with about 7 kg – and TC coring is both faster and cheaper. In soft clays, push-tube sampling or reverse-circulation coring are required. Large bulk samples from weathered or weak shales can be obtained down to depths of 20–30 m using truck-mounted bucket augers or cable tool (shell-and-auger) rigs with diameters around 600 mm. These bulk samples are highly disturbed and typically 100 kg or more in weight, and must be mixed and quartered to yield more manageable, but still representative, test specimens. A typical sampling grid has drillholes laid out in a square pattern on 50–100 m centres.

The principal features to be recorded when logging drillcore from shale deposits proposed for brick-making include the following:

- Detailed ply *lithology*, especially variations in sand and clay content (estimated using penknife scratch and thread rolling tests, for example). Distinctions should be drawn between the different fine-grained sedimentary lithologies, including claystone, siltstone, mudstone, shale, clay–shale and laminite.
- Apparent differences in *mineralogy*, indicated by colour, slaking or shrink–swell behaviour, and in degree of weathering or alteration.
- The presence of *hard bands* and non-clay beds (mainly sandstone or limestone), which might have to be mined separately.
- The presence of *deleterious minerals* such as gravel clasts; carbonate,

ironstone, silcrete and phosphate nodules; fossil fragments and calcite veins; disseminated sulphides; and gypsum lenses and veins.
- The presence of *fuels* such as coaly and carbonaceous bands, lignite and bituminous matter.

Some of these characteristics, such as the presence of siderite nodules, shrinking and swelling clays and disseminated pyrite along bedding, may only become apparent after the core has been exposed to the atmosphere for some time. Sampling can be carried out by splitting the core longitudinally to obtain both sublayer (ply) specimens and seam composites. The ply samples may be subjected to only simple index testing, but composites – representing potential working sections of the seams to be mined – should be run through a range of ceramic firing trials.

16.8   TESTING

Routine testing of brick clays is largely confined to firing small biscuit-like samples (briquettes) at varying moisture contents and temperatures, typically at 50°C intervals between 950 and 1200°C. Shrinkage, both oven-dried and fired, is measured and fired colour noted, with tensile strength assessed by simple flexural breakage (modulus of rupture) tests. Samples are prepared from drillcore by crushing and grinding to pass a 425 μm sieve. A number of simple tests can then be used to assess grain size distribution semi-quantitatively in the fine sand/silt/clay range and clay mineralogy. These include the following:

- *Linear drying shrinkage* (LDS) in which a moistened sample is oven dried for 24 h at 45°C and the percentage contraction of the clay is recorded. This differs from the linear shrinkage test used in soil mechanics, where the specimen is a slurry and is dried at 105°C. It also differs from the linear fired shrinkage (LFS), although the two are sometimes lumped together as total shrinkage.
- *Plastic limit* (PL), *liquid limit* (LL) and *plasticity index* (PI) are likewise indications of both clay mineralogy and clay content. LL is often plotted against PI, or PI is divided by clay content to estimate 'activity' (hence potential shrinkage) of the soil sample. PI can also be multiplied by the clay content ($-2\,\mu m$ fraction), since shrinkage is partly due to clay mineralogy and partly due to clay size percentage.
- *Moisture absorption* (MA) under standard conditions is related to specific surface area of the clay mineral platelets and hence to their composition. A pure kaolinite has a total surface area (TSA) of $100–200\,m^2/g$, while pure smectites are in the range $600–800\,m^2/g$; clay/silt/sand mixtures will have lower TSA values.

More elaborate – and therefore less frequently performed – tests are concerned with clay mineral determination, chemical composition and grain size distribution. Clay and accessory mineral composition can be determined using X-ray diffraction (XRD) and differential thermal analysis (DTA) techniques, while chemical analyses are expressed in terms of weight percentage of oxides present (with SiO_2, Al_2O_3, Fe_2O_3 and CaO being the main constituents). Particle size distribution is determined by hydrometer methods after drying, crushing and dispersing the 'clay' sample (which may contain mostly fine sand and silt). The coarser grains can be examined under a binocular microscope.

REFERENCES AND FURTHER READING

Bell, F.G. (1992) An investigation of a site in coal measures for brick-making materials. *Engineering Geology*, **32**, 39–52.

Bender, W. and Handle, F. (eds) (1982) *Brick and Tile Manual*. Bauer, Wiesbaden.

Brick Development Association (1974) *Bricks: Their Properties and Use.* Construction Press, Lancaster.

Brownell, W.E. (1976) *Structural Clay Products*. Springer, Berlin.

Clews, F.H. (1969) *Heavy Clay Technology*, 2nd edn. Academic Press, London.

Gillott, J.E. (1968) *Clay in Engineering Geology*. Elsevier, London.

Herbert, C. (1979) *The Geology and Resource Potential of the Wianamatta Group*. Geological Survey of New South Wales, Bulletin 25.

Keeling, P.S. (1963) *The Geology and Mineralogy of Brick Clays*. Brick Development Association, UK.

Prentice, J.E. (1990) Structural clay products, in *Geology of Construction Materials*, Ch. 5. Chapman and Hall, London.

Worrall, W.E. (1968) *Clays and Ceramic Raw Materials*, 2nd edn. Elsevier, London.

CHAPTER 17

Waste, by-product and synthetic materials

This group of materials includes a variety of rock-like and granular solids that are largely wastes, though some – notably blast furnace slags – are nowadays regarded more as industrial by-products. Their uses, in order of increasing value, include: as fill; as brick clay; as road sub-base and basecourse; as aggregate and fillers for asphalt and concrete; as cementitious raw materials, extenders and stabilizing agents; and in lightweight aggregate, insulation materials and non-polishing roadstone. The *in situ* value of most of these materials is low and sometimes negative, meaning that the user is paid to remove them. The chief sorts of waste and by-product materials with present-day or future construction potential comprise:

- *Slag*, mainly that derived from iron- and steel-making, but not excluding wastes from other metallurgical refining.
- *Boiler ash* from coal-fired power generation and incinerator residues, including furnace bottom ash (FBA) and pulverized fly ash (PFA).
- *Coal washery rejects*, both coarse shale fragments and fine tailings; burnt shale from waste rock spoil-piles can also be included.
- *Tailings*, fine sand and silt from alluvial gravel washing and mineral processing.
- *Demolition wastes*, mainly bricks, concrete and asphalt.

Typical grading curves for these materials are illustrated in Figure 17.1. Market demand for these materials varies greatly: hardly any slag goes to waste in Australia these days, and recycled demolition waste is displacing mineral (natural rock) aggregates in many low-value applications. On the other hand, only a tiny proportion of coal-mining wastes and tailings are used as construction materials. Many potential uses of wastes – such as ceramic clays from tailings – are presently at the experimental rather than commercial stage. Other applications, such as fly ash filler in concrete, are well established but consume only a small proportion of the available waste.

An excellent survey of waste and by-product utilization in

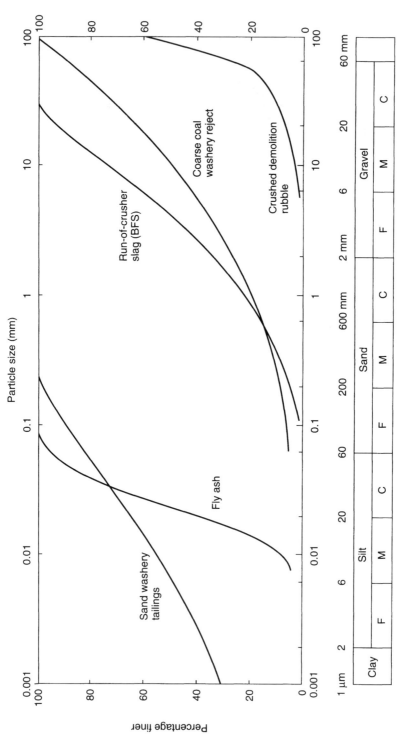

Figure 17.1 Typical waste material grading curves, fining towards the left. The steep curves indicate materials (like fly ash) with a narrow range of particle sizes (i.e. they are uniformly graded), while sloping curves indicate an even mixture of sizes. The coarse demolition rubble at the right is an open-graded material produced by single-stage crushing.

construction is provided by the OECD (1977), though slag and fly ash are reviewed in greater detail in Hotaling (1982), and colliery wastes in Rainbow (1987). The use of recycled construction materials in Australia is reviewed in Institution of Engineers Australia (1996).

There are powerful environmental arguments in favour of using these materials, despite difficulties in specifying, testing and handling them:

- Their use slows the depletion of natural aggregate resources, prolonging the life of existing quarries and postponing the opening of new ones.
- Many of these materials in their natural state – as waste dumps or tailings lagoons – are visually unattractive or impediments to development. Their use as construction materials can thus help to rehabilitate land previously derelict.
- Their utilization may reduce energy consumption in two ways: by minimizing inward haulage of natural materials from distant quarries; and by making use of energy 'invested' in their production (such as substituting slag for fired clinker in cement manufacture).

Conventional materials specifications have in the past ignored regional differences in the availability of natural construction materials, environmental priorities in resource use, and even the beneficial properties of some alternative materials – such as self-cementing in slag and rounded PFA particles. Many aggregates were 'overspecified', meaning that high-quality stone was stipulated where lower grades would suffice. For example, the same standard of roadbase is not required in residential streets as in freeways, nor is high-strength aggregate needed for mass concrete.

While the environmental advantages of using waste materials wherever possible for new construction are therefore now obvious, there are some serious drawbacks. These limitations include the following:

- *Suitability for end-use*   Most wastes are suitable only for the lowest-value end-use, generally as embankment material and backfill. It is uneconomic, and indeed wasteful of energy, to haul them far from spoil-piles, since natural fill materials of equal quality are nearly always available close to the construction site.
- *Variability of properties*   Most wastes have highly variable engineering properties, particularly their clay and moisture contents. This is partly due to waste products varying with changes in processing technology and raw materials, and partly a consequence of random tipping practices.
- *Assessment of deposits*   As a consequence of this, very careful sampling and testing are required to assess waste deposits and to specify those suitable for construction use. Furthermore, separation of waste components may be needed before they can be processed or used.

- *Unproven in service*   Because the interpretation of most engineering test results is largely empirical, waste and by-product materials suffer from their lack of a proven record in service. In particular, the long-term durability of these materials will be uncertain, and may lead to their rejection even when other properties appear favourable.
- *Hazardous nature*   Specific hazards are associated with some waste materials. Coal-mining refuse, for example, may catch fire by spontaneous combustion, while other wastes may discharge troublesome or even toxic leachate.
- *Localized occurrence*   Finally, most wastes and by-products occur in localized concentrations, for example adjacent to coal washeries or power stations. In such areas they can easily satisfy local demand with only a small proportion of the plant output, yet it is uneconomic to ship them to areas of construction material deficiency.

17.1   IRON- AND STEEL-MAKING SLAGS

Ferrous slags have been used as construction materials for over a century, making them the best known and most valued of the alternative materials. Current world output of ferrous slags is about 180 Mtpa, of which about 65% is *blast furnace slag* (BFS) and the remainder *steel furnace slag* (SFS). Because slag production is tied to steel output, the amounts generated in developed countries are static or declining. Furthermore, the proportion of slag to iron product is decreasing with improved blast furnace technology. At present about 0.25 t of slag is produced per tonne of pig iron, down from 0.75 t over the past few decades. On the other hand, now that slag is a saleable material rather than a waste, it is produced to a consistent quality, and may be further processed to obtain even more valuable end-products. Hence if all blast furnace slags were produced in vitrified form (granulated or pelletized), and thus suitable for grinding as a cement substitute, about 15% of world cement requirements could be met – but there would be none left for aggregate or roadbase!

17.2   BLAST FURNACE SLAG (BFS)

Blast furnace slag consists mainly of silica (SiO_2) and alumina Al_2O_3, emerging as a liquid at about 1500°C. Depending on the method of cooling, several products can result: air-cooled, granulated, expanded and pelletized BFS.

Air-cooled BFS

Air-cooled BFS is a synthetic igneous rock with a cellular or vesicular texture, similar to scoria. It is manufactured in three forms: as coarse uncrushed angular rubble up to 300 mm diameter ('rock slag', see Figure 17.2); –100 mm unscreened run-of-crusher material; and as crushed and screened dense-graded roadbase (DGB). As an unbound roadbase, the most attractive properties of BFS are the ease with which it can be broken down (by on-road grid or vibratory rolling), its self-cementing action and its good bonding with bituminous seals. Nevertheless, BFS has been more successful as bound basecourse than in meeting specification requirements for dense-graded but unbound base, owing to the lack of fines generated by crushing.

As a concrete aggregate, BFS offers excellent bonding with the cement matrix, reduced heat of hydration and high long-term strength. Although not a true lightweight aggregate, since its particle SG is only about 20% lower than mineral aggregates, BFS is attractive in some structural concretes because of the reduced density and high tensile strength that it imparts. Such concrete also tends to have low permeability and hence high resistance to chloride-ion penetration and sulphate attack in marine environments.

Granulated BFS

Granulated BFS is produced by rapid cooling or quenching of molten slag to form a coarse sand-sized mixture of glassy particles. It was

Figure 17.2 Stockpiled rock slag – coarse 'skulls' of cellular blast furnace slag about 0.3 m average size, generally used for open-graded macadam roadbase. A small proportion of fines is generated by on-road rolling, but the strength of these pavements is mainly due to interlocking blocks and self-cementation.

formerly used largely as a sand substitute (particularly in the coarse size range) in drainage layers and asphaltic concrete, and as a pipe backfill. As a filter material it is superior to quartz sand because of its microporosity; its particle angularity improves the skid resistance of asphaltic wearing courses; and its alkalinity retards pipe corrosion when used as a bedding material.

However, the most promising use of granulated slag is as a partial substitute for clinker in cement manufacture, a 'cement extender' in other words. For this purpose it must first be finely milled, becoming *ground granulated slag* (GGS). GGS is a pozzolan, a substance exhibiting cementitious properties in the presence of moisture and a lime activator. The chemical properties of slags and other cementitious materials are compared in Figure 17.3. Note the chemical similarities between slag and cement, and that between natural pozzolans and fly ash; cementitious properties seem to increase with the CaO/SiO_2 ratio.

Where GGS is used as a cement extender, it is mixed with Portland

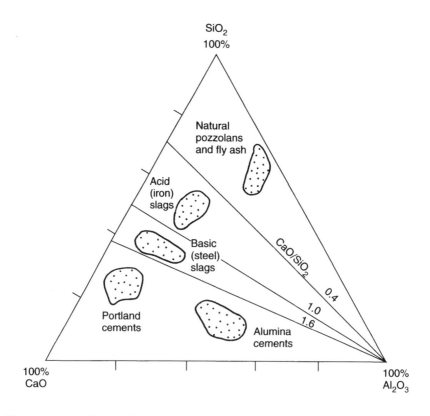

Figure 17.3   Chemical compositions of slags and other cementitious materials. (After OECD, 1977.)

cement and fly ash, in blends such as 'slagment'. Blended cement can include up to 85% ground slag, although 30–70% is more usual. It may also be used as a cement raw material, mixed in before firing, rather than later interground with the fused clinker. Blended cement concretes are not only cheaper: they are more sulphate-resistant, have good workability and may develop high strength. The pozzolanic properties of granulated slag are the basis of gravel–slag and sand–slag road pavement materials (see below and Figure 17.4).

Expanded BFS

Expanded BFS is produced by controlled air and water cooling, and is a true lightweight ($0.8\,t/m^3$) aggregate. The particles are either angular or rounded, depending on the process and whether or not they are crushed. 'Foamed slag' is one variety, produced by pouring molten BFS into shallow steel trays. The polishing resistance of crushed foamed slag is claimed to be superior to other types of BFS, owing to its vesicular fabric and hardness.

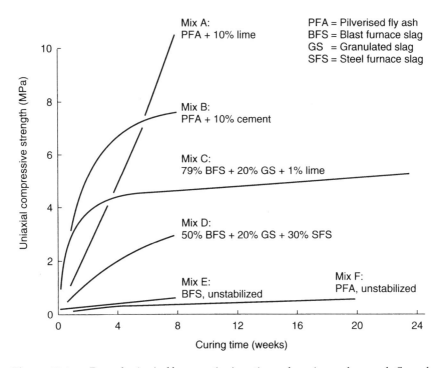

Figure 17.4 Pozzolanic (self-cementing) action of various slag and fly ash mixtures. The best pavement material is mix A, which offers rapid strength gain and adequate long-term strength, though D would be adequate. (Data from OECD, 1977, and Heaton and Bullen, 1982.)

Pelletized BFS

Pelletized slag is another variety of expanded BFS, made by cooling slag on a rotating drum and then flinging the semi-plastic fragments through the air. The aim is to produce rounded particles in which the cellular structure caused by expanding gases is closed off by a glassy skin. This reduces water demand in concrete, while improving workability and reducing thermal conductivity, although the original purpose was simply to seal in hydrogen sulphide gas and thereby meet air-quality regulations.

17.3   STEEL FURNACE SLAG (SFS)

The slag produced from steel-making operations is quite different from BFS in being denser, harder, less vesicular, and consequently having superior wear and polishing characteristics in pavement surfaces. However, SFS is also more variable in composition and properties, owing to different steel-making processes (the chief ones being basic oxygen (BOS) and electric arc furnace (EAF)), and the fact that it is produced in batches rather than continuously cast. Chemically, it is much lower in alumina and silica content, and richer in lime (some of which occurs as free lime, CaO). Some SFS contains so much free lime and iron that it is recycled as part of the blast furnace charge. Particle specific gravity is high, 3.25–3.50, and bulk densities of compacted BFS are up to $2.8 \, t/m^3$.

The main problems with SFS are its inconsistent properties, its tendency to swell on wetting and its high bitumen absorption. The swelling is caused by the hydration of free CaO; this process is both time-dependent and varies with particle size. Swelling can be reduced by acid treatment or weathering for several months, but it makes SFS generally unsuitable as a concrete aggregate. Its high bitumen absorption is a cost penalty in asphaltic mixes, although the rough microtexture of SFS chips enhances their skid resistance and the lime content inhibits stripping. The high SG relative to lighter mineral aggregates increases haulage costs for SFS roadbase, although this is less of a penalty with high-value skid-resistant wearing-course asphalt and a positive advantage in railway ballast due to increased pulldown forces on sleepers.

One promising outlet for steel slag is as a component of *gravel–slag* ('grave–laitier') bound roadbase. These mixtures balance the low particle SG of BFS with denser SFS and make use of its lime content as an activator for the GGS binder. They also provide a market for three separate steelworks by-products. Strength results for a typical blend (50% BFS, 30% SFS and 20% granulated slag) are included as mix D in

Figure 17.4. In some cheaper but lower-strength roadbase blends, most of the BFS is replaced by marginal-quality gravels and sand.

17.4 BOILER ASH

Burning pulverized coal in power station boilers at around 1500°C produces dust-sized fly ash and gritty bottom ash (Figure 17.5). Fly ash, which makes up about 80% of the combustion residues, is carried upwards with the flue gases and is trapped by cyclones or in electrostatic precipitators. It is transported to storage either pneumatically (in the dry state), or hydraulically (as a slurry) into ash dams and evaporation lagoons. It can also be 'conditioned' by the addition of a small amount of water, causing aggregation of the particles. In this state it can be moved to storage dumps by conveyor. The remaining unburnt mineral matter falls to the base of the furnace, where it sinters to form a glassy sand and fine gravel-sized bottom ash. Bottom ash can be cooled by air or water, and hence is described as 'dry bottom' or 'wet bottom' ash (or boiler slag).

Pulverized fly (or fuel) ash

Pulverized fly (or fuel) ash, generally known as PFA, ranges in particle diameter from about 1 μm (0.001 mm) to 0.2 mm – that is, from clay to fine sand – but is predominantly silt-sized (2–60 μm, 0.002–0.060 mm). This puts it in the same size range as GGS, to which it has similar, though weaker, cementitious properties. Chemically, it is much higher in silica and lower in lime, and close to the composition of natural pozzolans (Figure 17.3). Fly ash includes a proportion of spherical and glassy particles with specific gravities in the range 1.9–2.4; its *in situ* bulk density is only about $0.8 \, t/m^3$, indicating a high porosity and loose packing. Its compacted density is $1.1–1.5 \, t/m^3$ at an optimum moisture content of 18–35%, though this density is rarely achievable.

Bottom ash

Bottom ash (or furnace bottom ash), known as FBA, particles range in size from silt and fine sand to fine gravel, and in specific gravity from 2.4 to 2.7. Compacted densities are low, around $1.1 \, t/m^3$, and optimum moisture contents are high (up to 35%) due to the large silt content. Particles tend to be angular in shape and to have a glassy texture, with a high angle of internal friction (around 40°).

Uses of PFA and FBA

Perhaps one-fifth of fly ash production in Australia and the USA is used commercially, mainly in concrete and roadbase mixtures, or as fill. Most

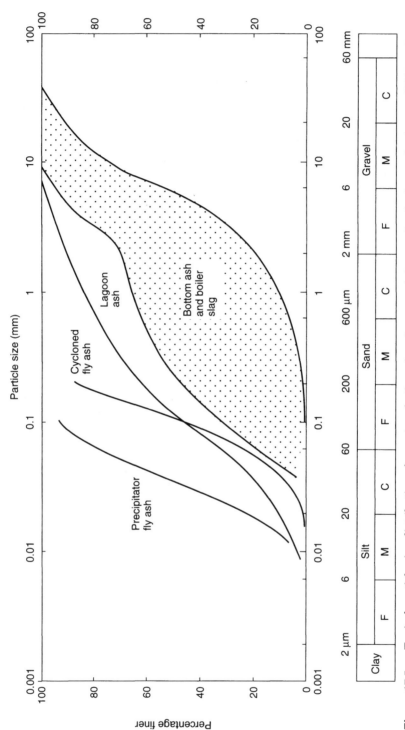

Figure 17.5 Typical particle size distributions for various boiler ashes. Lagoon ash is a randomly dumped mixture of PFA and FBA. Note the much greater grading variability for FBA compared with fly ash. (After OECD, 1977.)

of the remainder is deposited in ash dams near power stations. The marketing challenge with fly ash is finding new high-volume uses, since there is little scope for increased use in concrete.

Fly ash is a versatile material, as emphasized by its applications in *cement- and concrete-making*, where it is used variously as a raw material prior to firing; ground in with clinker as a cement extender; and mixed in wet concrete as a filler and set retarder. It can also improve the workability (and pumpability) of wet concrete, and reduce water segregation ('bleeding'). In addition, PFA reduces heat generation and shrinkage in setting concrete. Replacement of cement by fly ash to the extent of 30% is common, and lean mixes may contain up to 70%. These 'econocretes' are of low strength (10–20 MPa), often using marginal-quality aggregates and high stone/cement ratios (up to 11/1), which are cheap but adequate for applications such as pavement sub-bases.

The low bulk density makes fly ash (and bottom ash) suitable as *lightweight fill* on compressible subgrades such as peat (Figure 17.6). Settlement within these fills is negligible, partly because of their low density and partly through their self-cementing properties. Both the cohesion and friction angle increase with time, a factor that has also encouraged the use of PFA as backfill behind bridge abutments. Compacted fly ash is practically impervious to seepage, but it is very

Figure 17.6 Lightweight bottom ash (FBA) fill on freeway construction site, Minmi, New South Wales. This 12 m high embankment overlies 20 m of soft estuarine mud. Note one hazard of silty fill materials – dust movement on a hot, windy day.

susceptible to capillary rise and hence to frost heave. A free-draining underlayer, or capillary break, is required to prevent this. Ash embankments are prone to wind and water erosion (both on the surface, due to rill action, and internally, by piping), and may exude heavy-metal leachate.

Despite its self-cementing action, fly ash alone is unsuitable as a pavement material because of its fine grain size, uniform grading and moisture sensitivity. Nonetheless it is widely used in combination with lime or cement as a *stabilizing agent*. At normal air temperatures lime–PFA mixtures react more slowly with soil than cement, allowing time for spreading, compacting and shaping. Nonetheless, the long-term strength of lime–PFA stabilized soils may be close to that of cement-modified soil. An extreme example of a stabilized pavement is that of a haul road recently constructed at Eraring, New South Wales, using only PFA mixed with 4–10% cement, similar to mix B in Figure 17.4. The performance of this road section has been satisfactory to date, and if the trial is successful in the longer term it will open up an important market and means of disposal for fly ash.

An increasing amount of furnace ash (PFA and FBA) is being used as *mine backfill* in abandoned shallow mines. This is potentially another large-volume application, since power stations and coal-mines are generally close to each other, so two environmental nuisances can be eliminated at one stroke. The ash is used as dry bulk fill, as fine aggregate in cement grout and as an ingredient in 'rock paste', a viscous mixture of PFA with shale fragments and lime. Dry PFA can be blown down boreholes into underground caverns, a process known as *pneumatic stowing* (Figure 17.7), and subsequently moistened; a small proportion of cement may be added to accelerate hardening.

The uses of FBA are similar to but more limited than those of PFA: as lightweight fill and mine backfill. However, mixtures of bottom ash, fly ash and slag respond well to cement and lime stabilization and show promise as bound roadbase.

17.5 COAL-MINING WASTES

The three main waste products from coal-mining are coarse and fine washery rejects and overburden (Figure 17.8). For every tonne of coal mined in New South Wales, about 10 t of overburden, 0.2 t of coarse reject and 0.05 t of fine reject are produced. Nearly all of this material is presently disposed of in waste dumps.

Coarse refuse

Coarse refuse ('chitter', 'minestone' or coarse washery reject, CWR) is made up of angular and platy sandstone and siltstone rock fragments

Figure 17.7 Pneumatically stowed dry fly ash in shallow abandoned colliery tunnel adjacent to a new highway cutting, Swansea, New South Wales. The void above this was subsequently filled with a wet grout containing 88% fly ash. (Photo: C. Francis.)

ranging up to 75 mm in diameter. These clasts are coated with about 10% silt and clay fines of low plasticity, giving the mixture a dark grey colour and a muddy appearance. Particle specific gravity is 1.75–2.15 and is inversely proportional to the carbonaceous content. The parent rock may be quite strong – sometimes over 100 MPa – but its durability is variable, with shale, mudstone and claystone fragments subject to slaking.

CWR is mainly used for filling at the washery site and as rockfill for tailings dams, but the sounder varieties have been laid as unsealed haul road pavements. Its success here, despite high void ratios and heavy wheel loadings, has been attributed to interlock of the angular particles. It is important to note that such roads are constantly compacted by heavy vehicle traffic and watered, primarily to lay dust but also to maintain the pavement density and to inhibit slaking. They are also frequently reshaped by motor graders and resheeted whenever necessary. Macadam pavements of this type are not necessarily as successful in sealed urban road pavements, where a higher standard of surface finish is needed. Stabilized CWR mixtures have been used as sub-base, though some problems have been experienced due to sulphate ions in the waste.

Coarse reject is also used to a limited extent in the UK as *rock paste*, for filling large underground cavities in abandoned shallow coal and

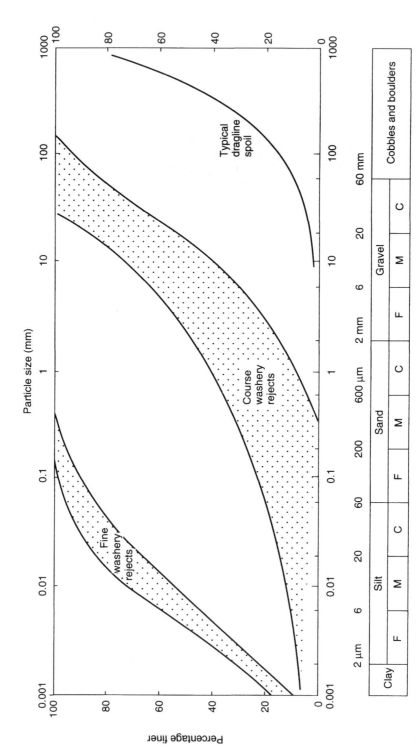

Figure 17.8 Approximate grading envelopes for coarse and fine coal washery rejects compared with particle size distribution for dragline spoil.

limestone mines. Screened (–50 mm) minestone is mixed to a porridge-like consistency with water, fly ash and about 1% lime activator, and poured down large-diameter boreholes into the old workings. Because of its viscosity, rock paste does not spread far, unlike wet PFA. If necessary, the smaller voids and roof spaces can be filled by pressure injection of cement–PFA grout.

Fine refuse

Fine refuse (washery tailings or 'slimes') is largely composed of coal and carbonaceous shale particles in the sand and silt size ranges. It is relatively easily dewatered and could provide a useful fuel for cement-making kilns, if the ash and moisture content could be kept uniform. Dried tailings have also been pelletized and sintered to form lightweight aggregate particles.

Overburden

Overburden is used for opencast mine backfilling, but even as fill it is inferior to CWR in being much coarser and less uniform, with a high proportion of weathered rock. *Burnt shale* is overburden that has been baked by spontaneous combustion to a brick-like consistency, and is therefore more attractive as a roadbase material. It has been used for this purpose in the UK, though supplies of the best quality are dwindling and are not being replaced, because modern spoil-pile dumping practice inhibits spontaneous combustion.

17.6   SAND WASHERY REJECTS

The waste products from sand washing are mixtures of fine sand, silt and clay. Where the fine sand content is dominant, it can sometimes be sold as mortar sand; and where clay of moderate plasticity is the main constituent, the mix may be suitable for brick-making blends. Despite satisfactory firing trials, however, there has been reluctance on the part of Australian brick producers to use this new material, mainly due to its high moisture content. The main present use for sand washery tailings is, therefore, as a reshaping material or subsoil in the restoration of abandoned pits.

Below a surface crust sand washery tailings generally contain only 30–40% of solids, even after years in settling/evaporation lagoons 3–6 m deep. The simplest method of dewatering is simply to allow the tailings to drain for a short time in shallow ponds, then scrape up the wet residues and spread them in layers about 0.5 m thick for air drying. Although this procedure requires rehandling two or three times and a relatively dry climate, it allows the tailings to be used for site rehabilitation within months. Hydrocycloning or electro-osmosis may be required for tailings dewatering in wet climates.

17.7   DEMOLITION WASTES

The demolition wastes of interest as recycled construction materials include building rubble, pavement materials and masonry.

Building rubble is mainly brickwork and concrete, which is fragmented by hydraulic breakers and often crushed on site as well, before being used as fill. The incentives here are economic as much as environmental – the shortage of dumping sites and the cost of haulage in urban areas. Masonry comprises relatively valuable stone blocks and intact bricks, which are salvaged and re-laid rather than being crushed.

The main problems associated with re-use of demolition wastes are the separation of softer inclusions (timber, plastics, glass) from the hard core and the removal of steel reinforcement from concrete. The latter requires two-stage crushing, with magnetic and hand sorting between the primary jaw crusher (Figure 17.9) and secondary cone or impact crusher. Rolls crushers are suitable for breaking up brick panels and unreinforced concrete slabs. Single-stage crushing by either jaws or rolls

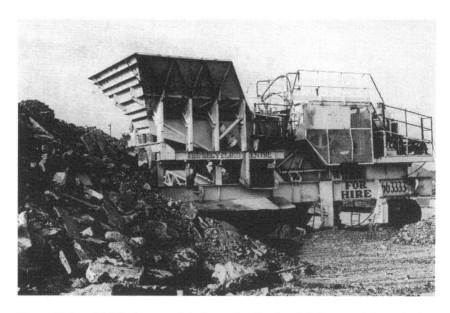

Figure 17.9   Mobile jaw crusher (note the flywheel) being used for reducing demolition rubble. The feedstock (left side) is a mixture of concrete slabs up to 0.63 m across, rock fragments and asphalt pavement chunks. A mobile impact crusher is used for secondary comminution to produce roadbase.

produces a harsh, open-graded rubble; production of dense-graded roadbase requires at least two stages of comminution.

Recycled concrete aggregate (RCA) is gaining acceptance for low-strength concrete, road sub-base and free-draining rubble. Indeed, some European countries require that most demolition rubble be recycled for these materials, while in Australia government specifications have been redrafted to ensure that RCA is not excluded. The quality of RCA depends on how effectively the adhering mortar has been removed, but particle water absorption tends to be high (up to 8%). Compressive strengths of RCA concrete are typically 10–20% lower and shrinkage somewhat greater than concrete made with new aggregate. However, aggregate blends with up to 20% of RCA are claimed to produce quality concrete.

Road pavement materials (asphalt, concrete slabs) are being crushed for use as unbound or stabilized sub-bases, or mixed with cement as lean-mix sub-base for rigid pavements. Wearing-course asphalt layers can be planed up and treated with hot bitumen, bitumen emulsion or bitumen rejuvenator and then relaid as recycled asphalt pavement (RAP). As with on-site crushing of demolition rubble, the use of RAP is motivated as much by cost savings and construction expediency as by environmental considerations. A proportion of polish-resistant stone can be added to the mixture, and the operation carried out in-place using a repaver or off-site through a hot-mix plant.

The recycling of unbound granular pavements by *in situ* stabilization is discussed in Chapter 18. Railway ballast has also been recovered from closed lines and cleaned for re-use, or crushed for road sub-base.

17.8 SYNTHETIC AGGREGATES

These are not wastes or even by-products, but high-value manufactured aggregates with especially low particle density or high polishing resistance not available in natural crushed rock ('mineral aggregates'). The former is used in lightweight structural concrete, in concrete blocks for partition walls, and in fireproof and insulating concrete; the latter finds use in skid-resistant asphalt. The raw materials for synthetic aggregates are principally clay, ground shale, slate wastes and coal washery rejects. Fly ash, bauxitic clay, alumina ('red mud') and waste glass are included in some products.

Lightweight aggregate

The desirable properties for lightweight aggregate (LWA) include the following:

• Particle SG less than 1.20 and sometimes below 0.75. Lightweight concrete has densities in the range 0.3–1.9 t/m³, with structural varieties at the high end and insulating concrete at the low end of the range.

- Adequate strength and toughness, approaching that of mineral aggregates. Note that it is tensile rather than compressive strength that is usually required of lightweight concrete. This is derived from the tenacity of the cement–aggregate bond, rather than from the intrinsic strength of the aggregate particles.
- Satisfactory particle shape (which, where concrete pumpability is important, can mean rounded as well as equidimensional chips). This also facilitates compaction, minimizes cement consumption and eliminates bridged voids.

LWA is manufactured from slag by foaming, or from shale by kiln firing ('bloating'). Where shale is the principal raw material, its desirable properties include a fairly wide pyroplastic temperature range, a significant carbonate mineral content, and the ability to retain gas within a swelling mass. In this way bubble formation can take place without complete vitrification and pore collapse. The firing is usually performed in rotary kilns similar to those used for cement-making. Sintering is an alternative process, carried out at temperatures somewhat above those employed in brick ovens, but lower than in cement kilns. Sintered aggregates tend to be more harsh and porous than fired products.

Ceramic roadstone

Ceramic roadstone is manufactured primarily as a non-polishing, and therefore skid-resistant, aggregate. It is used in wearing-course asphalt laid at high-risk sites, such as roundabouts and near pedestrian crossings. Resistance to particle rounding and surface smoothing is characteristic of rocks that are either finely vesicular or composed of hard minerals set in a softer matrix. Such lithologies, where they are available, are prone to wear rapidly; roads surfaced with these aggregates may have to be resheeted with asphalt every 2–3 years. Hence more expensive, but more consistent and harder-wearing, synthetic aggregates are sometimes used instead.

Ceramic aggregates are denser (SG 1.00–2.00) than LWA, and much harder. Polished stone values (PSVs) of 60–90 have been obtained, where 50–60 is typical of good-quality mineral aggregates. The standard material is calcined Guyana bauxite (RASC bauxite), but a number of other synthetic aggregates are claimed to perform as well.

REFERENCES AND FURTHER READING

Heaton, B.S. and Bullen, F. (1982) Properties of stabilized blast furnace slag roadbase. *Proceedings, 11th Australian Road Research Board Conference*, **11** (3), 168–75.

Hotaling, W. (ed.) (1982) *Extending Aggregate Resources.* ASTM Special Publication No. 774.

Institution of Engineers Australia (1996) *National Symposium on the Use of Recycled Materials in Engineering Construction*, May. Institution of Engineers Australia and Australian Geomechanics Society, Sydney.

Leventhal, A.R. and De Ambrosis, L.P. (1985) Waste disposal in coal mining – A geotechnical analysis. *Engineering Geology*, **22**, 83–96.

OECD (1977) *Use of Waste Materials and By-Products in Road Construction.* OECD Road Research Series. OECD, Paris.

Rainbow, A.K.M. (ed.) (1987) *Reclamation, Treatment and Utilization of Coal Mining Wastes.* Elsevier, Amsterdam.

Sharp, K.D. (ed.) (1993) *Fly Ash for Soil Improvement.* ASCE Geotechnical Special Publication No. 36.

Waller, H.F. (ed.) (1993) *Use of Waste Materials in Hot Mix Asphalt.* ASTM Special Technical Publication 1193.

CHAPTER 18

Stabilization and pavement renovation

The term 'stabilization', as it is used here, means enhancing the engineering properties of a soil by blending in a chemical or granular additive. The principal additives that are used are:

- Gravel, coarse crushed aggregate, grit and loam
- Portland cement and cement–slag blends
- Lime (quick-lime, hydrated lime) and gypsum
- Lime–pozzolan (lime plus fly ash or ground slag) mixtures
- Hot bitumen and cold bitumen emulsion

The first of these processes is referred to as *granular stabilization*, while the others are collectively known as *chemical stabilization*. Stabilization is employed when it is more economical to overcome a deficiency in a readily available material than to bring in one that fully complies with the specification requirements. For the purposes of this chapter, *modified* materials are lightly stabilized, say with less than 3% of chemical additive, while *bound* materials have more cement added and consequently behave in a more rigid fashion.

The improvements in engineering properties caused by stabilization can include the following:

- Increases in soil strength (shearing resistance), stiffness (resistance to deformation) and durability (wear resistance).
- Reductions in swelling potential or dispersivity (tendency to deflocculate) of wet clay soils.
- Other desirable characteristics, such as dustproofing and waterproofing unsealed roads.

The permeability of granular soils is generally decreased by stabilization, but the granulating action of lime makes clay soils pervious (it also dries them out).

Stabilization has had a chequered history in Australia. Like many innovations, its early successes were oversold and its subsequent failures were then magnified. Laboratory results, under closely controlled conditions, were invariably much better than those achieved on construction sites. Stabilization came to be regarded as a last resort for upgrading substandard materials where no economic alternative was available. These early problems were especially notable with cement stabilization, due to rapid setting in warm weather (before compaction could be completed) and longer-term pavement cracking, water entry and internal erosion by subgrade pumping. Lime was more successful, especially in combination with fly ash. However, all methods – both chemical and granular – suffered from poor mixing on the road, using the primitive equipment then available.

In recent years, stabilization has regained favour because of three developments:

- A better appreciation of the cyclic loading effects of heavy traffic has created a need for stronger pavements that often cannot be provided by realistic thicknesses of unbound granular materials.
- Purpose-built equipment is now available for *in situ* stabilization to improve mix homogeneity; an even higher standard of blending can be achieved in quarry batching plants.
- Extensive pavement reconstruction of the Australian rural highway system is becoming necessary as many of these roads approach the end of their 30-year design life.

Although road construction is the principal market for soil stabilization techniques, they have also been applied to a limited extent in soil foundation strengthening. Light buildings with strip footings on swelling clays and water-retaining structures have been the main users. Deep stabilization by lime injection has recently been used to strengthen old and poorly compacted railway embankments and to increase soil bearing capacity beneath piles.

The classic reference book on soil stabilization, with particular reference to Australian conditions, is that by Ingles and Metcalf (1972). Though the photographs in particular are more than a little dated, it covers the fundamentals of all the stabilization methods in common use today. A briefer, but more up-to-date, summary can be found in NAASRA (1986).

18.1 AIMS AND METHOD SELECTION

The aims of stabilization therefore include:

- *Construction expediency*, for example, improving the trafficability of wet, weak or loose material to carry vehicles as soon as possible.
- *Quality enhancement* of a marginal material, for example, sub-base-quality roadbase sufficiently improved in tensile strength to meet basecourse requirements.

- *Subgrade improvement* by drying and stiffening, to reduce the required thickness of expensive pavement courses above.
- *Reducing moisture infiltration* (especially via unsealed road shoulders) to the subgrade, where permeable bases such as fines-deficient crushed stone are used.
- *Increasing pavement stiffness*, to ensure strain compatibility with higher-modulus asphaltic wearing course (hence avoiding 'reflection cracks' propagating upwards from the base).
- *Reducing dispersivity* of sodic clay soils, thus preventing internal erosion of small dams and canal walls – a major cause of failure.
- *Pavement rehabilitation*, or *in situ* recycling of unbound granular pavements approaching the end of their fatigue life.

The selection of an appropriate stabilizing agent and construction procedures involves a number of considerations, including the following:

- The material property to be modified (early or ultimate strength, permeability, moisture sensitivity, dispersivity and so on).
- The nature of the material to be stabilized, mainly in terms of its particle size distribution, liquid limit and plasticity; the possibility of deleterious organic matter or sulphates is also sometimes investigated.
- The site environment (temperature, rainfall, built-up area or rural, sensitivity of surroundings to contamination from stabilizers, and so on).
- Chemical quality of the water available for compaction (salinity, temperature, pH), particularly if bore water has to be used.
- Availability of specialist plant and operators; although modified farm equipment was used in the past, the results were often disappointing, especially where mixing was incomplete.

Table 18.1 summarizes the mechanisms and applicability of various stabilizing agents, and Figure 18.1 illustrates their suitability with respect to the predominant soil particle size and plasticity. In broad terms, cement and bitumen are best suited to granular and non-plastic soils, while lime performs better in cohesive ones. Granular stabilization is much more feasible where coarse material is being added to fine; in practice, adding fines is impossible unless both the materials are quite dry.

From Figure 18.1 it can be seen that cement is not the preferred stabilizer for clay- to coarse silt-sized soils. It is nearly impossible to mix cement evenly through a stiff and impermeable clay soil. In such a case, if the strength required necessitates the use of cement, this can be done by first using lime to granulate the soil by clay particle aggregation into a 'pseudo-sand', which is more amenable to cement admixture. In other words, a sequence of stabilizers may be required to achieve the desired result.

Table 18.1 Mechanisms and applications of stabilization

Mechanism	Effects	Suitable soils
Granular Blending two poorly graded soils, usually coarse into fine (not clayey) soils	Higher compacted density, more uniform mixing, increased shear strength	Gap-graded or gravel-deficient (*gravel, sand* addition), or harsh FCR[a] (*loam* addition)
Cement Mixing small amounts (cement *modification*) or larger proportions (cement *binding*) into soil or FCR[a]	Improves shear strength, reduces moisture sensitivity (modification); greatly increases tensile strength and stiffness (binding)	Most soils, especially granular ones; large amounts of cement needed in clay-rich and poorly graded sands, hence expensive
Lime Mixing hydrated lime or quick-lime in small to moderate amounts into soil	Increases bearing capacity, dries wet soil, improves friability, reduces shrinkage	Cohesive soils, especially wet, high-PI clays
Lime–pozzolan Mixing lime plus fly ash or granulated slag into soil or FCR[a]	Similar to cement, but slower acting and less ultimate strength	As for cement, plus clayey soils that do not react with lime
Bitumen Agglomeration, coating and binding of granular particles	Waterproofs, imparts cohesion and stiffness	Granular, non-cohesive soils in hot climates

[a] FCR = Fine crushed rock roadbase.

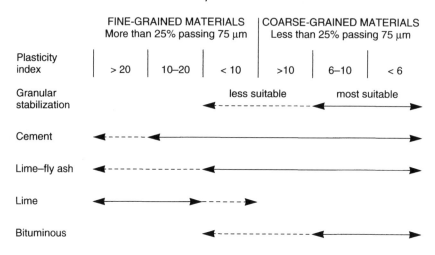

Figure 18.1 Methods of stabilization in relation to their suitability for soils of different grain size and plasticity.

Apart from the particle size distribution (grading) of the soil, its dominant clay mineral may be important to the success of the stabilization programme. This is particularly true of heavy montmorillonitic clays ('black soils'), which can have their shrink–swell characteristics dramatically reduced, owing to substitution of Na^+ by Ca^{2+} (from added lime). Adding common salt (a source of Na^+) to water in earth dams has the reverse effect – it flocculates colloidal particles, causing them to settle and seal off leakage cracks in the dam floor. This is why dams full of moderately saline bore water are so clear, while rivers draining dispersive clay catchments are cloudy. The effects of soil mineralogy on stabilization response are summarized in Table 18.2.

Cement, lime, lime–pozzolan mixtures and bitumen may be considered to form a stabilizing continuum based on setting time, with cement at one end and lime at the other. Some aspects of this continuum are illustrated in Figure 18.2, which shows the typical strength development characteristics of each member. The top curve represents a moderately well-graded gravel bound with 5% cement; this combination exhibits rapid strength gain in the first three days, and a steady increase to 100 days and beyond. The clay stabilized with an equal proportion of lime achieves a more modest strength increase and does so more slowly. The lime–pozzolan mixture, in this case granulated blast furnace slag with a lime activator, is a compromise that offers the slow setting of lime with the higher final strengths possible with cement.

Two considerations are important in construction: adequate early strength to allow traffic onto the stabilized material as soon as possible, and sufficient long-term bearing capacity. These appear to be satisfied by

Table 18.2 Effects of soil mineralogy on stabilization response[a]

Mineral	Recommended stabilizer	Remarks
Crushed gravels and FCR	Sandy or silty loam, crushed shale (must be dry)	Improves grading, workability, increases compacted density
Quartz sands	As above	Improves grading, increases density and imparts plasticity
	Cement	For density, shear strength, impermeability
	Bitumen, bitumen emulsion	For cohesion, water-proofing
Carbonate sands	Lime	Lowers PI, increases shear strength
Kaolinite, illite	Lime	For drying, friability and later strength
	Cement	For early strength, especially if lime previously applied
Montmorillonite and mixed-layer clays	Lime	For drying, friability and PI reduction
Dispersive (sodic) clays	Lime, gypsum	To resist deflocculation and internal erosion
Allophane	Lime–gypsum mixes	For strength
Halloysite	Drying	Causes irreversible granulation and shrinkage
Volcanic ash	Lime	Promotes pozzolanic setting

[a] Adapted from Ingles and Metcalf (1972).

the cement-stabilized gravel, but its rate of setting may be too rapid to allow full compaction and pavement shaping. Furthermore, high ultimate strength is not necessarily an advantage in road pavements, and the strength imparted by cement is much less for other types of soils, especially clay-rich ones. On the other hand, a compressive strength of 2 MPa is an excellent result for a clay subgrade, since it can substantially reduce the thickness of the overlying pavement layers. Unfortunately this laboratory result could not be consistently reproduced in construction, because of the difficulty of uniform mixing in clays.

18.2 GRANULAR STABILIZATION

Granular stabilization is a form of *mechanical stabilization* that involves the mixing of two natural soils, generally with the aim of improving grading and compacted density (though it may not reduce plasticity). This distinction is drawn because mechanical stabilization can also include compaction (discussed in Chapter 10), dewatering, soil grouting and even thermal treatment of clay subgrades. The common forms of granular stabilization include the following:

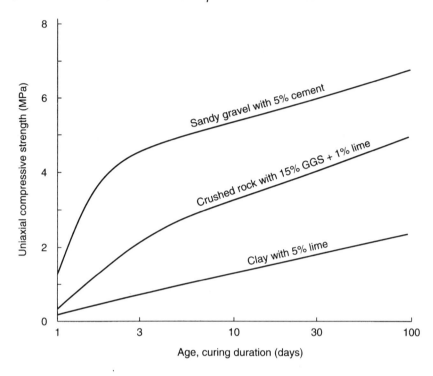

Figure 18.2 Typical strength gain characteristics for stabilized soils under laboratory conditions. See text for further explanation.

- Mixing soil materials within a borrow pit to meet specification requirements, or simply to improve the uniformity of the run-of-pit product. This is by far the most common form of granular stabilization. Usually the aim is to distribute gravel particles, which tend to occur in pockets, more evenly through the soil during stockpiling.
- Blending crushed but marginal-quality materials to generate a product that is just within specification, though often only for lightly loaded pavements (such as minor streets and parking areas).
- Adding dry fines (silt and very fine sand, but not clay) to 'harsh' crushed roadbase that is otherwise deficient in binder. Loam soils or crushed shale are generally used, but care is required to ensure that the resulting base remains within acceptable PI limits.
- Mixing crushed stone into well-graded but gravel-deficient sands on the road (Figure 18.3). Here the angular crushed gravel reinforces a marginal material, increases its shear strength and improves the homogeneity of the product. It also reduces the wear rate of unsealed pavements under traffic.

In-pit blending is achieved by bulldozer blading, linear stockpiling and

Figure 18.3 On-road granular stabilization, Darlington Point, New South Wales. An imported 30 mm crushed basalt is being blended into a well-graded but very silty alluvial sand of maximum size 5 mm from a local pit, in the proportion 1:4.

cross-pit pushing – basically any method that moves, turns, spreads and piles will do. It is sometimes accompanied by moistening ('conditioning') to bring the gravel up to OMC before loading. Quarry blending allows carefully proportioned but closely sized components to be drawn out of storage bins and thoroughly mixed by rotating paddles in a pugmill (Figure 18.4). On-road granular stabilization is uncommon these days because it often leads to size segregation and slows construction. Simple windrowing by graders or tining by rotary hoes has been superseded by specialist machines such as the pulvimixer (a small mobile pugmill) and the rockbuster (a small self-propelled impact crusher).

The major advantage of granular stabilization is that two inferior but locally available, and therefore cheap, materials can be combined to meet requirements that neither could satisfy alone. It is rarely possible to blend more than two components, or to stabilize clay-rich soils. The main disadvantage is that the improvement in material properties may be too small in many cases to justify the cost.

18.3 CEMENT STABILIZATION

Cement and water combine to form calcium silicates, which bond and waterproof granular soil materials, and can yield compressive strengths

Figure 18.4   Mobile stabilizing pugmill installed at a temporary freeway quarry, Mittagong, New South Wales. The vertical bin contains cementitious stabilizer; crushed roadbase is fed into the hopper (left, behind bin), mixed by a pugmill beneath the bin and loaded into trucks by the elevating conveyor.

more than twice those of similarly lime-treated soils. Higher early strengths in particular can be achieved, though in some cases these are excessive and lead to cracking. Cement-stabilized materials ('soil–cement') are sometimes arbitrarily divided into those which are simply *modified* with a few per cent of cement, which behave as conventional unbound flexible pavement courses, and those which are *bound* with up to 16% cement. Bound materials have greatly enhanced elastic moduli (2–20 GPa, or 10–50 times the stiffness of untreated granular basecourses). Their compressive strengths exceed 2 MPa and they have measurable tensile strengths, and generally behave as semi-rigid pavements. Very heavily stabilized soils are sometimes referred to as *lean mix* or soil concrete. A typical stabilizing tractor is illustrated in Figure 18.5.

One of the main problems of cement-stabilized soils is their tendency to crack. Cement-modified soils are assumed to have networks of fine cracks, which facilitate their flexible behaviour under traffic loading, or at least do not adversely affect this. Cement-bound materials, on the other hand, develop widely spaced but more continuous cracks due to tensile strain concentrations, hydration shrinkage and subgrade movement. The most successful materials used in cement-bound pavements are crushed rock and well-graded natural gravel–sand mixtures; plastic or poorly graded sands gain less strength at the same cement content. Cement setting may be retarded by the presence of organic matter in the soil (peat, lignite or organic clay), and hardened soil–cement may be weakened by sulphate-bearing groundwater.

Figure 18.5 On-road stabilizing plant, Bathurst, New South Wales. This tractor is being used to cement-modify granular base to a depth of 150 mm during reconstruction of a rural highway. The shroud in front covers a rotating drum fitted with mixing picks; dry cement and water are fed in by different hoses in controlled quantities.

18.4   LIME AND LIME–POZZOLAN STABILIZATION

Lime stabilization

Lime plus water is not in itself cementitious, but reacts with soil particles to form calcium silicate. Silica for the reaction is drawn from the clay minerals present, ultimately producing the same bonding agents as Portland cement. This is why lime can successfully be mixed with a high-plasticity clay, whereas cement is not so effective. The lime also modifies the soil's texture, changing it to a more friable (granular) consistency typical of low-plasticity materials.

Soil stabilization using lime is broadly similar in technique and result to cement modification, but differs in three important ways:

- Lime reacts well with heavy clays but poorly with granular soils (the opposite of cement).
- It sets slowly, allowing more time for the stabilized material to be spread, compacted and shaped.
- While the ultimate strength achieved by soil–lime is usually less than that of soil–cement, cracking is also greatly reduced.

Strengths comparable with those of cement-bound pavement courses can be produced by lime–pozzolan mixtures, but not by lime alone in any proportion. In fact, strength gain ceases at about 8% lime, but it is still increasing at 10% cement with some well-graded gravels.

Calcium hydroxide (slaked or hydrated lime) is generally used nowadays in preference to calcium oxide (quick-lime), though both are effective stabilizers. Quick-lime is caustic and hence dangerous to handle, is susceptible to moisture uptake in storage, and gives off much heat during hydration. In civil engineering terminology 'lime' therefore generally means slaked lime.

Lime is most often used as a construction expedient, to support traffic on wet subgrades. It may also be a prelude to cement stabilization, owing to its immediate granulating or clay-aggregating action. Lime reacts fastest with wet, high-plasticity montmorillonitic clays and can cut PI by nearly half, while greatly improving workability and ease of compaction. In fact, this can be such an asset on soft, wet earthworks and foundations that the accompanying strength increase is regarded as a bonus.

Lime–pozzolan stabilization

Lime–pozzolan stabilization combines the slow setting of soil–lime with the higher strengths, especially tensile strength, possible through cement stabilization. If the soil contains little clay mineral, then fly ash (PFA) is used as the source of finely divided silica. It is a *pozzolan* – a material that is not itself cementitious, but which when mixed with lime reacts to form a cement. Lime–fly ash mixes often give better stabilization results than cement, particularly at high temperatures. In addition, the ultimate strength of the stabilized soil can be controlled by varying the proportion of lime to fly ash. The most commonly used lime–pozzolans have lime and PFA in ratios between 1:2 (for maximum strength) and 1:7 (for maximum economy), averaging about 1:4.

18.5   BITUMEN STABILIZATION

Bituminous stabilization is the least used of the soil modification methods discussed here, mainly due to greater cost. Its main virtue is that bitumen works best in the same granular soils most amenable to cement stabilization, but is not limited by ambient temperature or setting time. In fact, hot conditions are preferred – a great advantage in summer – and the stabilized material can even be temporarily stockpiled during its curing stage. This material is a lean form of asphaltic concrete, though it still requires a sprayed seal or asphalt wearing course to protect it from wheel abrasion. Bitumen stabilization adds cohesion to non-plastic soils, reduces the sensitivity of some cohesive soils to moisture uptake, and binds aggregate particles into a stiffer mass. It is best suited to fines-deficient or slightly dirty (PI < 6) sands and gravels, but like cement may give satisfactory results in a wide range of soils.

Bituminous stabilization may be effected using hot bitumen, cut-back bitumen or cold bitumen emulsion.

- *Foamed bitumen* is produced by mixing hot bitumen, water and additives in a special chamber, such that the resulting vapour expands to about 10 times its original volume. In this hot, low-viscosity state the bituminous foam has great wetting and penetrating power. It rapidly disperses through the soil mass, effectively coating and binding particles. The advantages of this method are that the product is more homogeneous and does not need the curing period that either cut-back bitumen, or especially emulsions, require before they can carry traffic.
- *Cut-back bitumen* stabilization involves diluting bitumen with a thinner ('cutter'), generally kerosene, which allows a longer mixing time. The disadvantage of this method is that the mixture must be aerated to evaporate the cutter oil prior to laying.
- *Bitumen emulsions* are the most convenient, and therefore most widely used, agents of bitumen stabilization. Bitumen is emulsified with about 40% water by dispersing it as fine droplets in the presence of various surface-active agents ('surfactants'). 'Anionic' emulsions are

negatively charged (alkaline), whilst the 'cationic' emulsions are positively charged (acidic). Emulsions may be further diluted in the field with water but, owing to their charge, the emulsion can be 'broken' – that is, the droplets agglomerate – if water with an opposite charge is used.

Because the emulsion must dry back before the stabilized mixture stiffens sufficiently to bear traffic, such a process is really only suited to areas with long periods of warm, dry weather. In dry soils the emulsified bitumen tends to form blobs, and hence pre-wetting is needed. In damp soils the *in situ* moisture plus the weight of the emulsion must be taken into account, since both are included in the compaction fluid and should together equal OMC.

18.6 PAVEMENT RENOVATION

Pavement renovation or recycling by *deep-lift in situ stabilization* offers great promise for strengthening and extending the life of rural highways. The cost is around one-fifth that of full reconstruction, with the next cheapest alternative, strengthening by means of granular or asphaltic overlays, about half that of a new pavement. These two-lane roads are still adequate in terms of their geometric design – grade, width, curve radius and so on – but their pavements are wearing out due to accumulated traffic-induced strains. The requirement in New South Wales alone, where much of the trunk road network was constructed using natural gravel pavements during the 1960s, could be for several hundred kilometres of rehabilitation per year.

The method involves ripping up the existing pavement and subgrade to a depth of 300–400 mm, mixing in 4–6% of cementitious stabilizer and water or 3–5% of foamed bitumen, and recompacting the pavement as a single layer in a continuous operation. The equipment used is similar in principle to the tractor shown in Figure 18.5, but larger and more powerful. The renovated section can be re-opened to traffic within 24–48 hours, a considerable bonus since it eliminates the need for temporary side tracks. Typically, several hundred metres along one half of the road are stabilized one day and the adjacent lane treated on the next day. A sprayed bituminous prime seal is generally applied as a curing membrane on the third day, though full resurfacing may be delayed for some weeks. The cementitious stabilizer is typically a mixture of Portland cement with one-third to one-quarter fly ash or ground slag, but foamed bitumen and cement–emulsion treatment are offered by some contractors.

The result is a bound pavement with a target resilient modulus of about 5 GPa, which represents a six- to fifteenfold increase in stiffness compared with that of the previous flexible granular base. This in turn can halve the pavement thickness required over weak subgrades, say those with CBR values around 2, though the improvement is less for strong subgrades (CBR 10 or more) and for less-trafficked roads. Although deep-lift stabilization offers a lot of promise, not least of which is the elimination of the quarries and gravel pits otherwise required for the new pavement materials, it does present some problems, namely that:

- Even with stabilization, many older flexible pavements are too thin, sometimes 200 mm or less, to provide sufficient strength for the anticipated traffic volumes (typically 10^6 to 10^8 ESA over 20–30 years). These pavements require overlays in addition to stabilization.
- Thin granular pavements also mean that a proportion of clay-rich subgrade is mixed in with the bound layer, making it unsuitable for cement stabilization. Lime–pozzolans would then be the logical choice, but at a penalty of longer curing time and lower ultimate strength.
- The energetic compactors (impact rollers and vibratory padfoot rollers operating at maximum amplitude) required for satisfactory compaction to depths of 300–400 mm in a single layer are likely to cause vibration complaints in built-up areas. Hence deep-lift stabilization is in practice limited to rural sites.
- Heavily patched pavements and those with multiple asphalt wearing courses can damage the mixing tines. Pavements with a proportion of coarse gravel and rock fragments, or subgrades with shallow rock bars, will also cause severe pick loss. These situations require a two-stage approach (ripping followed by mixing).

Procedures for specifying and testing deep-lift-stabilized pavements have not yet been standardized. The main issues are: the type and percentage of stabilizer; mixing, moisture and spread rate control (Figure 18.6); and the degree of compaction. At present cement–slag blends are the most commonly used stabilizing agents. Ordinary Portland cement is equally effective, but more expensive. Mix uniformity is assessed using the heat of neutralization test or X-ray fluorescence to measure calcium content. Stabilization trials indicate that mixing and pulverization with present equipment are satisfactory, and that cement contents within 1–2% of target (usually about 5%) are achievable.

Compaction to full depth presents problems, since the density of the treated layer (target: at least 100% of standard Proctor throughout) diminishes with depth. Hence a much greater compactive effort is required than is usual in earthworks. This can be accomplished with about 20 passes of a heavy (15–18 tonne) vibratory padfoot roller. Another approach to transmitting compactive effort through a thick

Figure 18.6   Measuring cement spreading uniformity during a deep-lift stabilization trial. The samples enclosed by the steel frames were subsequently tested for calcium content by their heat of neutralization. Note the purpose-built stabilizer at the rear, with its shrouded tine array behind the front wheels, and the adjacent highway lane still open to traffic. (Photo: I.R. Wilson.)

layer is to use impact ('square wheel') rollers. However, a smooth drum or pneumatic tyred roller is needed to finish the process and remove the pock-marked surface left by the pad feet; alternatively, this surface can be trimmed off by a grader. *In situ* density has to be assessed using neutron backscattering devices, as sand replacement tests are difficult to perform at these depths and do not produce results speedily enough. The test frequency is about four per $1000\,m^2$ of pavement.

At present, pavement acceptance is based on unconfined compressive strength (UCS) of the stabilized material; resilient modulus is the preferred criterion, but the test is much more difficult to perform. Target UCS is 2.5 MPa at 28 days, though pavement strength may increase to 5–8 MPa over a year; this is approximately equivalent to a resilient modulus of 4–5 GPa. Typically, at least four pairs of UCS test specimens are cored for each kilometre, say one per $2500\,m^2$.

Pavement rehabilitation using foamed bitumen technology is also now being offered in Australia. The advantages claimed for this method over cement stabilization are superior fatigue characteristics (a more flexible pavement that is less prone to cracking), less sensitivity to hot or wet weather during reconstruction, and better control over binder content and mixing. The bituminous mixture remains workable for some hours, yet like cement-treated materials the renovated pavement can be re-opened to traffic within a day.

18.7 INVESTIGATIONS FOR PAVEMENT RENOVATION

Generally, only unbound granular pavements with thin sprayed seals are suitable for renovation. Crushed bases are preferred to natural gravels because of their lesser clay content, homogeneity and superior grading, though the latter are much more common in New South Wales rural highways. Asphalt wearing courses thicker than about 75 mm have to be stripped ahead of stabilization. Ideally, the old pavement will be at least 300 mm thick, such that no subgrade material has to be incorporated in the recycled pavement. This is rarely the case, so a proportion of subgrade has to be tolerated. Especially thin pavements – say those of 200 mm or less – or expansive clay subgrades may require an overlay of new granular base to compensate for these deficiencies. A proportion of coarse aggregate may also be added during stabilization to improve the material grading. Other problems with old pavements that complicate renovation include wet and low-density sections, and the presence of coarse gravel in the base or rock bars in the subgrade. Ground-penetrating radar shows much promise for detecting these inhomogeneities and for mapping differences in layer thicknesses.

Test holes are excavated to below subgrade level at about 100 m intervals in alternate lanes to investigate the old pavement. These are logged and sampled to record each pavement course and its moisture content. *In situ* density and dynamic cone penetrometer tests are carried out in some holes. Initially the samples are subjected to classification tests and CBR is measured for some of the subgrade samples. Later, pavement/subgrade composites will be tested with varying proportions of different binders to determine the most appropriate stabilizer and dosage. It must be remembered that laboratory testing is carried out under ideal conditions, which cannot be fully replicated during construction. Factors that will generally differ markedly between field and laboratory include:

- The thoroughness of soil pulverization before the stabilizer is added and the homogeneity of the mixture afterwards (this means that more stabilizer is usually added on the road than the laboratory optimum, to ensure that the whole of the recycled layer receives the minimum needed).
- The degree of compaction achieved (generally field compactive effort is greater than standard Proctor, but is not uniformly distributed).

- The curing temperature, which in summer may be higher than the laboratory (hence more rapid strength gain), and in winter may be lower.

Compressive strength increases in direct proportion to the density achieved, but may diminish with cementitious stabilizers if there are delays between mixing and compacting. Achieved field results will almost always be inferior to laboratory trial results, but may nonetheless be quite adequate for the purpose.

REFERENCES AND FURTHER READING

Ingles, O.G. and Metcalf, J.B. (1972) *Soil Stabilization – Principles and Practice.* Butterworths, Sydney.

Kedzi, A. (1979) *Stabilized Earth Roads.* Elsevier, Amsterdam.

Lay, M.G. (1981) *Source Book for Australian Roads.* Australian Road Research Board, Melbourne.

NAASRA (National Association of Australian State Road Authorities) (1986) *Guide to Stabilization in Roadworks.* NAASRA (now Austroads), Sydney.

Wood, L.E. (ed.) (1978) *Recycling of Bituminous Pavements*, ASTM Special Technical Publication 662.

Youdale, G.P., *et al.* (1994) Deep-lift recycling of granular pavements. *Road and Transport Research*, **3** (3), September, 22–35.

CHAPTER 19

Environmental planning and management

The establishment of a large quarrying operation almost invariably attracts controversy because of its location: in order to serve the market, it generally has to be situated in a semi-rural area, yet close to urban centres. It represents a major industrial impact on the local landscape and is resented accordingly by the existing residents, who can see little advantage to themselves – and much inconvenience over an indefinite period – arising out of the development. Much of this opposition is misinformed, or based on observations of derelict extractive sites abandoned decades previously, with no attempt at (or requirement for) pit rehabilitation. Paradoxically, where rehabilitation is most successful, the former quarry sites cannot be recognized as such by the public, while examples of dereliction are immediately obvious.

Most of the substantive objections to quarrying – as opposed to the merely emotional – can be satisfied, or at least placated, by careful location, design and operation of these sites. The tendency for quarries to become fewer, larger and longer-lived may increase their environmental impact locally, but economies of scale mean that resources – in the form of specialist professional advice and equipment – will be available to minimize the damage. Furthermore, a few large and long-term operations are more easily regulated by government authorities than many small-scale temporary sites. Nevertheless, this policy risks creating supply monopolies or at least oligopolies.

Where a major new quarry is being planned, the regulatory authorities will usually provide facilities for public comment on the proposal, as described in the *Environmental Impact Statement* (EIS). If the objections are particularly numerous or vocal, a semi-judicial public enquiry is sometimes conducted. Using the information contained in the EIS as a baseline, the regulators may impose '*Conditions of Consent*' on the development – restrictions on air and water quality, blast vibration amplitude, quarry operating hours and so on. Progressive restoration work is generally stipulated and monitored, and quarry operators can be penalized for non-compliance if necessary. Usually, however, self-regulation is preferred, with the onus on the quarry operator to meet environmental guidelines and satisfy complaints from local residents. It is very much in the operator's interest to practise good public relations by meeting these obligations, since the alternative is the heavy hand of the regulatory bureaucracy.

19.1 ENVIRONMENTAL OBJECTIONS TO QUARRYING

Geological considerations

Environmental objections to new quarry sites fall into a number of categories, of which the foremost is that 'better sites are located elsewhere'. Therefore, the primary aim of the geological input to an extractive-industry EIS is to demonstrate that the proposed site is a good one – preferably the best available in terms of reserves and quality – by reference to the regional geology. The EIS should also point out why the likely alternatives are less attractive than they seem to objectors.

Consider, for example, a proposal for a large *hard rock quarry* in the Sydney Basin. In an area of sedimentary rocks such as this, only a few very localized formations (such as dolerite sills and basalt flow remnants) are likely to be suitable as aggregate sources. Because the region has been intensively studied and geologically mapped over many decades, there is little chance of large new igneous bodies being discovered at or near the surface. Hence it can be presumed that all potential hard rock quarry sites have been identified, even if most have not yet been properly investigated. Furthermore, the regional resources inventories are public documents, available to both quarry developers and their opponents.

Next, consider a commodity like *brick clay*. In gross geological terms, the Sydney Basin has a number of shale formations covering thousands of square kilometres, so the 'better site elsewhere' argument is much more pertinent in this case. Hence a shale pit developer has to demonstrate that:

- There is a real need for further shale mining, caused by diminished reserves in existing pits, increasing brick demand or a requirement for a particular seam that is not found elsewhere.
- Much of the volume of these widely distributed shale formations consists of sandstone or other unsuitable ceramic materials.
- Alternative deposits of equal quality are now either built upon, locked up in national parks, or otherwise inaccessible to extraction.
- Other potential deposits that meet the above criteria are too remote from markets, main roads or processing works to be economically developed at present.

Environmental considerations

It is not sufficient for the developer simply to seek a commercial advantage over competitors by abandoning a pit that is relatively expensive to work in favour of a new site that is cheaper. Assuming, however, that a good case has been made for a new quarry and a site has been selected as the best available on geological grounds, the main environmental objections that can be expected include the following:

- *Visual impact* The visual impact of the quarry landform and its associated facilities must be considered. This may, for example, be a broad pit devoid of grass and trees, or a steep benched face in unweathered rock, both alien to the surrounding landscape. The facilities within the quarry may be equally unattractive crushers, conveyors, stackers, machinery sheds, haul roads, stockpiles and spoil heaps.
- *Loss of land* Land can often be lost, usually from agricultural production. The 'land take' for various extractive operations is illustrated in Figure 19.1, which demonstrates that in this regard hard rock quarrying is much more defensible than gravel mining. In addition, sand and gravel pits are often located beneath highly productive floodplains and alluvial terraces, while hard rock terrains are usually too rugged for agriculture.
- *Dust* Heavy vehicle traffic on unsealed haul roads, drilling, blasting, crushing, conveying and stockpiles will all cause dust. This problem is particularly severe in dry and windy conditions (Figure 19.2), or where fine grinding (for example, limestone dust manufacture) is part of the production process.
- *Ground vibrations* Ground vibrations and air overpressure will result from blasting, and may extend far beyond the range of other nuisances. These rarely cause structural damage, but impose discomfort or irritation on surrounding residents (see Chapter 20).
- *Noise* Trucks, crushing and conveying machinery all produce noise. This category of nuisance (other than traffic noise) normally extends for a few hundred metres at most.
- *Water pollution* This may result mainly from suction-cutter dredging, but also from sand and gravel washing, and from turbid storm runoff. An associated problem with dredging is that of upstream and downstream erosion, as the stream seeks to equalize its gradient by infilling the worked-out riverbed cavity. Groundwater pollution may also result from disposal of lubricants and industrial chemicals into water-filled pits, or into open joints in hard rock quarries.
- *Traffic volume* There will be problems associated with increased traffic on rural roads and suburban streets, particularly since most of this is likely to be heavily loaded trucks. Apart from vehicular congestion, safety problems, engine noise and mud from wheels, the

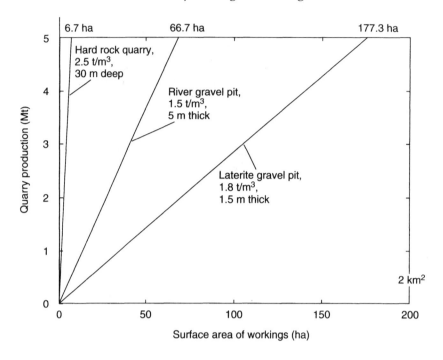

Figure 19.1   Comparison of land area consumed against aggregate production, indicating the relative economy of hard rock quarrying. Note that the most wasteful of all is extraction from thin duricrust deposits, though production from such a pit would never exceed 0.5 Mt.

increased traffic loadings. are likely to cause extensive pavement failures along the lightly built rural roads near the quarry. Hence they are usually opposed by both nearby residents and the local council.

19.2   REMEDIAL MEASURES

The first and most fundamental measure that can minimize conflict between quarrying interests and the public is to ensure that maximum production is obtained from existing sources, or from extensions to these. 'High grading' by removing only the best-quality or most cheaply worked reserves (and, conversely, premature closure due to resident pressure) should be resisted by the regulatory authorities. Future quarry sites should be designated as such in regional planning schemes and provided with surrounding buffer zones to avoid urban encroachment.

Existing producers may be required to contribute to rehabilitation levies for abandoned quarries and to pursue agreed restoration plans within their own pits, in return for what sometimes amounts to a near-monopoly on production within a particular area. One interesting trend

Figure 19.2 Dust nuisance at a temporary quarry site on a windy day, southeastern Queensland. Most of the dust is from open vibratory screens and conveyors rather than from the crusher. (Photo: R.A. Bathurst.)

is for developers to offer to rehabilitate old pits as part of a new and larger quarry (see the example of Calga quarry in section 19.5). Another trend is conjunctive use, where a small quarry is developed in parallel with landfilling (Figure 19.3), or as a prelude to industrial or residential subdivision. In this case the extractive phase is generally subordinate to these more profitable activities.

Visual impact

The visual impact of quarrying can be reduced by careful screening. In hilly sites this can be achieved by using ridges and gullies to hide the workings, by avoiding cliff face or hilltop mining (even where these would be cheaper), and by facing away from the direction of urban growth (Figure 19.4). In flat terrain simply mining downwards will be sufficient to conceal the quarry after a short time, provided all quarry facilities are located within the pit. In both cases the exposed working area, between unmined and restored ground, should be kept as small as possible.

All sites can be improved by tree planting, which should ideally commence some years before the quarry is developed. Large plantations not only improve a company's environmental credit, they act as visual, noise and dust barriers, and may even provide a cash crop. Similarly,

Figure 19.3   Conjunctive use of a quarry site, Reading, UK. As Eocene sands are extracted to the right (outside the photograph), the pit is being backfilled with waste (left). (Photo: J.H. Whitehead.)

Figure 19.4   Quartzite quarry face oriented so as to be unseen from the eastern suburbs of Adelaide, South Australia; this arrangement also radiates airblast away from most residents. This site is located along the sensitive Hills Face zone, where quarrying is now subject to severe environmental controls.

spoil can be shaped into natural-looking barriers with grass, shrub or tree cover, depending on which is the dominant vegetation in the area. Native species are preferred to exotics, but fast growth may be the main consideration initially.

Buildings, production machinery and stockpiles should, wherever possible, be located inside the pit and out of the public gaze. Where these are visible, they should be painted to match the surrounding land colours. Exposed rock faces can be spray-painted with bitumen emulsion or other pigments to match naturally weathered cliffs, and bench outlines can be broken up by backfilling and revegetation, as illustrated in Figure 19.5. This also shows the effect of gradual barrier removal after worked faces have been reshaped and revegetated.

Dust

Dust pollution is mainly reduced by water sprays and dust suppressant chemicals directed onto unsealed haul roads and stockpiles, and particularly by keeping the area stripped of topsoil as small as possible at any given time. In many cases frequent spraying is a convenient method of disposal for silt-laden washery water, since haul roads act as huge evaporation pans. In the longer term, dust generation can be minimized by replanting of overburden dumps and benches. Tree screens also act as traps for the coarser dust particles, and conveyors can be covered or even fully enclosed to inhibit dust movement. Dust from blasthole drilling is collected using shrouds and cyclones attached to the drilling rigs, while bin storage and fully enclosed screening decks eliminate most airborne dust from product processing and storage (see Figure 19.2).

Noise

Machinery noise (and also dust generation) is tackled by housing crushers, screen decks and associated materials-handling equipment in large insulated and airtight sheds. A small pressure differential (interior below atmospheric pressure) ensures that no dust will escape. Hard rubber and polyurethane screens minimize screening noise to a dull rumble, which is perceptible only in the immediate vicinity. The enclosures also provide a visual barrier, which can be camouflaged by painting to match the prevailing countryside tones. Unfortunately the shape and height of these multi-storey buildings cannot so easily be disguised!

Water pollution

The principal countermeasure used to prevent water pollution from dry pit quarrying, as distinct from dredging, is to require that operators

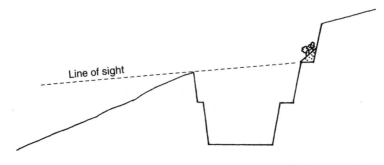

(a)   Development of quarry slot and upper bench rehabilitation (years 1–10)

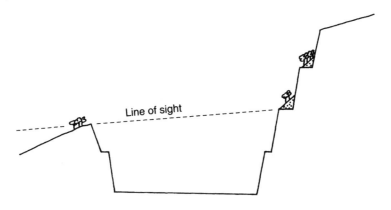

(b)   Maximum quarry development and removal of sight barrier (years 30–50)

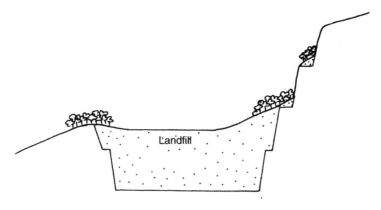

(c)   Final site rehabilitation (year 60)

Figure 19.5   How to hide a quarry – concealment by topographic barriers, staged development and rehabilitation, and face camouflage.

recycle all contaminated water within their holdings (Figure 19.6). This can have the added benefit of reducing demand on sometimes scarce local water resources. Recycling is carried out by decanting from a series of settling ponds, with the best-quality water reserved for the washery and the worst for dust suppression. Clarified water along with a small amount of natural flow is stored in a recycling dam, while flood runoff is routed along a diversion channel from a small upstream weir. This avoids overtopping the tailings lagoons, which could cause a mudflow downstream.

Control of water pollution caused by *onstream dredging* is more difficult. The simplest solution is to restrict dredging to enclosed lagoons away from the mainstream of the river (*offstream dredging*). Another remedy, where suction-cutter dredges are used, is to pump the slurry through a floating pipeline to a processing plant and settling ponds on land, or at least to discharge muddy water into barges for subsequent treatment. Suction-cutter dredges are preferred to draglines, since less muddy water is generated in the digging process.

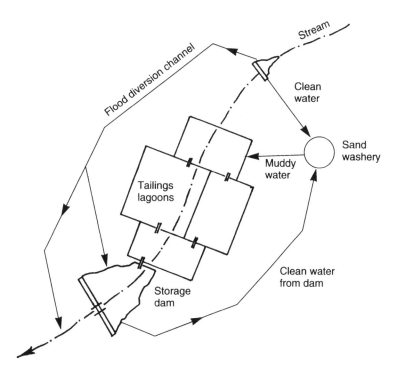

Figure 19.6   A typical water management system for a sand quarry. The chief aims here are to recycle as much water as possible and to prevent storm flows from inundating the tailings lagoons. See text for further explanation.

19.3   STATUTORY EIS REQUIREMENTS

An EIS for an extractive-industry development, such as a soil or rock quarry, is required to address a number of issues of general public concern, plus others that are peculiar to the site in question. Usually one or two issues – for example, the number of heavy truck movements per day or the protection of a particular habitat – will loom large in the public mind and demand a specific reply in the EIS. The information supplied should be objective and succinct (frequently it is neither), so that the public and the consent authorities can form a clear judgement as to the environmental consequences of the proposed development. Persons and communities likely to be especially affected by the project should be able to identify their situation in the documents. Failure to supply full details can result in a climate of even greater mistrust than usual between local residents and intending quarry operators, and can lead to expensive and prolonged court battles and – ultimately – to refusal of the application.

In broad terms, the content of an extractive-industry EIS should therefore include the following:

- *Summary*   A summary of the significant features of the project, the investigations carried out, the main environmental impacts expected and the measures proposed to ameliorate these (in not more than three pages).
- *Project description*   A description of the project, providing the economic and technical background to the proposal, the site location and its size, present land-use and ownership, zoning and planning constraints.
- *Objectives*   The objectives of the development, emphasizing its place in regional planning schemes for the orderly development and conservation of geomaterial resources.
- *Methods*   Specific information on the extraction and processing methods to be used, including types of machinery, blasting and crushing procedures, plant and stockpile layouts, traffic arrangements and hours of operation.
- *Physical environment*   A full description of the existing physical environment of the site and adjacent areas, emphasizing its geology (especially quarryable rock reserves), its soils (especially those with agricultural or site restoration potential), and its climate (especially those factors relevant to blasting, dust movement and site regeneration).
- *Environmental impact*   An analysis of likely environmental impacts resulting from the development, such as: changed river flow and water quality; increased noise and vibration relative to pre-extraction land-uses; and possible disturbance to plant and animal habitats, archaeological or historical remains.

- *Protection measures*   Proposed environmental protection measures (operational safeguards) to be carried out as part of the project, such as: monitoring blast vibration; measuring windblown dust and downstream water turbidity; progressive site rehabilitation; noise muffling; and protecting sensitive locations within the development area.
- *Project justification*   A community justification for the project, such as: reduced transport costs to users of construction materials; replacement of deteriorated, dwindling or worked-out deposits; or supply of superior-quality materials. This would also include a discussion of alternative materials and extraction sites, with reasons for preferring the proposed development.
- *Possible objectors*   A list of public and private organizations approached for comments, and an indication of how their objections are being addressed. The specific needs of residents closest to the site – say within 500 m – should receive particular attention.

The scope and contents of a typical major quarry EIS are summarized in Table 19.1.

19.4   EXAMPLE: DIAMOND HILL QUARRY ENQUIRY

This is an example of a hard rock quarry application that was refused development approval largely on geological grounds, following strong local opposition, which eventually led to a public enquiry in 1979. The proposed quarry site is a small, poorly outcropping volcanic breccia and basalt intrusion – a volcanic neck about 220 m across and 5.5 ha in area. It is situated in rural land of considerable beauty near Kurrajong, New South Wales, in the foothills of the Blue Mountains about 80 km west of Sydney. The enquiry threw much light on the different attitudes of opponents and supporters of quarry development, and on the social and engineering issues involved. It also showed that in this case the protesters were much better prepared with their technical arguments than were the developers.

The application was refused primarily because the deposit is small and would therefore have had little effect on the Sydney aggregate market, but would have greatly inconvenienced nearby residents and caused environmental impacts out of proportion to the postulated economic benefits. Other contributing factors were that:

- The applicant company had done little geological investigation, but relied on surface mapping, magnetometer traverses and a few drillholes to define a poorly exposed intrusion.
- Because of the poor geological information available – which necessitated much speculation about the shape and size of the

Table 19.1 Scope and contents of a large quarry EIS

Summary

Introduction
Description of proposal and objectives
Applicant company profile
Site location, extent and boundaries
Scope and format of EIS

Characteristics of the resource
Site geology (lithology, weathering, structure)
Available materials and reserves
Exploration performed (drillholes, pits, testing, etc.)

Existing physical environment
Topography, landforms and drainage
Regional geology and relation to site geology
Soils (types, distribution, fertility)
Climate (especially rainfall and prevailing winds)
Hydrology (surface and groundwater)
Air quality and acoustic environment
Flora and fauna (especially rare and endangered species)
Visual environment, especially proximity to sensitive sites

Existing social and economic environment

Description of proposed development
Product types, reserves and quality categories
Stages and timing of development, expected quarry life
Possible future expansion or product diversification
Quarrying equipment and processes to be used
Pit layout, haul roads, stockpiles and facilities
Site access roads, rail facilities, overland conveyors
Water supply, storage and treatment facilities
Workforce size and skills, hours of operation, safety provisions
Nearby quarries: present capacity, products and reserves

Environmental impact assessment
Demands on existing surface and groundwater resources
Water discharge (volume, quality, locations)
Erosion hazards upstream and siltation hazards downstream
Effects on flora and fauna (especially rare or endangered species)
Effects on heritage sites, protected areas
Visual catchment and quarry visibility
Likely changes in air quality and prevailing noise levels
Land-use and socio-economic changes expected

Proposed environmental safeguards
Environmental management plans and audit procedures
Water conservation, treatment and recycling measures
Stormwater routing, temporary storage, silt traps
Catchment tree planting and erosion control measures
Specific sites to be preserved, and protective measures
Visual, acoustic and dust screens to be provided
Dust/noise monitoring procedures and control measures
Blast vibration monitoring procedures and control measures
Progressive site rehabilitation plan and timing
Proposed quarry end-use and decommissioning plan

Project evaluation
Socio-economic justification (environmental costs/benefits)
Assessment of alternative quarry sites
Alternative pit layouts and advantages/disadvantages
Alternative transport modes and assessment
Consequences of not proceeding with development

intrusion – estimates of probable reserves down to a depth of 50 m varied widely, from 20 Mt (applicant) to 5 Mt (opponents). The maximum extractable reserves were estimated by the regulatory authority at about 7 Mt, with possibly 1–4 Mt of waste and overburden.

- Despite the fact that volcanic breccia pipes such as the Diamond Hill deposit are notoriously variable in both size and rock quality, and generally suitable only as sources of medium- to poor-quality roadbase, no proper assessment of the proportion of first-grade basalt aggregate to second-class breccia roadbase was made.
- The existing secondary road access to the site was narrow and winding, and would have required much upgrading to handle quarry trucks. The company had no definite plans for this and it was doubtful if the size of the deposit was sufficient to fund the work.
- The proposed development offered no real competition, much less a replacement, for the two main operating roadbase quarries in the region, at Hornsby (30 km northwest of Sydney) and Prospect (30 km west). Both are larger and closer to markets, other than the small local one. Nor could superior quality be demonstrated for the Diamond Hill product on the basis of the small amount of drilling and test data available.
- The site had previously been rejected by three other companies, and two earlier development applications had been turned down by the local council.

19.5 EXAMPLE: CALGA SANDSTONE QUARRY EIS

This EIS was prepared to support a development application to the local council for the re-opening of a sandstone quarry near Calga, New South Wales, some 70 km north of Sydney. A previous operator had allowed large volumes of sand and silt to be washed down the adjacent creek system (similar to Figure 19.7), and the pit had been closed by an order of the Land and Environment Court. The new owner proposed to generate a variety of construction materials from the weakly cemented sandstone, including dry screened bricklaying sand, washed concrete sand, crushed sandstone roadbase and brick clay from shale lenses. About 90% of the output would be sand products, and the new quarry has an estimated working life of 25–30 years.

The main environmental issue here was the prevention of further runoff and downstream siltation by construction of a dam, which would also be used for storing and recycling wash water. Lesser issues included: the protection of a small area containing rare plants (proposed solution: exclusion of this land from the mining area); drawdown of the groundwater table, possibly causing a neighbour's wells to dry up

Figure 19.7   Sand wave deposited by storm flows through an abandoned dune sand pit about 500 m upstream from photo location, Wamberal, New South Wales. The depth of the sand is almost one metre.

(proposed solution: the affected bores to be deepened or replaced); and the return of land to agricultural production after early restoration.

Approval of the project, it was claimed, would provide an engineering remedy to the existing downstream siltation problem and restore land mined previously to cultivation. It would also allow a substantial sand resource to be developed close to a major transport route (the Sydney–Newcastle freeway) and close to a growing market (the Gosford–Wyong district and the northern suburbs of Sydney). Rejection, on the other hand, would leave the site derelict and postpone downstream remediation. In addition, Gosford residents might suffer a cost penalty through lack of competition and haulage from distant existing sources, or alternative pits would have to be opened. These arguments were evidently persuasive, since the pit has now been re-opened and re-equipped.

19.6   EXAMPLE: MOUNT MISERY AND MOUNT FLORA QUARRY PROPOSALS

These are two potential hard rock quarry sites located only 12 km apart, west and north respectively from Mittagong, New South Wales, about 100 km southwest of Sydney. Both projects have finally been approved after many years of legal and administrative delays. They are discussed together because they are geologically alike – both are large microsyenite intrusions – and their investigation and planning have proceeded in parallel, even though their developers are rival quarrying companies. They are the largest potential sources of high-quality aggregate within economic hauling distance of Sydney. Probable reserves within the initial quarry limits are 23 Mt at Mount Misery and 32 Mt at Mount Flora, but inferred reserves run into hundreds of millions of tonnes at both sites. In short, they are both potential 'superquarries'.

The strategic importance of these deposits to the Sydney construction-materials market was recognized in a 1970s land-use planning investigation by Department of Mineral Resources (DMR) geologists, and quarry development applications were submitted for both sites in 1976. These were refused by the local council because of resident opposition, but a subsequent Commission of Inquiry recommended that no land-use incompatible with quarrying, such as residential subdivision, be allowed nearby; this at least kept the option of future extractive use open.

A Working Party made up of representatives from the DMR, the council and other relevant authorities was set up to define more precisely the limits of the igneous bodies and possible buffer zones, and to consider how local objections to quarrying could best be met. Revised development applications along with new EISs were submitted by the applicant companies in 1989, and after lengthy delays both appealed to the New South Wales Land and Environment Court on the grounds of 'deemed refusal' by the consent authority – at that time the shire council. The Court handed down a decision favouring the companies in 1991.

This was a clear conflict of interest between Mittagong area residents, who saw little benefit to themselves in either project, and those of the citizens of Sydney, who are dependent on aggregate resources (at Prospect quarry and the Penrith Lakes scheme) that are likely to be exhausted by 2010 at the latest. The opposition was probably intensified by the likelihood that the eventual quarry or quarries would be both large and long-lived, since either site has sufficient reserves to supply the Sydney market for decades.

The technical issues, unlike the political ones, were fairly straightforward; they involved visual screening, blast vibration limits and monitoring procedures, hours of operation and so on. One interesting argument put forward by the quarry opponents was that the top 20 m of the microsyenite intrusions is hydrothermally altered, and therefore slightly water-absorbent and unsuitable as aggregate for high-strength concrete. This was conceded, but the applicants countered by designating this material as a source of roadbase, for which demand is much greater. An important concession to local residents around the Mount Flora site will be an 8 km private haul road between the quarry and the Mittagong–Sydney freeway, to keep trucks off shire roads. A conveyor and rail transport to Sydney were also considered, but have been rejected on cost grounds for the time being.

REFERENCES AND FURTHER READING

Culshaw, M.G., *et al.* (eds) (1987) *Planning and Engineering Geology.* Geological Society of London, Engineering Group Special Publication No. 4.

Holmes, P.J. (1988) The environmental impact of quarrying. *Quarry Management,* April, pp. 37–43.

Knight, M.J., *et al.* (eds) (1983) *Collected Case Studies in Engineering Geology, Hydrogeology and Environmental Geology.* Geological Society of Australia, Special Publication No. 11.

Poulin, R., Pakalnis, R.C. and Sinding, K. (1994) Aggregate resources: production and environmental constraints. *Environmental Geology,* **23**, 221–7.

Roberts, D.I. (1981) Construction aggregates and the cost to conservation. *Quarry Management and Products,* **8** (March), 167–74.

Selby, J. (1984) *Geology and the Adelaide Environment.* South Australia Department of Mines and Energy, Handbook No. 8.

Tomlinson, R.V. (1983) The public impact of quarrying. *Quarry Management and Products,* **10** (February), 85–90.

CHAPTER 20

Blast monitoring and control

Blasting is, after heavy vehicle traffic, the source of most public complaints received during quarry operations. The principal nuisances associated with blasting, illustrated in Figure 20.1, are caused by seismic waves transmitted through the earth (ground vibrations) and airborne overpressure (airblast). Flying rock fragments and airborne dust are lesser hazards. Most of these objections can be satisfied by monitoring and progressive improvements in blast design, though vibration can never be completely suppressed. Monitoring serves both the public interest and that of the quarrying company, since explosive energy lost as vibration detracts from fragmentation as well as being an environmental nuisance.

Ground vibrations are propagated outwards by surface waves, though 'body' waves (compressional and shear) predominate close to the

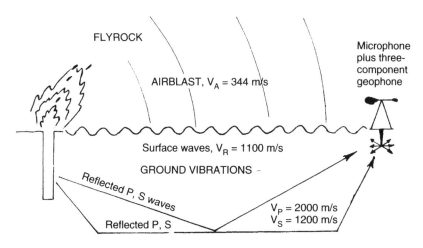

Figure 20.1 Environmental nuisances connected with blasting: ground vibration, airblast and flyrock. A monitoring station equipped with a buried three-component geophone and tripod microphone is shown at the right. Note the connection between blasthole venting and flyrock generation.

blasthole. Although these seismic waves are potentially damaging, airborne vibrations ('airblast' or overpressure) are usually more annoying to the public. Airblast consists of energy with a range of frequencies, some audible but mostly below the hearing threshold. There is no definite relationship between the size of a blast and the vibrations felt by neighbours; site geology, weather conditions and degree of explosive confinement are also important factors.

Airborne debris (*flyrock*) can also be a problem in blasting, especially where airblast is severe. This occurs where explosive gas energy is vented prematurely into the atmosphere, propelling rock fragments upwards and outwards at high velocity. Flyrock is, nevertheless, more a hazard to nearby workers and quarry buildings rather than to distant residents, and should be treated as a safety issue rather than an environmental nuisance.

20.1   GROUND VIBRATIONS

Peak particle velocity

Surface waves are characterized by relatively low velocity, low frequency and high amplitude (compared to the compressional and shear waves produced by the same event). The particle motion associated with their outward propagation is roughly elliptical, so to define ground vibration at any point it is necessary to measure this motion in three mutually perpendicular directions. This may be done in terms of particle displacement, velocity or acceleration, although velocity is considered the best indicator of potential damage to structures. The velocity signal is recorded by a three-component geophone at the monitoring point, and the maximum vector sum of these orthogonal vibrations is defined as the *peak particle velocity* (PPV) or peak vector sum.

Note that particle velocity at a point on the ground is distinct from, and much less than, the wave *propagation velocity*. Furthermore, PPV diminishes rapidly with distance from the shot point, while wave velocity is constant. Most statutory limits on blast-induced vibrations are set in terms of permissible PPVs, quoted in millimetres per second (mm/s). The present regulatory trend is to fix these far below structurally damaging velocities or even human discomfort ranges, down to where blasting is scarcely perceptible to distant residents.

Amplification

Damage can also be caused by signal *amplification* within a structure, where the ground vibration frequency spectrum contains the natural (resonant) frequency of the structure, as shown in Figure 20.2. At the resonant frequency, severe annoyance to residents and cracking can occur even at low PPVs, particularly in tall buildings. Most houses have fundamental resonant frequencies around 5 Hz, while for walls and floors this may be higher (10–20 Hz). The frequency content of the blast vibrations is therefore of great importance, and higher PPVs can be tolerated at higher frequencies.

Foundation conditions can also have a considerable influence on

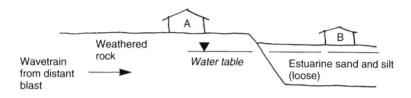

(a)   Houses on different foundations

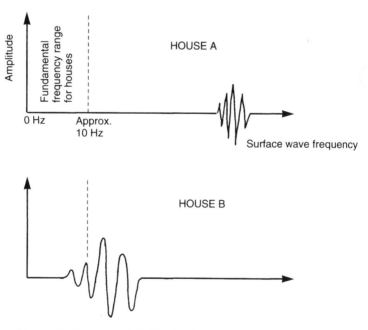

(b)   Ground vibrations recorded at the two houses

Figure 20.2   Effect of geology on blast vibration levels in nearby houses located on different foundation materials. At house B the dominant frequencies are lower, overlapping its resonant frequency, and amplitudes are greater – hence ground vibrations will be much more alarming than at A.

damage caused by signal amplification, and may provide the explanation for damage recorded at large distances from the blast site. Soft rock and soil filter out high frequencies, and the remaining vibrations can resonate within certain types of ground. Amplitude enhancement is greatest where structures are founded upon saturated and soft sediments (Figure 20.2), followed by those on loose dry soil, weak rock and hard bedrock, in decreasing order of risk.

20.2   AIRBLAST

Airborne vibrations are generated by radiation from surface waves moving along the ground, by heaving around blasthole collars, by violent release of gas pressure from blastholes, and by firing of uncovered detonating cord. The higher-frequency (noise) component is perceived as a sharp whipcrack or as a dull rumble. Lower airblast frequencies (concussion), in the range 10–20 Hz, cause windows and lightweight building panels to vibrate at audible frequencies.

These emissions differ from ground vibrations in a number of ways. They are lower in amplitude; propagate much more slowly (around 340 m/s); and can be recorded using a single transducer, since air pressure is equal in all directions. Paradoxically, because they vibrate at (or can be re-radiated at) audible frequencies, they are more apparent to observers than ground waves. Most people find this merely intrusive or annoying, but some are alarmed.

Quite small blasts can generate complaints if meteorological or blasthole conditions are unfavourable. Temperature inversions – warm air above cold – can cause airblast energy to be refracted downwards at distances of several kilometres from the quarry (Figure 20.3), and for overpressures to be intensified two- to threefold there. Strong winds may also channel the overpressure in a particular direction. At the same time, areas closer to the blast may remain quiet. In addition, though terrain barriers may attenuate the audible frequencies, sub-audible vibrations may simply propagate over these obstacles.

20.3   STATUTORY LIMITS

Environmental regulatory bodies set limits on levels of vibration and airblast from quarries and mines, but often decline to become involved unless complaints are received. This self-regulation policy means that the quarry operator has an incentive to minimize blast vibrations and, by regular monitoring and consultation with residents, to be seen to be doing this.

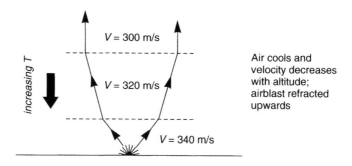

(a)   Normal weather

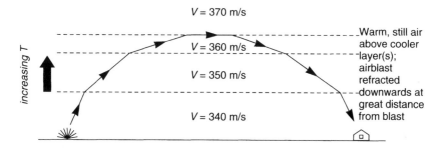

(b)   Temperature inversion

Figure 20.3   Effects of temperature inversions on airblast energy propagation, causing the nuisance to be spread to distant residences where blasting is normally imperceptible. Much of this energy is generally directed upwards, but on still sunny mornings or cloudy days it may be refracted downwards several kilometres from the quarry.

Vibration limits

Vibration limits are set in terms of PPV and vary with the time of day. The maximum acceptable PPV for blasting in New South Wales on Monday to Saturday between 9 a.m. and 3 p.m. is 5 mm/s, with this limit not to be exceeded in more than 5% of blasts, and 10 mm/s never to be exceeded. Outside these hours – the busiest of the day, when blasting is least noticeable – PPVs must be even less and in some cases blasting is prohibited. The aim is to set vibration limits that do not cause concern to the majority of people. These are termed 'comfort' limits and are well below damaging PPVs, except for historic buildings.

Airblast limits

Airblast limits are set as levels of air pressure in decibels, dB(linear). These are measured by means of a microphone with a lower cut-off frequency of 2 Hz (to exclude 'ground roll' caused by wind). The limit set in New South Wales for Monday to Saturday between 9 a.m. and 3 p.m. is 115 dB, to be achieved in at least 95% of blasts, and 120 dB never to be exceeded. This is approximately equivalent to a particle velocity of 1 mm/s, well below damaging overpressure levels, and highlights the fact that nowadays comfort limits for airblast dictate blast design rather than ground vibration limits.

20.4 MONITORING OF BLAST VIBRATION

The equipment used at a blast vibration monitoring station is illustrated in Figures 20.1 and 20.4. Ground vibration and airblast are recorded simultaneously, peak vector sum is calculated continuously, and all digitized data are stored. Printed outputs can generally be obtained, and some instruments can store data from a number of events for later analysis. Alternatively, this can be stored in an FM tape recorder and later passed through a signal analyser.

Usually the station is set up near the closest residence to the blast, since the highest amplitudes can be expected there. However, a more distant but particularly sensitive site, such as a hospital or school, may have to be monitored as well.

Measuring ground vibration

Three practical problems are encountered when measuring ground vibration:

- Inadequate coupling of the geophone to the ground, which may result in signal amplification due to resonance.
- A monitoring period that is too short to record both ground vibration and airblast (for example, at 1 km the airblast arrives approximately 3 s after the ground vibration).
- 'Noise' due to swaying trees, passing vehicles and nearby pedestrians.

Particular care is needed when small PPVs are expected and the 'trigger level' (at which the instrument starts to gather data) must be set very low. In this situation the monitor should not be left unattended, to ensure that the recorded event was in fact produced by the blast and not by random noise or ground roll.

Figure 20.4 Typical blast monitoring station set-up. In this case the geophone is surface-mounted on a kerb and is rather close (about 2 m) to the adjacent building, to avoid road traffic. The two cases house the seismograph and the tape recorder. (Photo: G.W. Won.)

Measuring airblast

The chief requirements when measuring airblast are that the microphone is checked with a portable calibrator and placed at least 3 m from any reflecting structure at 1.2–1.5 m above the ground. The main problem encountered when measuring airblast is that wind often registers higher than the set limits. Wind gusts have a frequency of up to 2 Hz and are similar to airblast, except for a slower 'rise time'. Hence background levels of wind should be recorded for comparison with airblast results.

20.5 CONTROL OF GROUND VIBRATION

Many blast vibration problems can be solved to the mutual benefit of the objecting residents and the quarrying company, since explosive energy lost as seismic or air disturbance diminishes that available for fragmentation. With careful seismograph monitoring, including stations much closer to the blast than the affected residences, even the offending holes can be identified. In addition, seismic wavetrain analysis can be supplemented by high-speed photography and numerical modelling to build up a complete blast history, almost millisecond by millisecond. There are two chief ways in which ground vibration can be reduced: by limiting the weight of explosive, and by decreasing the degree of confinement.

Weight of explosive

The first way in which ground vibration can be reduced is by limiting the weight of explosive detonated at any moment (the maximum instantaneous charge, or MIC). This is achieved by the use of *delays* between holes, and is the most effective single technique for reducing PPV. It is also useful in decreasing airblast, since this is partly related to the amplitude of surface waves. The disadvantage is that short delays may improve fragmentation, but long delays are required to reduce ground vibration amplitude. Furthermore, at large distances – the 'far field' in blasting jargon – the benefit of delays is lost due to wavetrains overlapping; fortunately, PPVs are also greatly diminished here.

Degree of confinement

Secondly, ground vibration can be controlled by decreasing the degree of confinement of the charges. This can be accomplished in a number of ways: by reducing the stemming length, bench height and burden distance, and by decoupling the charge from the surrounding blasthole walls. Decoupling involves the use of decked or low-density charges, which limit the amount of energy input to the ground (and, in particular, the proportion of strain energy to heave energy). Collectively, these measures will reduce the proportion of low-frequency, high-amplitude (hence high-PPV) surface waves produced. The drawback is that some – in particular, shorter stemming and narrower burden – increase the likelihood of airblast complaints.

20.6 TRIAL BLASTS AND SCALED DISTANCE

Achieving the most economic blast design to maximize fragmentation and keep vibration and airblast within acceptable limits is largely a matter of trial and error. When blasting at a new location, a series of trials are usually conducted, with monitoring stations at varying distances from single small charges.

Safe charge weights for a particular site may be estimated using an empirical relationship developed by the US Bureau of Mines (USBM) and known as the *scaled distance formula*. The input data are a series of PPVs recorded at varying distances from the shot point, usually a single blasthole with a small and constant charge weight. The PPV obtained for each blast is plotted on log–log graph paper against the distance (*d*) between the shotpoint and the recording station, as shown in Figure 20.5. In this example five 1 kg rounds were fired and PPVs up to 13 mm/s were recorded at distances of 19–55 m from the shot point. The result is an empirical *site law* of wave propagation through the rockmass at this location, which can be used to predict vibration levels from future larger production blasts.

Figure 20.5 is derived and used as follows:

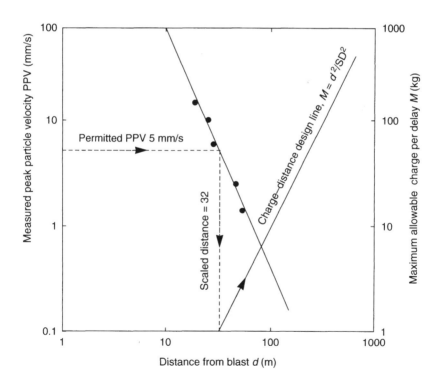

Figure 20.5 Nomogram for estimating the maximum charge that can be detonated at a particular site. The dots are the PPVs measured with small charges set at different distances from the monitoring station, while the right-hand diagonal represents the empirical 'site law' for meeting statutory PPVs at this location. See text for further explanation.

- Measured PPVs (left hand y-axis) are plotted against distance (x-axis). The scaled distance (SD) corresponding to the maximum allowable PPV is then read from the x-axis (in this case that PPV is 5 mm/s, hence the SD is 32). This means that the PPV for a nominal charge of 1 kg drops below 5 mm/s at 32 m from the shot point.
- A second line is plotted for distance (x-axis) versus maximum charge weight per delay (M, right hand y-axis), using the formula:

$M = d^2/(SD)^2$
$M = d^2/1024$ when SD = 32

For example, the nomogram predicts that the maximum allowable charge at a distance of 300 m on this site is 88 kg.

The first line plotted is the mean line of best fit, with 50% of points above the line and half below. Consequently, the maximum charge per delay obtained from this plot will result in 50% of blasts exceeding 5 mm/s. When only 5% of the blasts are allowed to exceed this PPV, a regression line corresponding to this probability is plotted parallel to the mean line. In general, the line corresponding to 5% exceedance will move closer to the mean line as more data points are obtained and the degree of statistical certainty increases.

Some caution is needed in using scaled distance predictions, since the USBM formula does not allow for the overlapping effects of multiple-hole wavetrains at great distances from the shot point (i.e. the duration of the waveform increases in the 'far field', as shown in Figure 20.6). Hence MICs should be set low initially and then progressively increased, with the resulting vibrations carefully monitored.

20.7   AIRBLAST CONTROL

The rattling of windows caused by airblast is commonly, but mistakenly, attributed to ground vibration. However, airblast from quarry blasting is most unlikely to cause damage to residential structures, with window cracking at high overpressures (greater than 140 dB, well above the statutory limit in New South Wales) being the only real concern. Unfortunately, overpressure results show much more scatter and attenuate more slowly with distance from the source than do PPVs, partly due to weather conditions.

Measures to reduce airblast are broadly similar to those for minimizing ground vibration, although sometimes the two requirements conflict. In addition to reducing the MIC, they include the following:

- Ensuring that *stemming* is adequate in length and tightness. Using crushed aggregate in place of drill cuttings as backfill is recommended. Longer stemming columns can, however, result in poor fragmentation of rock around the blasthole collar and intensified ground vibration.

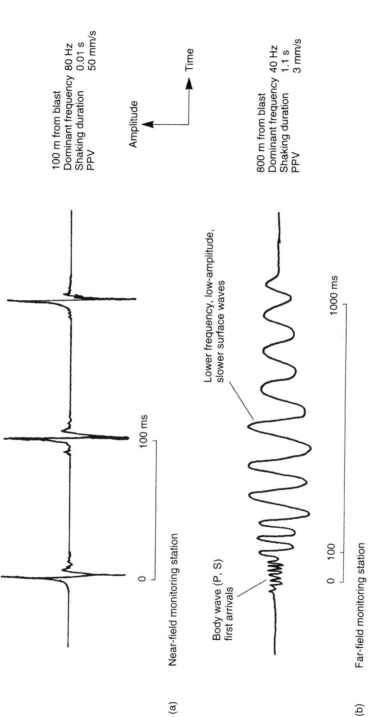

100 m from blast
Dominant frequency 80 Hz
Shaking duration 0.01 s
PPV 50 mm/s

800 m from blast
Dominant frequency 40 Hz
Shaking duration 1.1 s
PPV 3 mm/s

Amplitude

Time

100 ms

0

100 ms

(a)   Near-field monitoring station

Lower frequency, low-amplitude,
slower surface waves

Body wave (P, S)
first arrivals

0   100

1000 ms

(b)   Far-field monitoring station

Figure 20.6   Comparison of ground vibration wavetrains recorded close to a blast pattern ('near field') and at a distance of about 800 m away ('far field'). In the near field individual blastholes can be distinguished and misfires identified (note the inter-hole delay of 100 ms). The far field wavetrain is of much longer duration, with a higher proportion of low-frequency energy. The benefits of delays are lost here due to multiple reflections, refractions and interference, but amplitudes and PPVs are also much diminished.

- Covering surface *detonating cord*, which generates a high-pitched crack due to its high velocity of detonation. This is in any case being replaced by non-electric (Nonel) charge initiation, which not only eliminates the noise but also allows each blasthole to be separately delayed.
- Avoiding *unfavourable weather conditions*, such as heavy cloud cover or strong winds blowing towards nearby residences. In particular, blasting should be suspended on days of temperature inversion (still cold air overlying warmer near-surface layers, as shown in Figure 20.3); these are indicated by smoke haze, fog and low cloud.
- *Orienting benches* where possible so that they do not face towards the nearest residences. After venting, the main source of overpressure is bench face vibration at high PPVs. Quarry layouts where blasting faces are below ground level, behind ridges or screened by earth bunds are also helpful in deflecting overpressure.
- Exercising strict control over *blasthole deviation*. Wandering holes lead to variations in burden and hole spacing. Where these distances are too small, venting and sympathetic detonations (adjacent blastholes firing almost simultaneously) can occur; where too large, ground vibration amplitudes increase and fragmentation suffers.
- Infrequent, *larger multi-row blasts,* rather than daily single- or double-row patterns, reduce face PPV and improve fragmentation overall. The disadvantage here is that the blast, when it comes, will be unexpected and therefore more startling to some people!

Another very useful technique for improving blast design and reducing airblast is to record the blast on videotape, and analyse it frame by frame in slow motion. Stemming ejection and venting can be seen on the video image as emerging puffs of smoke prior to face move-out. Flyrock and delay misfires can also be detected. The latter may be even more obvious on near-field seismic records, which should be compared with the frame-by-frame analysis to build up a time history of the blast.

REFERENCES AND FURTHER READING

Dowding, C.H. (1985) *Blast Vibration Monitoring and Control.* Prentice-Hall, New York.

Dowding, C.H. (1992) Suggested methods for blast vibration monitoring. *International Journal of Rock Mechanics and Mining Sciences*, **29** (2), 145–56.

McKenzie, C. (1990) Quarry blast monitoring. *Quarry Management*, December, pp. 23–9.

McKenzie, C. (1993) Quarry blast diagnosis – Can it help? *Quarry Management*, July, pp. 17–23.

Morhard, R.C. (1987) *Explosives and Rock Blasting*, Ch. 11. Atlas Powder Co., Dallas, TX.

Siskind, D.E., *et al.* (1980) *Structure Response and Damage Produced by Ground Vibration from Surface Mine Blasting.* USBM Report of Investigations 8507.

CHAPTER 21

Quarry reclamation

Quarrying is only a temporary land-use, and hence it is essential that the condition in which these sites are abandoned should not prejudice their future uses. In the past, when land and mineral resources were considered inexhaustible, and quarrying operations were small and scattered, public opinion tolerated the presence of derelict pits in return for the cheap materials and jobs that they provided (Figure 21.1). This consensus has been reversed by urban encroachment, by the greatly increased scale of modern extractive operations and – not least – by changed public attitudes towards the natural environment, particularly that part of it close to their homes.

Figure 21.1   Quarry abandonment, old style – a former slate pit in north Wales. Note planar slope failures along a persistent joint set oblique to cleavage and the large proportion of wastage (up to 90% in some slate pits). (Photo: J.H. Whitehead.)

The extractive industries are facing greatly increased environmental regulation world wide, and quarry reclamation is a now mandatory in most cases. This involves either creating a natural-looking landform, or converting the site to industrial or residential land-use. How this reclamation is achieved depends on whether the quarry is at the planning stage, is being actively worked, or has already closed.

- *Proposed quarry sites* These days proposed quarry sites are mostly planned as large-scale operations, with working lives of 20–100 years. Reclamation work is planned from the outset and implemented in parallel with extraction. A final landform can be aimed at, and any changes to this geometry can be programmed well ahead of site completion. From the quarry operator's viewpoint, progressive site rehabilitation reduces double handling of waste and overburden; makes use of idle plant during slack production periods; and, not least, presents a good public image. From the regulator's viewpoint, it reduces the amount of land exposed at any time; yields a more predictable result; and minimizes the damage should the operator abandon the site prematurely.
- *Operating quarry sites* In the case of operating quarry sites, the considerations are similar, the aim being to integrate extraction and restoration as closely as possible. The principal difference is that many active quarries have been operating for 50 years or more, most of that period with no long-term plan or even a requirement for one, and under a number of different owners. The present management is expected simultaneously to plan for the future, to remedy past mistakes and to meet current environmental standards.
- *Abandoned quarries* These will generally have to be restored at public expense, although this may be financed through an environmental rehabilitation levy imposed on all current producers. Many smaller worked-out pits have nonetheless been recycled at private expense as housing subdivisions, factory sites, storage areas, testing facilities and even as shopping malls. In these cases the increased value of the land due to urban envelopment can easily cover the cost of remedial work.

  Public utilities such as pumping stations, water treatment facilities and freeway cuttings have also been located in former quarries, so that the bill for restoration was included in the cost of construction. Occasionally this type of reclamation can be partly self-financing, since the construction may generate a temporary demand for the re-opened quarry's products. In addition, the extraction of these materials provides the opportunity for reshaping the mined-out void into a more acceptable landform.

There is a considerable literature on land restoration, including quarry reclamation, mainly from the viewpoint of landscape architecture and revegetation problems (such as Haywood, 1974; Bradshaw and

Chadwick, 1980; Coppin and Bradshaw, 1982). The geotechnical remediation of derelict and filled land, including properties of waste products and dump reconstruction techniques, are dealt with in Fleming (1991) and Fell *et al.* (1993).

21.1 END-USES AND RECLAMATION STRATEGIES

The end-use of a quarry will depend on public priorities, climate, site geology and prevailing land values. Generally, quarries go through an intermediate stage as waste disposal sites between their initial extractive phase and their final use, and this is discussed in the next section. The final use categories include:

- *Agriculture* This use is largely confined to extensive and shallow workings in high-rainfall areas, such as sand and gravel pits within alluvial floodplains. Some flooded pits have been used for fish farming, but recreational fishing and boating are much more common.
- *Nature conservation* Sites that have been left derelict for a long period regenerate naturally to some degree, and are often dedicated as nature reserves with minimal modification. Former limestone and hard rock quarries can be especially suitable, since they contain a variety of habitats and often provide a refuge for wildlife from surrounding urban or agricultural land.
- *Forestry* Commercial tree farming may be viable on more level and well-watered ground where the soil quality is not adequate for agriculture, such as former sand pits. However, most former quarry sites are too small and topographically uneven for this purpose.
- *Recreational uses* Most abandoned quarries can find some recreational use, including as parklands, golf courses, ornamental lakes and playing fields, since the activity chosen can be fitted to the geometry of the site.
- *Industrial estates* These are often located on former quarried land, largely because of its industrial zoning and access to transport facilities. Industrial users are better able than home owners to afford the expensive foundations and slope modification required in these situations, and are more tolerant of their lack of visual appeal.
- *Housing estates* Houses have been built in former quarries, but many problems have resulted. The most widespread is foundation settlement, but fear of toxic wastes, leachate springs and methane emissions have limited recycling for residential purposes.

There is in fact no limit to the uses that former quarries may be put, as Figure 21.2 demonstrates. In Australia, recreational facilities are by far the most common end-uses, with industrial estates or single factories often located in former hard rock quarries.

Figure 21.2 An unusual example of quarry rehabilitation – as a tennis court, near Mount Gambier, South Australia. The sunken aspect also provides protection from winter winds! The Gambier Limestone here is a soft, non-abrasive rock that is cut as dimension stone.

21.2 QUARRY LANDFILLS

Disposal of domestic and industrial wastes is the most common intermediate use for disused quarries, and generally precedes final restoration. In fact, many abandoned pits close to urban centres are now more valuable as landfill sites than they were for construction materials because of the high cost of tipping. About 90% of waste in the UK, for example, goes into past quarry workings, and the proportion is probably similar in other developed countries. Municipal domestic waste (MDW) provides a bulk material that is not only free but profitable for backfilling, levelling and reshaping the quarried void. Further incentives for dumping in operating pits are that visual, noise and dust screening measures are already in place, backloading aggregate trucks may be possible, and any additional traffic generated is less noticeable than it would be near a disused site.

The main geotechnical problems are common to all waste disposal sites and not just to quarry landfills: leachate generation, gas emission and ground settlement. In most instances these can be controlled by careful design and fill management, and sometimes methane emissions can be burned as a fuel. The significance of these problems also varies a lot with the geological environment and the proposed end-use. For example, leachate formation in nearly impermeable clay pits is less alarming than it is in highly pervious gravel excavations, which are in hydraulic connection with potable groundwater. Similarly, settlement is much more of a consideration in landfills that are to be built upon rather than those intended for open recreation spaces.

Leachate generation

Leachate generation is minimized by preventing infiltration of underground or surface waters into the fill, mainly by placing it well above the water table and enclosing it within compacted soil covers. Leachate movement is controlled by selecting quarry sites with tight walls, or by making these impermeable by means of clay or synthetic liners. The use of liners and leachate drains is illustrated in Figure 21.3. From the quarry floor upwards these comprise moist compacted clay, a sand/filter fabric drainage blanket, and high-density polyethylene (HDPE) geofabric sheeting. The purpose of this elaborate layering is not just to prevent downward and outward percolation of liquid pollutants into the local groundwater system, but also to direct it towards collection points (wells and sumps) from which it can be pumped for treatment.

Landfill gas emission

Landfill gas migration has been recognized as a problem following methane explosions above old dumps and observations of its toxic effects on some vegetation. Although methane is the main constituent, and is explosive in 5–15% mixtures with air, carbon dioxide and other gases are usually also present. Because of the slow rate of anaerobic decay in modern compacted landfills, emissions may continue over 20–30 years. Control measures such as gas barriers, gas-permeable layers and gas extraction (vacuum) wells are therefore usually installed in large landfills.

The collected gas may simply be vented or flared-off to the atmosphere, although flows from major dumps are increasingly being put to use for on-site power generation and kiln-firing. The generators are typically 1 MW diesel units burning a minimum 30% methane mixture. Where operating quarries and clay–shales are located adjacent to landfills, the waste gas has also been used for aggregate drying and kiln fuel, although it is unsuitable – without treatment – for feeding into a town gas supply.

Ground settlement

Settlement of made ground is the major obstacle to the redevelopment of quarry landfills as building sites. Downward movements may range

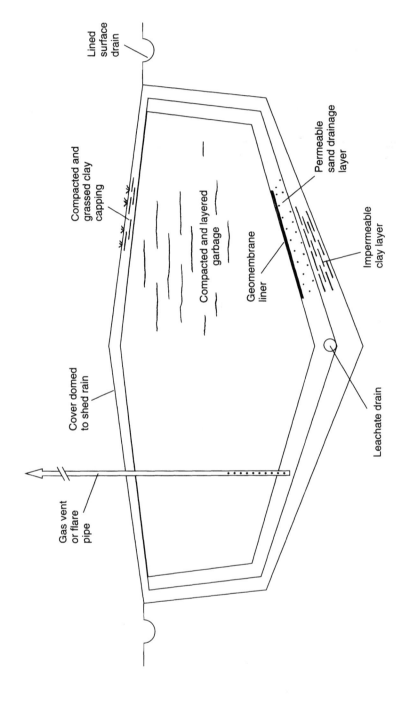

Figure 21.3 Design features of a simple landfill (not to scale). See text for further explanation.

between 3% and more than 10% of fill thickness – more than a metre in some cases – and take decades to be completed. An example of an initially uncompacted landfill is illustrated in Figure 21.4. Loose MDW has a density of less than $0.8\,t/m^3$ and a compacted density of 1–$1.5\,t/m^3$, depending on the degree of rolling and the proportion of cover soil thickness (about 1:10 is typical). Compaction is performed by tracked loaders and bulldozers or, increasingly, by cleated or spiked steel-wheeled spreader/compacters. The degree of initial compaction achieved is very low by the standards of road earthworks; most fill densification results from self-weight and long-term secondary consolidation (creep). An average long-term rate of settlement is about 0.8% per log cycle of time. For a typical 10 m thick fill layer this would represent about 80 mm in the first 10 days, 160 mm after three months and 240 mm over three years.

The reality, however, is that landfill settlement is generally neither uniform nor predictable. *Differential settlement* is normal because of

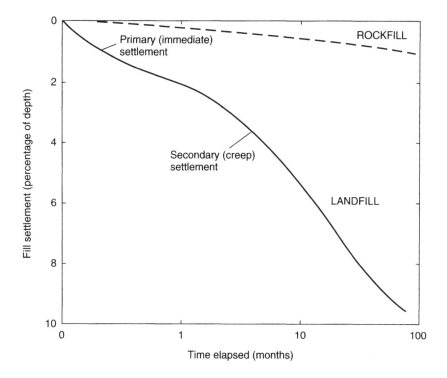

Figure 21.4 Self-weight compaction of a domestic waste landfill compared to that of a rockfill dam embankment. The cumulative settlements (in terms of fill height) were 1.1% at 10 years for the rockfill and about 12% at 20 years for the landfill. Note, however, that about half the settlement of this loose landfill occurred in the first year.

variations in waste composition, moisture content, degree of compactive effort applied and rate of decay. *Collapse settlement* can result from saturation due to a rising water table, or from sudden closure of a large cavity within the fill (caused by, for example, a rusted-out tank). High fills placed against benched hillside quarries are subject to downslope creep and landslides, particularly where porewater pressures can develop due to infiltration through inadequate surface seals or open joints in the adjacent rockmass.

21.3   PLANNING, PROCEDURES AND MATERIALS

Shape, size and layout

Planning for reclamation should begin with the present and expected final shape, size and layout of the quarry (in other words, its 'geometry'). This will largely determine the eventual landform to be created, since – unlike opencast coal-mining – the volume of product extracted will be much greater than that of the pit rejects. Effective planning requires detailed topographical plans at scales around 1:500 and contour intervals of 1 m or less, which should be updated every two to five years. This can most effectively be done using aerial photographs, preferably in colour, and photogrammetric contouring. These plans and photographs also provide an excellent base for geological mapping and geotechnical investigations.

Available materials

The second consideration should be the materials available for reclamation. These are of two types, quarry rejects and imported wastes. *Quarry rejects* comprise overburden, inferior aggregate, crusher grit, clay lumps, washery tailings and (sometimes) stockpiled topsoil. *Imported wastes* include domestic garbage, building rubble and solid industrial wastes. An important task – particularly in the case of abandoned workings – is to make an inventory of the materials available, since some will be more useful than others for restoration, while others may be noxious (offensive smelling, flammable, corrosive) or even toxic (poisonous to plants and animals). The inventory should indicate the location, volume and physical properties of each material class, so that a selective burial plan can be devised. Broadly speaking, three types of restoration materials are required:

- *Bulk fill*, mostly garbage but including the cover soil on daily lifts. This replaces most of the mined-out stone, gravel or shale and provides the general shape of the final landform. At most sites waste represents about 90% of the backfill volume and its proportion is increasing as cover soil is used more economically.

- *Subsoil*, generally only about 1 m thick, which is used for landfill capping and final shaping. Washery fines, clayey overburden and other quarry wastes are typical subsoil materials. Where bulk filling is not carried out – for lack of suitable waste or because of environmental restrictions – subsoil is used mainly for minor landscaping and to save topsoil for essential purposes. It is also used, after copious applications of lime and fertilizer, as a topsoil extender.
- *Topsoil* is the critical element for plant restoration, although it is usually only about 0.5 m thick and constitutes only about 1% of the backfill volume. In operating quarries this is stripped and stockpiled before the main overburden removal begins, and should be replaced within a few months on rehabilitated ground.

Hydrogeological regime

Another major consideration is the hydrogeological regime operating in the subsurface around the quarry. This is a simple task where groundwater occurs far below the quarry floor and is unused because of salinity or low rockmass permeability, but where there is any risk of pollution more rigorous investigations will be required. These could involve falling-head or pump-out permeability tests, extensive piezometer (observation well) installation, water level and water quality monitoring. Where leachate migration is suspected, electrical geophysics and chemical tracers are used to map groundwater movement away from the quarry. Simplified hydrogeological models for three different types of unsealed landfilled quarries are illustrated in Figure 21.5. Groundwater down-gradient from the gravel pit is obviously most at risk, though some degree of leachate cleansing is possible due to cation exchange with clay particles in the alluvium. Leachate moving along open joints below the hard rock quarry is not attenuated in this way, and may travel rapidly towards groundwater discharge points in a highly permeable aquifer.

Slope stability

Long-term slope stability of the quarry walls can be a consideration where public access to the old workings is to be allowed. Bench and rock face stability that is adequate during mining will probably not satisfy safety requirements for recreational spaces. A survey of joint, bedding and foliation orientation, spacing and shearing resistance may be required, together with an assessment of possible long-term face deterioration due to weathering. Considerable face scaling, overhang removal and artificial support of loose blocks could be required. The

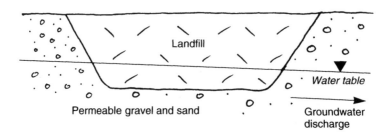

(a) Sand and gravel pits

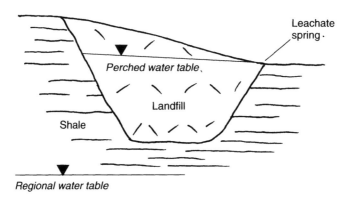

(b) Brick shale pits

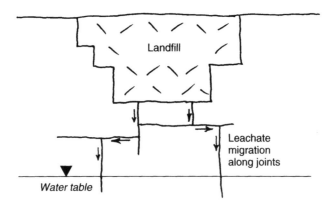

(c) Hard rock quarries

Figure 21.5 Hydrogeological models of three different types of unsealed quarry landfill. See text for further explanation.

most common types of support used are rockbolts, cable anchors, steel meshing, shotcreting and underpinning. A safer and cheaper solution in most cases is to limit public access to steep rock faces by means of fences and dense vegetation.

21.4 RECLAMATION OF SAND AND GRAVEL PITS

Former sand and gravel workings are the showcases of quarry reclamation, since they are relatively easy to reshape into natural-looking landforms and generally flood on abandonment. They can therefore be converted into ornamental lakes, which may be used for water sports, for passive recreation, or as the centrepiece of a nature reserve. They may also have a flood control function, providing short-term detention capacity during high stages of the adjacent river. The main geotechnical considerations in rehabilitating sand and gravel pits are as follows:

- The risk of *groundwater pollution*, particularly where the pit floor lies close to, or below, the water table. However, leachate migration may pose a risk even where the landfill base is some metres above the water table.
- *Erosion protection* of the lake rim from surface runoff and wave action. This can be tackled by careful surface drainage design, grassing and rock facing ('beaching stone') over geotextile matting. Slopes of about V:H = 1:3 (18°) are generally satisfactory. The risk of bank collapse due to seepage pressures following rapid pond lowering should also be considered.
- Provision for *flushing* stagnant lakewater. Usually this is done by means of a high-level connection to a nearby river, allowing floodwaters to spill through the lake and be discharged at lower elevation downstream.

Reclamation materials, in the form of unused alluvium, silt and clay tailings from sand washing, clay lumps and oversize cobbles are likely to be readily available. Washery tailings are used primarily for reshaping and as fill, but can be fertilized and seeded where no better topsoil is available to provide a growth medium for plants. Oversize is used for wave protection stone.

Example: Penrith Lakes scheme, Castlereagh, New South Wales

Alluvial sand and gravel deposits beneath the floodplain of the Nepean River near Castlereagh, 60 km west of Sydney, supply about half of Sydney's coarse aggregate requirements. By the time extraction ceases early next century, roughly 230 Mt of aggregate will have been mined. The former pits are being replaced by a complex of lakes, foreshore recreation areas and residential subdivisions, covering 1900 ha along an 11 km reach of the Nepean.

This is the largest quarry reclamation project in Australia and is being carried out by the Penrith Lakes Development Commission (PLDC), a private company set up by the three quarrying companies working the deposit. The PLDC was established in 1980 to plan, design and coordinate extraction in what had previously been a patchwork of pits operated since the late nineteenth century. The total cost of the scheme was estimated at A$188 million in 1988, and is financed by production levies on PLDC members and by sales of restored land.

About half the area of the scheme comprises seven lakes and an Olympic-standard rowing course, with 40 km of shoreline and a maximum depth of 5 m. One aim of the multiple lakes is to separate users (swimmers, sail boaters, wildlife) but it also allows for scenic diversity and for water movement through the system, with water levels falling from south to north. Trees are used to screen operating quarries, for shelter belts and in amenity plantings.

Geotechnical aspects of the scheme design include: prevention of shoreline erosion and sedimentation; preservation of the underlying alluvial aquifers; and provision of adequate nutrient flushing and flood control capacity. Some soils are susceptible either to surface erosion by waves or runoff, or to internal erosion (tunnelling) by piping. This arises from their light texture, their lack of soil structure following excavation and the presence of dispersive clays. Potentially erodible foreshore soils are protected by grassing, geotextile mats and stone-pitched banks, and power boats are forbidden. Channels are lined and spaced so as to prevent scouring, ponding and interbank erosion.

21.5   RECLAMATION OF HARD ROCK QUARRIES

Hard rock quarries present a number of geotechnical problems, making them the most difficult category of all to rehabilitate. There is likely to be little waste material available for filling and much of this will be coarse, slow-weathering rock fragment scree; the rock faces may be high, steep and impossible to reshape without extensive blasting; and the presence of benches, flat floors and obvious excavation scars detracts from any illusion of natural origins. Therefore the first decision required is whether to reclaim the quarry as part of an industrial landscape – in which case straight faces and benches are quite appropriate – or to attempt to recreate a natural panorama by selective trimming (Figure 21.6) and tree screening. The second decision is whether it is feasible and desirable to use the quarry as a landfill site prior to either end-use.

The industrial end-use requires that the quarry faces be trimmed to remove projections and overhangs, and that loose rock be scaled off or

Figure 21.6 A former quartzite quarry so successfully reshaped that former benches are almost invisible, near Adelaide, South Australia. Grasses have been planted, but some sheet erosion and sedimentation (at pit floor) are visible, and the native treescape has not yet been reproduced. The site was recycled for housing and recreational uses.

pinned back with rockbolts and mesh. Some additional small-scale extraction may be necessary to make best use of the void, and to level the floor. The aim is to create a regular excavation shape with stable and straight walls, similar to a road cutting.

Rehabilitation of a quarry as a nature reserve or public recreation area may range from simply tidying up the site and replanting trees in soil pockets, to blasting down faces and reshaping using imported fill prior to topsoiling, mulching and planting (Figure 21.7). Selective planting in sheltered niches allows soil to be used most economically, but where finer-grained wastes such as sewerage sludge are available in quantity some backfilling to create more natural-looking slopes is possible. Given time and non-interference, native plants will recolonize even bare rock; the cavities, rubble piles and overhangs left by extraction provide refuges for a variety of small wildlife. In these situations the geotechnical input is minimal compared to floral and faunal considerations.

Hard rock quarries are generally suitable for most types of landfill, except where open joints are present and in hydrological connection with unconfined groundwater. This may prevent filling of some limestone and closely jointed basalt quarries, but even in these cases it

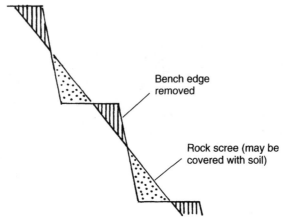

(a) Benches blasted down

(b) Benches fractured and planted

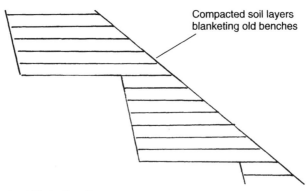

(c) Benches backfilled with soil

Figure 21.7   Remedial works on hard rock quarry benches. The type of treatment depends on slope steepness, bench height and width, and the availability of backfill soil.

may be possible to provide a clay floor and wall seal, or to fill with inert wastes. Deep quarries where the floor is below the water table should preferably remain unfilled (other than with clean granular wastes), but may present rehabilitation difficulties due to their cold, de-oxygenated and layered stagnant water.

21.6 RECLAMATION OF BRICK-PITS

Abandoned brick clay and shale workings are probably the best quarry sites for landfilling, owing to their low permeability and the availability of clay cover material. Nonetheless, some groundwater movement into and out along joints is possible, particularly in lithified shale pits, and confined leachate may simply ooze out at the surface if the fill is allowed to become saturated with no provision for drainage (see Figure 21.5).

The long-term stability of near-vertical clay and shale faces also presents different problems to hard rock exposures. A shale mass will dilate and weather along joints and bedding planes loosened by stress relief. Filling will provide lateral support for these slopes, but if the pit is to remain unfilled, some cutting-back, benching and slope drainage will be necessary.

Example: Sydney Park, St Peters, New South Wales

Sydney Park is the largest of several former brick-pits in the Sydney region that have been or are being landfilled prior to final restoration as public reserves. As landfill sites they offer a number of advantages: they are large ($1-5\,Mm^3$) present-day eyesores; they are situated at inner city locations, where there is a demand both for waste disposal space and for parkland; the surrounding shale is nearly impermeable and its groundwater saline; and if left unfilled the weathered shale slopes will deteriorate.

The Sydney Park site covers 35 ha formerly occupied by seven brick-pits in the industrial suburb of St Peters. Shale extraction has long been in decline here and several tips operated on the site between 1952 and 1985. Filling ceased at the northern end – Stage I of the park – in the early 1970s and redevelopment commenced in 1984. Up to the present about 20 ha has been fully rehabilitated. The brick-pits were located in Ashfield Shale, a lithified Triassic mudrock that was formerly the largest single source of brick and tile shale in Australia.

In the completed portion of the park, the depth of landfill varies from 15 m to 45 m, with 1.3–1.5 m of soil cover. Leachate movement is monitored by observation wells, and a pumping system has been installed to prevent pollutant migration laterally into the adjacent Botany Sands aquifer (Quaternary). Total settlement at the site is believed to have been at least 1 m, but no records were kept. Methane emission remained considerable even 15 years after the completion of filling, and affects about 87% of the initial park area. This first became obvious when a number of trees died after reaching 2 m height, although grass and shallow-rooted shrubs were unaffected. Methane concentrations averaging 42% at depth 2 m have been measured, while at another Ashfield Shale landfill in the Sydney area (Merrylands brick-pit) flows have been sufficient to run a small electric power station.

REFERENCES AND FURTHER READING

Allen, D.F. (1983) Quarry site rehabilitation and after use, in *Surface Mining and Quarrying*. Institution of Mining and Metallurgy, London.

Bradshaw, A.D. and Chadwick, M.J. (1980) *The Restoration of Land*. Blackwell, London.

Coppin, N.J. and Bradshaw, A.D. (1982) *Quarry Reclamation*. Mining Journal Books, London.

Fell, R., Phillips, A. and Gerrard, C. (eds) (1993) *Geotechnical Management of Waste and Contamination*. Balkema, Rotterdam.

Fleming, G. (ed.) (1991) *Recycling Derelict Land*. Institution of Engineers, London.

Haywood, S.M. (1974) *Quarries and the Landscape*. British Quarrying and Slag Federation.

Selby, J. and Hiern, M.C. (1984) *Mine and Quarry Reclamation in South Australia*. South Australia Department of Mines and Energy, Special Publication No. 4.

Whiffen, P. and Walker, D. (1984) Rehabilitation at Stoneyfell Quarry. *Quarry Management*, May, pp. 283–91.

Wilkinson, D.J. (1984) Quarries and waste disposal: techniques and equipment. *Quarry Management*, March, pp. 159–62.

Index

Page numbers in *italics* refer to Tables, and those in **bold** refer to Figures.